Lecture Notes in Mathematics

Edited by A. Dold and B. Eckmann

1182

Saharon Shelah

Around Classification Theory of Models

Springer-Verlag
Berlin Heidelberg New York Tokyo

Author

Saharon Shelah
Institute of Mathematics and Computer Science
The Hebrew University, Givat Ram
91904 Jerusalem, Israel

Mathematics Subject Classification (1980): 03Cxx, 03Exx

ISBN 3-540-16448-0 Springer-Verlag Berlin Heidelberg New York Tokyo
ISBN 0-387-16448-0 Springer-Verlag New York Heidelberg Berlin Tokyo

Printing and binding: Beltz Offsetdruck, Hemsbach/Bergstr.
2146/3140-543210

Introduction

Though our focus is extending classification theory in various directions, there is here material which, I think, will interest researchers in quite different directions: general topology, Boolean algebras, set theory (mainly on coding sets) monadic logic and the theory of modules.

The only paper dealing directly with first order theories is a) of the Notes. A long time ago, in solving a problem of Keisler, we showed that if a (say countably first order) theory T, has only homogeneous models in one $\lambda > \aleph_0$ then this occurs in every $\mu > Min\{\lambda, 2^{\aleph_0}\}$. However this was done for sequence homogeneous models and here we prove the parallel theorem for model homogeneous models.[1]

In 2) we continue the classification of theories over a predicate. Here amalgamation properties over finite diagrams of models play a prominent part, and the combinatorics involving the non-structure theorems becomes much harder. Hence we do it by forcing. We also restrict ourselves to theories without two-cardinal models. This will be continued and the classification (for countable theories in a convenient set theory) will be completed in a paper together with Brad Hart.

Another generalization is the classification of first-order T under any first-order definable quantifiers. We know much on this by previous works of

---

[1] Some further light is thrown on the proof by the following theorem.
Th: If T is countable superstable unidimensional, then one of the following occurs:
(1) T is categorical in every uncountable cardinal
(2) T has the maximal number of models in every uncountable cardinals
(3) The number of models of T of power λ is $Min\{2^\lambda, 2^{2^{\aleph_0}}\}$

Baldwin together with the author and of the author. An important case left open in the latter are the pairs (T, Q^{mon}) (monadic logic) for T unstable; it was known that if some monadic expansion of T has the independence property the pair was "complicated" e.g. has Hanf numbers like second-order logic. Here we prove that the other pairs in this case, are all similar and have smaller Hanf numbers.

A more basic question is whether we can classify generalized quantifiers which are definable in stronger logics. For this we deal with the quantification on a family K of relations over a fix universe U, closed under isomorphism. Surprisingly, we get a reasonable picture when we classify them under the suitable equivalence relation of binterpretability or biiexpressability; essentially, equivalence relations are enough and they are well understood. Also a nice surprise is the use of techniques from classification of first-order theories.

Next let us turn to non-structure results. For infinitary languages we show that for singular cardinals, the infinitary logic not only does not provide us with Scott sentences, we even do not have a nice dichotomy for $no(M)$.

We use a development of the non-structure techniques to get Boolean algebras with few endomorphisms in most cardinals (in fact, satisfying the countable chain condition.)

Other results on Boolean algebras (or, essentially equivalently, on compact spaces) are on the possible number of ideals. Those numbers, though they do not actually have to be powers, have to satisfy strong (such) conditions. We also obtain strong restrictions on the cofinalities on the cardinal invariants $s(X), z(X), h(Z)$.

Many times non-structure theorems have required set theoretic investigations, as e.g. in "classification of non elementary classes I" the non-saturatedness of a natural ideals is used. We prove here e.g. that the ideal of non-stationary subsets of ω_1 is not $\aleph_1$-dense, assuming WCH (in (a) of the Notes).

Zwicker has recently introduced stationary coding. His reason was generalizing the theory of usual stationary sets, to stationary subsets of $P_\kappa(\lambda)$. It seems reasonable that such sets could be used in uniformizing and strengthening non-structure theorems, but existence theorems were lacking. In the two papers dealing with stationary codings we get various existence theorems. Mainly for this we deal again with the club filter of λ^+ or $P_\kappa(\lambda^+)$ concentrating on the 'wrong' cofinality getting non λ^+-saturatedness and generalizations. In e) of the Notes we prove in ZFC existence of weak variants of squares. Two other papers deal with modules. In one we prove (in ZFC) the existence of a non-standard uniserial module over some uniserial domains. This was a serious problem in the manuscript of Fuchs and Salce and many theorems were easier for standard such models. In the other (b) of the Notes) we investigate when abelian groups can be represented as the union of few free ones. Lastly one note deals with finite models.

ACKNOWLEDGEMENTS

The author thanks the United States - Israel Binational Science Foundation and the NSF for partially supporting this research.

Parts of this research were done while the author was at the following institutes:

Institute of Advanced Study and
Institute of Mathematics and Computer Science
The Hebrew University, Jerusalem, Israel

EECS and Mathematics Department
University of Michigan, Ann Arbor, MI, USA

University of California, Berkeley, Ca, USA

Department of Mathematics
University of New York, Buffalo, NY, USA

CONTENT

CLASSIFYING GENERALIZED QUANTIFIERS

Abstract : Finding a universe $\mathcal{U}$ we prove that any quantifier ranging on a family of n-place relations over $\mathcal{U}$, is bi-expressible with a quantifier ranging over a family of equivalence relations, provided that $V=L$. Most of the analysis is carried assuming *ZFC* only and for a stronger equivalence relation, also we find independence results in the other direction.

Notation :

1) $\bar{b}\approx_A\bar{c}$ means $\bar{b}=\langle b_i:i<n\rangle,\bar{c}=\langle c_i:i<n\rangle$, and: a) $b_i\in A$ iff $c_i\in A$, b) $b_i\in A$ implies $b_i=c_i$, c) $b_i=b_j$ iff $c_i=c_j$.

2) For a set Δ of $\varphi(\bar{x})$ (φ a formula, $\bar{x}$ a finite sequence of variables including all variables occurring freely in φ),

$$tp_\Delta(\bar{b},A,M)=\{\varphi(\bar{x},\bar{a}):\ \varphi(\bar{x},\bar{y})\in\Delta,\ \bar{a}\subseteq A \text{ and } M\models\varphi[\bar{b},\bar{a}]\}$$

We omit M when its identity is clear, and when $M=(\mathcal{U},R)$ write R instead of M. Replacing Δ by bs means $\Delta=\{\varphi(\bar{x}):\varphi$ atomic or negation of atomic formula$\}$. We write φ instead $\{\varphi\}$, and Δ will be always finite.

3) $S^m_\Delta(A,M)=\{tp_\Delta(\bar{b},A,M):\bar{b}\subseteq M,l(\bar{b})=m\}$

Introduction

In [Sh3] we gave a complete classification of a class of second order quantifiers: those which are first-order definable (see below an exact definition). We find that for infinite models up to a very strong notion of equivalence, biinterpretability, there are only four such quantifiers: first order, monadic, one-to-one partial functions, and second-order.

Our aim here is to see what occurs if we remove the restriction that the quantifier is first order definable. As we do not want to replace this by a specific $\mathcal{L}$-definable ($\mathcal{L}$-some logic) we restrict ourselves to a fix infinite universe $\mathcal{U}$. If we then want to restrict ourselves to $\mathcal{L}$-definable quantifiers, we will be able to remove the restriction to a fix universe $\mathcal{U}$.

Let us now make some conventions and definitions.

0.1 convention : 1) $\mathcal{U}$ will be a fix infinite universe

2) K will denote a family of n-place relation over $\mathcal{U}$, (for a natural number $n = n(K)$), closed under isomorphism, i.e. if R_1, R_2 are n-place relations on $\mathcal{U}$, $(\mathcal{U}, R_1) \cong (\mathcal{U}, R_2)$ then $R_1 \in K$ iff $R_2 \in K$.

3) Let $\bar{K}$ denote a finite sequence of such K's.

$$\bar{K} = \langle K_\ell : l < \ell(\bar{K}) \rangle, \quad \bar{K}_l^i = \langle K_{i,l}^i : l < l(\bar{K}_i) \rangle$$

4) Dom $R = \bigcup\{\bar{a} : \models R(\bar{a})\}$, $n = n(R)$ if R is an n-place relation (or predicate; we shall not strictly distinguish).

0.2 Definition : For any K, $\exists_K$ (or Q_K) denote a second order quantifier, intended to vary on members of K. More exactly, $L(\exists_{K_1}, \cdots, \exists_{K_m})$ is defined like the first order logic but we have for each $l = 1, m$ (infinitely many) variables R which serve as $n(K_l)$-place predicates, and we can form $(\exists_{K_l} R)\varphi$ for a formula φ. Defining satisfaction, we look only at models with universe $\mathcal{U}$, and $\models (\exists_{K_\ell} R)\varphi(R, \cdots)$ iff for some $R^0 \in K_\ell$, $\varphi(R^0, \cdots)$.

Remark : Note that quantifiers depending on parameters are not allowed. e.g. on automorphisms; on such quantifiers see [Sh4], [Sh5], [Sh6].

0.3 Definition : We say that K (or Q_K) is $\mathcal{L}$- definable ($\mathcal{L}$-a logic) if there is a formula $\varphi(R) \in \mathcal{L}$, R the only free variable of φ, and is appropriate, i.e. an n (K)-place predicate, such that for any n-place relation R on $\mathcal{U}$

$$(\mathcal{U}, R) \models \varphi(R) \text{ iff } R \in K$$

0.4 Definition : We say that $\exists_{K_1} \leq_{int} \exists_{K_2}$ ($\exists_{K_1}$ is interpretable in $\exists_{K_2}$) **iff** for some first-order formula $\vartheta(\bar{x}, \bar{S}) = \vartheta(x_0, \cdots, x_{n(K_1)-1}, S_0, \cdots, S_{n-1})$, (each S_l an $n(K_2)$-place predicate) the following holds:

(*) for every $R_1 \in K_1$ there are $S_0, \cdots, S_{n-1} \in K_2$ such that $(\mathcal{U}, S_0, \cdots, S_{m-1}) \models (\forall x)[R_1(x) \equiv \vartheta(x, S_0, \cdots, S_{m-1})]$

Remark : We can define $\exists_{K_1} \leq_{int}^{\mathcal{L}} \exists_{K_2}$ similarly, by letting $\vartheta \in \mathcal{L}$, but we have no need.

A weaker notion is

0.5 Definition : 1) We say that $\exists_{K_1} \leq_{\exp} \exists_{K_2}$ ($\exists_{K_1}$ is expressible by $\exists_{K_2}$) if there is a formula $\vartheta(x, S_0, \cdots, S_{m-1})$ in the logic $L(\exists_{K_2})$ such that:

(*) for every $R_1 \in K_1$, there are $S_0, \cdots, S_{m-1} \in K_2$ such that $(\mathcal{U}, S_0, \cdots, S_{m-1}) \models (\forall x)[R_1(x) \equiv \vartheta(x, S_0, \cdots, S_{m-1})]$.

2) We say that $\exists_{K_1} \leq_{inex} \exists_{K_2}$ ($\exists_{K_2}$ is invariantly expressible by $\exists_{K_2}$) if there is a formula $\vartheta(x, S_0, \cdots, S_{m-1})$ in the logic $L(\exists_{K_2})$ such that:

(*) for every $R_1 \in K_1$ there are $S_0, \ldots, S_{m-1} \in K_2$ such that for every K_3 which extends K_2, letting ϑ' is ϑ when we replace $\exists_{K_2}$ by $\exists K_3$:

$$(\mathcal{U}, S_0, \ldots, S_{m-1}) \models (\forall x)[R_1(x) \equiv \vartheta'(x, S_0, \ldots, S_{m-1})]$$

0.6 Definition : 1) We say $\exists_{K_1} \equiv_{int} \exists_{K_2}$ ($\exists_{K_1}, \exists_{K_2}$ are biinterpratable) if $\exists_{K_1} \leq_{int} \exists_{K_2}$ and $\exists_{K_2} \leq_{int} \exists_{K_1}$.

2) We say $\exists_{K_2} \equiv_{\exp} \exists_{K_2}$ ($\exists_{K_1}, \exists_{K_2}$ are biexpressible) if $\exists_{K_1} \leq_{\exp} \exists_{K_2}$ and $\exists_{K_2} \leq_{\exp} \exists_{K_1}$. Similarly for $\equiv_{inex}$: $\exists_{K_1} \equiv_{inex} \exists_{K_2}$ ($\exists_{K_2}, \exists_{K_1}$ are invariantly biexpressible) if $\exists_{K_1} \leq_{inex} \exists_{K_2}$ and $\exists_{K_2} \leq_{inex} \exists_{K_1}$.

3) We can define $\exists_{K_1} \leq_{int} \{\exists_{K_0}, \cdots, \exists_{K_{k-1}}\}$ as in Def. 0.4, but $S_0, \cdots, \in \bigcup_{i=1}^{k} K_i$, we let $\exists_K$ stand for $\{ \exists_{K_0}, \cdots, \exists_{K_{k-1}} \}$ where $K = \langle K_0, \cdots, K_{k-1} \rangle$; we define $\exists_{K^1} \leq_{int} \exists_{K^2}$ if $\exists_{K_l^1} \leq_{int} \exists_{K^2}$ for each l ; we also define expressible , invariantly expressible, biinterpratable and (invariantly) biexpressible similarly.

0.7 Notation : 1) If R_l is an n_l-place relation let $\sum_{l=0}^{n-1} R_l = \{\bar{a}_0 {}^\frown \cdots {}^\frown \bar{a}_{n-1} : \bar{a}_l \in R_l\}$.

2) Let $\sum_{l=0}^{n-1} K_l = \{ \sum_{l=0}^{n-1} R_l : R_l \in K_l \text{ for } l < n \}$.

3) $\exists_R$ stand for $\exists_K$ where $K = \{R_1 : (\mathcal{U}, R^1) \cong (\mathcal{U}, R \cdots)\}$.

0.8 Lemma : 1) $\leq_{int}, \leq_{inex}$ and $\leq_{\exp}$ are partial quasi orders, hence $\equiv_{int}$, $\equiv_{inex}$, $\equiv_{\exp}$ are equivalence relations.

2) $\exists_{\bar{K}_1} \leq_{int} \exists_{\bar{K}_2}$ implies $\exists_{\bar{K}_1} \leq_{inex} \exists_{\bar{K}_2}$ which implies $\exists_{\bar{K}_1} \leq_{\exp} \exists_{\bar{K}_2}$

3) $\exists_{\bar{K}}$, and $\exists_{K}$ are biinterpratable if $K=\sum_i K_i$ or $K=\bigcup_i K_i$ ($n(K_i)$ constant in the second case).

0.9 Lemma : 1) If $\bar{K}_1$, $\bar{K}_2$ are $\mathcal{L}$ -definable (i.e. each $K_{l,i}$ is) and $\exists_{\bar{K}_1} \leq_{\exp} \exists_{\bar{K}_2}$ then we can recursively attach to every formula in $\mathcal{L}(\exists_{\bar{K}_1})$ an equivalent formula in $\mathcal{L}(\exists_{\bar{K}_2})$.

2) If $\bar{K}_1$, $\bar{K}_2$ are $\mathcal{L}$-definable, $\exists_{\bar{K}_1} \leq_{\exp} \exists_{\bar{K}_2}$ then the set of valid $\mathcal{L}(\cdots\exists_{\bar{K}_1})$-sentences in recursive in the set of valid $\mathcal{L}(\exists_{\bar{K}_2})$ -sentences.

Remark : The need of " $\mathcal{L}$-definable " is clearly necessary. Though at first glance the conclusions of 0.9 may seem the natural definition of interpratable, I think reflection will lead us to see it isn't.

0.10 Definition : 1) We say that $\exists_{\bar{K}_1} \leq_{int} \exists_{\bar{K}_2}$ for a family of pairs $(\bar{K}_1, \bar{K}_2)$, uniformly, if the formulas $\vartheta_l\,(l<l(\bar{K}_1))$ depend on the $n(K_{1,l})$, $n(K_{2,j})$, $(i<l(\bar{K}_1), j<l(\bar{K}_2))$ only. (Clearly if we have only finitely many candidates for ϑ_l , it does not matter).

2) We use similar notions for $\leq_{\exp}$, $\leq_{inex}$,$\equiv_{int}$, $\equiv_{\exp}$, $\equiv_{inex}$.

§1 On some specific quantifiers.

1.1 Definition : 1) Let $K^{mon}_{\lambda} = \{A\subseteq\mathcal{U}: |A|=\lambda\leq|\mathcal{U}-A|\}$

2) but we write Q^{mon}_{λ} for $\exists_{K^{mon}_{\lambda}}$, and similarly for the other quantifiers defined below.

3) $K^{1-1}_{\lambda} = \{f : f$ is a partial one-to-one function.
$|\mathrm{Dom}\,(f)| = \lambda \leq |\mathcal{U}-\mathrm{Dom}\,(f)-Rang\,(f)|\}$.

4) $K^{eq}_{\lambda,\mu} = \{E{:}E$ is an equivalence relation on some $A\subseteq\mathcal{U}$, with λ equivalence classes, each of power μ, and $|\mathcal{U}-A| = |\mathcal{U}|\}$.

5) For $\lambda^{+}\geq\mu$, we let

$K^{*eq}_{\lambda,<\mu} = \{E{:}E$ is an equivalence relation, every equivalence class of E has power $<\mu$, for each $\kappa<\mu$, E has exactly λ equivalence classes of power κ, and $|\mathcal{U}-\mathrm{Dom}\,(E)| = |\mathcal{U}|\}$.

6) $K^{eq}_{\lambda,<\mu}$ $=\{E{:}E$ is an equivalence relation, with λ equivalence classes, each of power $<\mu$ and $|\mathcal{U}-\mathrm{Dom}\,E|=|\mathcal{U}|\}$.

7) $K^{mon}_{<\lambda}=\bigcup_{\mu<\lambda}K^{mon}_{\mu}$, $K^{1-1}_{<\lambda}=\bigcup_{\mu<\lambda}K^{1-1}_{\mu}$ and $K^{eq}_{<\lambda,<\mu}=\bigcup_{\chi<\lambda}K^{eq}_{\chi,<\mu}$ and $K^{*eq}_{\lambda,\leq\mu}=\bigcup_{\substack{\chi<\lambda\\ \chi^+\geq\mu}}K^{*eq}_{\chi,\leq\mu}$

of course, $K^{eq}_{<\lambda,\mu}=\bigcup_{\chi<\lambda}K^{eq}_{\chi,\mu}$, $K^{eq}_{\lambda}=K^{eq}_{\lambda,\lambda}$, $K^{eq}_{<\lambda}=K^{eq}_{<\lambda,<\lambda}$.

Remark : Of course, always $|\mathrm{Dom}\,E|\leq|\mathcal{U}|$.

1.2 Claim : Let $\lambda\leq\chi$. All results are uniform.

1) $Q^{mon}_{\lambda}\equiv_{int}Q^{mon}_{<\lambda^+}$ and $Q^{mon}_{<\lambda}\leq_{int}Q^{mon}_{<\chi}$; Q^{mon}_{χ} is $\exists_R$ for some R ; and $Q^{mon}_{<\mu}\equiv_{int}Q^{eq}_{1,<\mu}$.

2) $Q^{1-1}_{\lambda}\equiv_{int}Q^{1-1}_{<\lambda^+}$; $Q^{1-1}_{<\lambda}\leq_{int}Q^{1-1}_{<\chi}$, and $Q^{mon}_{<\lambda}\leq_{int}Q^{1-1}_{<\lambda}$; Q^{1-1}_{λ} is $\exists_R$ for some R, and $Q^{1-1}_{<\mu}\equiv_{int}Q^{eq}_{2,<\mu}$.

1.3 Claim : Let $\lambda\leq\chi$, $\mu\leq\kappa$, all results are uniform.

1) $Q^{eq}_{\lambda,\mu}\leq_{int}Q^{eq}_{\chi,\kappa}\equiv_{int}Q^{eq}_{<\chi^+,<\kappa^+}$ and $Q^{eq}_{<\lambda,<\mu}\leq_{int}Q^{eq}_{<\chi,<\kappa}$.

2) If $\lambda^+\geq\mu$, $Q^{*eq}_{\lambda,\leq\mu}\equiv_{int}Q^{eq}_{<\lambda^+,<\mu}$.

3) $\exists_K\leq_{int}Q^{eq}_{<\lambda}$ if $(\forall R\in K)[|\mathrm{Dom}\,R|<\lambda]$ (when λ is infinite, for λ finite $|\mathrm{Dom}\,R|^{n(R)}\leq(\lambda-1)^2$ is needed).

4) $Q^{eq}_{<\lambda,<\mu}\leq_{int}\{Q^{eq}_{<\chi_l,<\kappa_l}:l<k\}$ *iff* for some l, $\lambda\leq\chi_l\wedge\mu\leq\kappa_l$ or $\lambda<\aleph_0\wedge\chi_l>1\wedge\mu\leq\kappa_l$ or $\mu<\aleph_0\wedge\lambda\leq\chi_l\wedge\kappa_l>1$ or $\lambda<\aleph_0\wedge\mu<\aleph_0$ (but in the last three cases the interpretation is not uniform.)

Proof : Left to the reader.

1.4 Lemma : The following holds uniformly:

1) $Q^{eq}_{\chi,\chi}\leq_{inex}Q^{eq}_{\lambda,\mu}$ *if* $\lambda\geq\chi$, $\mu\geq\aleph_0$ and $\chi\leq2^{\mu}$

2) $Q^{eq}_{2^{\mu},\lambda}\leq_{inex}Q^{eq}_{\lambda,\mu}$ if $\lambda\geq2^{\mu}$, and $\mu\geq\aleph_0$

3) $Q^{1-1}_{<\lambda}\equiv_{inex}Q^{eq}_{<\lambda,\aleph_0}\equiv_{inex}Q^{eq}_{<\lambda,2}$ for $\lambda>\aleph_0$

4) $Q^{1-1}_{<\lambda}\equiv_{inex}Q^{eq}_{<\lambda,<\lambda}\equiv_{inex}Q^{eq}_{<\lambda,2}$ for $\lambda=\aleph_0$

Remark : 1) Clearly in (1) we get biinterpretability.

2) Because of the uniformity e.g. (2) implies $Q^{eq}_{2^{\mu},<\lambda}\leq_{inex}Q^{eq}_{<\lambda,\mu}$ if $\lambda>2^{\mu}$, $\mu\geq\aleph_0$.

Proof : Repeat the proofs in [Sh1], [Sh2].

1.5 Lemma : 1) For any K consisting of equivalence relations for some n, λ_l, $\mu_l\,(l<n)$, $\exists_K\equiv_{int}\{Q^{eq}_{<\lambda_l,<\mu_l}\;:\;l<n\}$.

2) For any $n, \lambda_l, \mu_l (l<n)$ for some equivalence relation E, $\exists_E \equiv_{int} \{Q^{eq}_{\lambda_l,\mu_l} : l<n\}$.

Remark : This lemma enables us to concentrate on analyzing quantifiers of the form $\exists_R$.

1.6 Lemma: For infinite cardinals λ,μ,χ,κ: $Q^{eq}_{\chi,\kappa} \leq_{inex} Q^{eq}_{\lambda,\mu}$ *iff* $Q^{eq}_{\chi,\kappa} \leq_{\exp} Q^{eq}_{\lambda,\mu}$ *iff* $\chi \leq \lambda \wedge \kappa \leq \mu$ or $\chi+\kappa \leq \lambda \wedge \chi + \kappa \leq 2^\mu$.

Proof: The first condition implies the second trivially the third implies the first by 1.3(1) (if $\chi \leq \lambda \wedge \kappa \leq \mu$) 1.4(1), (if $\chi + \kappa \leq \lambda, 2^\mu$) 1.4(2) (if $2^\mu \leq \lambda$, $\kappa \leq \lambda$ and $\chi \leq 2^\mu$). Now we assume the second is exemplified by $S_0, \dots, S_{m-1} \in K^{eq}_{\lambda,\mu}$ and suppose $E \in K^{eq}_{\chi,\kappa}$ is definable by an $L(Q^{eq}_{\lambda,\mu})$-formula (with $S_0, \dots, S_{m-1}$ the only non logical symbols, w.l.o.g. the elements were absorbed). The first case will be $\lambda \geq \mu$. Let E^* be the transitive closure of $\bigvee_{l<m} x\ S_l y$ (with domain $\bigcup_{l<m}$ Dom S_l). Then E^* is an equivalence relation with $\leq \lambda$ equivalence classes, each of power $\leq \mu$, hence $\bigcup_{\ell}$ can be represented as the disjoint union of $A_i (i < \alpha \leq \lambda)$ such that $S_l = \bigcup_{i<\alpha} (S_l \restriction A_i)$. Hence a permutation f is an automorphism of $(\mathcal{U}, S_0, \dots, S_{m-1})$ iff for some permutation h of α for each i ,$f \restriction A_i$ is an isomorphism from $(A_i, S_0 \restriction A_i, \dots, S_{m-1} \restriction A_i)$ onto $(A_{h(i)}, S_0 \restriction A_{h(i)}, \dots, S_{m-1} \restriction A_{h(i)})$.

Let $A_i = \{a_{i,j} : j < j_i \leq \mu\}$ and define E^+: $a_{i_1,j_1} E^+ a_{i_2,j_2}$ iff $j_1 = j_2$ and for some automorphism f of $(\mathcal{U}, S_0, \dots, S_{m-1})$, $f(a_{i_1,j_1}) = a_{i_2,j_2}$. Clearly E^+ is an equivalence relation on $\bigcup_i A_i$ with $\leq 2^\mu$ equivalence classes, and if B is an E^+-equivalence class then every permutation of it can be extended to an automorphism of $(\mathcal{U}, S_0, \dots, S_{m-1})$. Let B_i $(i < \gamma \leq 2^\mu)$ list the E^+-equivalence classes.

So if $(\exists x \neq y \in B_i)\ xEy$ then $(\forall x,y \in B_i)\ xEy$.

Let $B^* = \bigcup\{B_i : (\exists x \neq y \in B_i) xEy\}$, $B^{**} = \bigcup B_i : B_i \not\subset B^*, |B_i| > 2\}$, $B^{***} = \bigcup\{B_i : |B_i| \leq 2\}$. So on B^* $(\forall xy \in B^*)(xE^+y \rightarrow xEy)$ i.e. E^+ refine E,,and so E has $\leq 2^\mu$ equivalence classes, each of power $\leq |B^*| \leq |\bigcup_i A_i| \leq \lambda$.

Next on B^{**}, E^* refine E: for suppose xEy but $\neg xE^*y$, let $i<\gamma$ be such that $y\in B_i$, as $y\in B^{**}$ clearly there is $y'\in B_i$, $\neg xE^*y', \neg yEy'$ by a suitable automorphism necessarily xEy' but E is transitive and symmetric contradiction to the definition of B^{**}. So $E\restriction B^{**}$ has $\leq\lambda$ equivalence classes each of power $\leq\mu$. Thirdly on $B^{***}, E\restriction B^{**}$ has $\leq 2^\mu$ equivalence classes each of power $\leq 2^\mu$, (as $|B^{***}|\leq 2^\mu$. Lastly on $\mathcal{U}-B^*\cup B^{**}\cup B^{***}=\mathcal{U}-\bigcup_i A_i$, E is the equality.

By 1.3(4) we finish.

§2 Monadic analysis of $\exists_R$

Our aim is to interpret Q^{mon}_λ in $\exists_R$ for a maximal λ and show that except on λ elements R is trivial. So continuing later the analysis of $\exists_R$, we can instead analyze $\{Q^{mon}_\lambda, \exists_{R_1}\}$ where $|\text{Dom } R_1|\leq\lambda$. This is made exact below.

2.1 Definition : For any relation R let

$\lambda_0=\lambda_0(R)=Min\{|A|: A\subseteq\mathcal{U}$, and for every sequences

$\bar b,\bar c\in\mathcal{U}$ (of length $n(R)$) . $\bar b\approx_A\bar c$ implies $R[\bar b]\equiv R[\bar c]\}$.

where $\bar b\approx_A\bar c$ iff $tp_{bs}(\bar b,A,=)=tp_{bs}(\bar b,A,=)$.

Note that $\lambda_0(R)\leq|\text{Dom } R|$.

2.2 Theorem : 1) Uniformly $Q^{mon}_{\lambda_0(R)}\leq_{int}\exists_R$.

2) Uniformly $\exists_R\equiv_{int}\{\exists_{R_1}, Q^{mon}_{\lambda_0(R)}\}$ for some R_1, $|\text{Dom } R_1|=\lambda_0(R)$, $n(R_1)=n(R)$.

Proof : 1)

Case I : $\lambda_0(R)$ is an infinite regular cardinal.

Let $\bar R^m=\langle R^m_l: l<m\rangle$ denote a sequence of $n(R)$-place predicates or relations, $(\mathcal{U},R_l)\cong(\mathcal{U},R)$, and $\Delta=\Delta(\bar R^m)$ denote a set of formulas of the form $\varphi(\bar x,\bar R^m)$ closed under permuting the variables and identifying them. Let $k=k(\Delta(\bar R),\bar R)$ be the minimal natural number such that:

(*) there is a formula $\varphi=\varphi(\bar x,\bar y,\bar R)\in\Delta$ with $l(\bar x)=l(\bar y)=k$, and sequence $\bar a, l(\bar a)=l(\bar y)$ such that for every $A\subseteq\mathcal{U}$, $|A|<\lambda_0(R)$ there are sequences $\bar b,\bar c$ of length k , such that $\varphi(\bar b,\bar a)\wedge\neg\varphi(\bar c,\bar a,\bar R)$ but $\bar b\approx_A\bar c$.

Let $k(\Delta(\bar R^m))$ be the minimal $k(\Delta(\bar R^m,)\bar R^m)$

By the definition of $\lambda_0(R)$, $k=n(R)$, $\varphi=R(\bar x)$ satisfies (*) for $\bar a$ the empty sequence. By the minimality of k we can assume that $\bar b,\bar c$ are disjoint to A , and with no repetitions. Clearly as $|A|<\lambda_o(R)\le|\mathcal{U}|$, $\mathcal{U}$ infinite, for any such $A,\bar b,\bar c$ we can find $\bar b_l$ $(l=o,2k)$ such that $\bar b_o=\bar b,\bar b_{2k}=\bar c$ and $\bar b_l,\bar b_{l+1}$ differ at exactly one coordinate each $\bar b_l$ disjoint to A and without repetition. So w.l.o.g. in (*) $\bar b=\langle b\rangle^\frown\bar d,\bar c=\langle c\rangle^\frown\bar d$ (and $\bar d^\frown\langle b,c\rangle$ is disjoint to A and with no repetition, and let $\bar x=\langle x\rangle^\frown\bar z$, so $\varphi=(x,\bar z,\bar y)$.) Possibly $\bar z$ is empty (i.e. $k=1$) and then our conclusion is immediate as $\{b:\models\varphi[b,\bar a]\}$ and $\{c:\models\neg\varphi(c,\bar a)\}$ has power $\ge\lambda_o(R)$.

By the choice of $k=k(\Delta,\bar R)$ for every $e\in\mathcal{U}$ there is $A_e\subseteq\mathcal{U}$, $|A_e|<\lambda_o(R),e\in A_e$ such that for every $\bar d_1\approx_{A_e}\bar d_2$ ($\bar d_1,\bar d_2$ of length $k-1$) $\varphi(e,\bar d_1,\bar a,\bar R)\equiv\varphi(e,\bar d_2,\bar a,\bar R)$.

Now we define by induction on l, for $l=0,n(R)-1$ a set of formulas $\Delta_l=\Delta_l(\bar R_l)$ where $\bar R_l=\langle R_{l,i}:i<2^l\rangle$:

$\Delta_o(\bar R_o)=$ the closure of $\{R_{o,o}(x_o,\cdots,x_{n-1})\}$ under permuting and identifying the variables.

$\Delta_{l+1}(\bar R_{l+1})=$ the closure of

$\{(\forall z)[\varphi(z,\bar x,R_{l+1,0},\cdots,R_{l+1,2^l-1})\equiv\varphi(z,\bar x,R_{l+1,2^l},\cdots R_{l+1,2^l+2^l-1})]:$

$$\varphi(z,\bar x,R_{l,0},\cdots,R_{l,2^l-1})\in\Delta_l(\bar R_l)\}\cup\{\varphi(\bar x,R_{l+1,0},\cdots):\varphi(\bar x,R_{l,0},\cdots)\in\Delta_l(\bar R_l)\}$$

under permuting and identifying the variables.

Now we shall prove by induction on l that

$(**)_l\, k(\Delta_l(\bar R_l))\le n-l$.

For $l=0$, as we have mentioned above, this follows from the definition of $\lambda_o(R)$.

So we assume $(**)_l$ and prove $(**)_{l+1}$. As $(*)_l$ holds there are relations R^i $(i<2^l)$, $(\mathcal{U},R^i)\cong(\mathcal{U},R)$ and $k(\Delta_l(\bar R_l))=k(\Delta_l(\bar R_l),\bar R)$, where $\bar R=\langle R^i:i<2^l\rangle$, and let $\bar a_l^*$, $\varphi(\bar x,\bar y,\bar R)$ exemplify (*) for $k=k(\Delta_{l+1}(\bar R_l))$. If $k=1$ we finish of course, otherwise we shall prove that $k>k(\Delta_{l+1}(\bar R_{l+1}))$; this suffices of course.

Now for every $\psi=\psi(\bar u,\bar v,\bar R)\in\Delta_l(\bar R)$, $l(\bar u)<k$, and $\bar a\in\mathcal{U}$, there is a set $A_{\psi,\bar a}\subseteq\mathcal{U}$ of power $<\lambda_0(R)$ such that: if $\bar b\approx_{A_{\psi,a}}\bar c$, $l(\bar b)=l(\bar c)=l(\bar u)-l(\bar a)$ then $\models\psi[\bar b,\bar a,\bar R,]\equiv\psi[\bar c,\bar a,\bar R]$. We can assume $\bar a\subseteq A_{\psi,\bar a}$.

Now we define by induction on $\alpha<\lambda_0(R)$ $\bar d_\alpha$, b_α, c_α as follows. First let $A^0_\alpha=\bigcup_{\beta<\alpha}\bar d_\beta{}^\frown\langle b_\beta,c_\beta\rangle\cup\bar a^*$, and $A_\alpha=\bigcup\{A_{\psi,\bar a}:\bar a\subseteq A^0_\alpha,\ l(\bar a)<k$ and $\psi\in\Delta_l(\bar R)\}$. Now by the discussion after (*) there is $\bar d_\alpha{}^\frown\langle b_\alpha,c_\alpha\rangle$ disjoint to A_α and without repetitions, such that $\models\varphi[b_\alpha,\bar d_\alpha,\bar a^*]\wedge\neg\varphi[c_\alpha,\bar d_\alpha,\bar a^*]$

What are the truth values $t_{\alpha,\beta}$ of $\varphi[b_\alpha,\bar{d}_\beta,\bar{a}^*]$ and $s_{\alpha,\beta}$ of $\varphi[c_\alpha,\bar{d}_\beta,\bar{a}^*]$? Clearly if $\alpha=\beta$ then $t_{\alpha,\beta}$ is truth $s_{\alpha,\beta}$ is false. If $\alpha>\beta$, then we should remember that $A_{\varphi,\bar{d}_\beta\frown\bar{a}^*}\subseteq A_\alpha$, hence $b_\alpha,c_\alpha\notin A_{\varphi,\bar{d}_\beta\frown\bar{a}^*}$. Hence $t_{\alpha,\beta}=t^+_\beta=s_{\alpha,\beta}=s^+_\beta$. If $\alpha<\beta$ then let $\vartheta(\bar{z},x,\bar{y},\bar{R})=\varphi(x,\bar{z},\bar{y},R)$, and remembering that $A_{\vartheta,<b_\alpha>\frown\bar{a}^*}\subseteq A_\beta,A_{\vartheta,<c_\alpha>\frown\bar{a}^*}$, and $\bar{d}_\alpha$ is disjoint to A_β, it is clear that $t_{\alpha,\beta}=t^-_\alpha$, $s_{\alpha,\beta}=s^-_\alpha$.

As we can replace $\langle \bar{d}_\alpha\frown<\beta_\alpha,c_\alpha>:\alpha<\lambda_0(R)\rangle$ by any subsequence of length $\lambda_0(R)$ w.l.o.g. $t^+_\alpha=t^+$, $t^-_\alpha=t^-$ and $s^-_\alpha=s^-$ for every $\alpha<\lambda_0(R)$ we can assume t^+ is truth (otherwise interchange φ and $\neg\varphi$, b_α and c_α in the rest).

Now let h be the following permutation of $\mathcal{U}$: $h(c_{3\alpha+1})=c_{3\alpha+2}$, $h(c_{3\alpha+2})=c_{3\alpha+1}$ and $h(c)=c$ for any other element. Next, let for $2^l\le i<2^{l+1}$, $R^i=h(R^{i-2^l})$ and let $\bar{R}_{l+1}=\langle R^i:i<z^{l+1}\rangle$, $\bar{R}^1=\langle R_{i+2^l}:i<2^l\rangle$. Now

(a) $\psi(\bar{z},\bar{y},\bar{R}_{l+1})=(\forall x)[\varphi(x,\bar{z},\bar{y},\bar{R})\equiv\varphi(x,\bar{z},\bar{y},\bar{R}^1)]$ belong to $\Delta_{l+1}(\bar{R}_{l+1})$.

(b) $\models\psi(\bar{d}_{3\beta},\bar{a}^*,\bar{R}_{l+1})$ for $\beta<\lambda_0(R)$. This is equivalent to saying that h maps $\{e\in\mathcal{U}:\models\varphi[e,\bar{d}_{3\beta},\bar{a}^*,\bar{R}]\}$ into itself (as $h^{-1}=h$). i.e. we should prove $\models\varphi[e,\bar{d}_{3\beta},\bar{a}^*,\bar{R}]$ implies $\varphi[h(e),\bar{d}_{3\beta},\bar{a}^*,\bar{R}^1]$. If $e=h(e)$ this is trivial. Otherwise, $e=c_{3\alpha+i}$, $i\in\{1,2\}$; and $h(e)=c_{3\beta+(3-i)}$; if $\beta\ge\alpha$ this follows from $s_{3\alpha+i,3\beta}=s_{3\alpha+(3-i),\beta}=t^+$ (as $3\alpha+i$, $3\alpha+(3-i)>3\beta$) ; if $\beta<\alpha$ this follows from $s_{3\alpha+i,3\beta}=s_{3\alpha+(3-i),3\beta}=s^-$ (as $3\alpha+i$, $3\alpha+(3-i)<3\beta$).

(c) $\models\neg\psi[\bar{d}_{3\beta+1},\bar{a}^*,\bar{R}]$ for $\beta<\lambda_0(R)$ Just substitute $x=c_{3\beta+2}$ for the $(\forall x)$ in ψ's definition.

(d) The sequences $\{\bar{d}_\beta:\beta<\lambda_0(R)\}$ are pairwise disjoint. This is because for $\gamma<\beta$, $\bar{d}_\beta\subseteq A^0_\beta\subseteq A_\alpha$.

Now (a), (b), (c), (d) together show that $k(\Delta(\bar{R}_{l+1}),\bar{R}_{l+1})<k$ Hence $k(\Delta_{l+1}(\bar{R}_{l+1}))<k\le n-l$ (or $k=1$ and then $h(\Delta_{l+1}(\bar{R}_{l+1})=k((\Delta_l)(\bar{R}_l))$. So we have done the induction step in proving (**).

Now $(**)_{n(R)-1}$ show that for $\bar{a}^*$, $\bar{R}$ ($(\mathcal{U},R_l)\cong(\mathcal{U},R)$) and $\varphi(x,\bar{y},\bar{R})$, the powers of $\{c\in\mathcal{U}:\models\varphi[c,\bar{a}^*,\bar{R}]\}$ and of $\{c\in\mathcal{U}:\models\neg\varphi[c,\bar{a}^*,\bar{R}]\}$ are at least $\lambda_0(R)$, and we get the required interpretation.

Case II : $\lambda_0(R)<|\mathcal{U}|$ (and in particular $\lambda_0(R)$ finite)

Let $A\subseteq\mathcal{U}$ be a set of power $\lambda_0(R)$, such that $\bar{b}\approx_A\bar{c}$ implies $R[\bar{b}]\equiv R[\bar{c}]$. As $\mathcal{U}$ is infinite, we can find distinct $d_i\in\mathcal{U}-A$ $(i<n(R)^2)$. Define $\bar{d}=\langle d_i:i<n(R)^2\rangle$, $\varphi^*(x,\bar{d},R)=\bigwedge\{(\exists y_0,\dots,y_{k-1})$ [the elements $y_0,\dots,y_{k-1},x$ are pairwise distinct and if the elements $y_0,\dots,y_{k-1},d_m,x$ are pairwise distinct then

$\varphi(x, y_0, \ldots, y_{k-1}) \equiv \varphi(d_m, y_0, \ldots, y_{k-1})] : \varphi = \varphi(z_0, \ldots, z_k, R)$

is an atomic formula in $L(R)$ (so $k+1 \leq n(R)$) and $m < n(R)^2$, m, k are natural numbers $\}$. By the choice of A, $x \notin A \Longrightarrow \neg\varphi^*(x, \bar{d})$, hence $B = \{x \in \mathcal{U}: \mathcal{U} \models \varphi^*[x, \bar{d}]\}$ is a subset of A. Clearly $Q^{mon}_{|B|} \leq_{int} \exists_R$ (uniformly) hence it suffices to prove $|B| = \lambda_0(R)$ which follows from

(*) if $\bar{b} \cong_B \bar{c}$ then $R[\bar{b}] \equiv R[\bar{c}]$

for this it suffices to prove:

(**) if $\varphi(\bar{x}, R) \in L(R)$ is atomic, $\bar{b}, \bar{c}$ are sequences of length $l(\bar{x}) \leq n(R)$ without repetition then $\bar{b} \cong_B \bar{c}$ implies $\varphi(\bar{b}, R) \equiv \varphi(\bar{c}, R)$.

Let $\bar{b}\frown\bar{c}_0, \bar{b}\frown\bar{c}_1$ be sequences from $\mathcal{U}$, without repetition, $\bar{b} \subseteq B, \bar{c}_0, \bar{c}_1$ disjoint to B ; by the transitivity of $\equiv$, w.l.o.g. $\bar{c}_1$ is disjoint to $\bar{d}$, so for some i, $\langle d_i, d_{i+1}, \ldots, d_{i+k-1} \rangle$ (where $k = l(\bar{c}_0)$) is disjoint to $\bar{c}_0$ (and obviously to $\bar{c}_1$).

Now we shall prove that for every atomic $\varphi(\bar{x}, \bar{y}, R), l(\bar{x}) = k$, $l(\bar{y}) = l(\bar{b}) \models \varphi(\bar{c}_l, \bar{b}, r) \equiv \varphi(\langle d_i, \ldots, \rangle, \bar{b}, R)$ thus finishing. For this we define $\bar{c}_{l,m} (m \leq k)$ such that each $\bar{c}_{l,m}$ is with no repetitions, disjoint to B, b, $\bar{c}_{l,0} = \bar{c}_l$, $\bar{c}_{l,k} = \langle d_i, \ldots, d_{i+k-1} \rangle$, $\bar{c}_{l,m+1}, \bar{c}_{l,m}$ are distinct in one place only. By the definition of B (and φ) for every atomic $\varphi(\bar{x}, \bar{y}, R)$, $\models \varphi(\bar{c}_{l,m}, \bar{b}, R) \equiv \varphi(\bar{c}_{l,m+1}, \bar{b}, \bar{R})$ so we finish easily.

Case III. $\lambda_0(R)$ a singular cardinal.

We fix the relation R; now for every atomic formula $\varphi(\bar{x}, \bar{y}, R) \in L(R)$ and $\bar{b} \in \mathcal{U}$, $\varphi(\bar{x}, \bar{b}, R)$ define an $l(\bar{x})$-place relation on $\mathcal{U}$, let $\lambda_0(\varphi(\bar{x}, \bar{b}), R))$ be as defined as in Def. 2.1. Clearly the number of atomic $\varphi(\bar{x}, \bar{y}, R)$ (with no dummy variable, $\bar{x} \frown \bar{y} \subseteq \{x_i : i < m(R)\}$) is finite, and we can find φ and $\bar{b}$ such that $\lambda_0(\varphi(\bar{x}, \bar{b}, R)) = \lambda_0(R)$ and (under this restriction) $l(\bar{x})$ is minimal. Clearly $l(\bar{x}) > 0$ (as $A = \phi$ would serve), if $n=1$ we finish trivially. So assume $l(\bar{x}) > 1$, let $\bar{x} = \bar{z} \frown \langle x \rangle$. By the choice of $\varphi, \bar{b}$, for every $c, \lambda_0(\varphi(\bar{z}, c, \bar{b}, R)) < \lambda_0(R)$ and let $A_c \subseteq \mathcal{U}$ be such that

(i) $|A_c| = \lambda_0(\varphi(\bar{z}, c, \bar{b}, R)) \leq |\mathcal{U}|$

(ii) $\bar{d}_1 \sim_{A_c} \bar{d}_2$ implies $\varphi(\bar{d}_1, c, \bar{b}, R) \equiv \varphi(\bar{d}_2, c, \bar{b}, R)$

So $|A_c| = \lambda_0(\varphi(\bar{z},c,\bar{b},R)) < \lambda_0(R) \leq |\mathcal{U}|$.

For each c, by case II there are an atomic $\psi_c(x,\bar{y}_1,R)$ and $\bar{d}_c \in \mathcal{U}$ such that $|\{a : \models \psi_c(a,\bar{d}_c,R)\}| = \lambda_0(\varphi(\bar{z},c,\bar{b},R)) = |A_c|$; note there are only finitely many possible ψ_c's

Subcase III a: $\operatorname{Sup}_{c\in\mathcal{U}} |A_c| = \lambda_0(R)$.

Let $\lambda_0(R) = \sum_{\zeta<\kappa} \mu_\zeta$ where $\kappa = cf(\lambda_0(R))$, and each μ_ζ is regular $< \lambda_0(R)$. So assume $|A_{c_i}| \geq \mu_i$, so by Case II applied to $\varphi(\bar{z},c_i,R)$ we can interprete unformly $Q^{mon}_{<\mu_\zeta}$ and even $Q^{mon}_{<\lambda_0(R)}$ and moreover in this case, we have $\bar{a}_\zeta$ $(\zeta<\kappa)$ such that $\mu_\zeta \leq |\{e \in \mathcal{U}: \models \varphi[e, \bar{a}_\zeta, \bar{R}]\}| < \lambda_0(R)$ ($\bar{a}_\zeta$ is $\bar{d}_\alpha{}^\frown\bar{a}^*$ for some α). In particular we can interprete Q^{mon}_κ. Let $R = \bigcup_{\zeta<\kappa} \bar{a}_\zeta$ and E be the following equivalence relation on $\mathcal{U}$: bEc iff for every $\bar{a} \subseteq P$, $\varphi[b,\bar{a},\bar{R}] \equiv \varphi[c,\bar{a},\bar{R}]$. Let $\langle A_i : i<\chi\rangle$ be a list of the equivalence classes of E . If $\{i: |A_i| \geq 2\}$ has power $\geq \lambda_0(R)$, we get our conclusion easily; this holds also if there are at least two A_i of power $\geq \lambda_0(R)$, or even if $\operatorname{Sup}_{i<\alpha} |A_i| = \lambda_0(R)$. By the choice of P the only case left is $|\{i: |A_i|=1\}| \geq \lambda_0(R)$. So let $a_\alpha (\alpha<\lambda_0(R))$ be pairwise non E-equivalent $a_i \notin P$. Define a permutation :$h(a_{3\alpha+i}) = a_{3\alpha+\beta-i}$ for $i=1,2$ and $h(e)=e$ otherwise. Define $\bar{R}$, $\bar{R}_{l*}$ as in case I and $\varphi^*(x, P, \bar{R}) = (\forall x_0, \cdots, x_{k-2}) [\bigwedge_{i<k} P_i(x_i) \rightarrow \varphi(x, x_0, \cdots, x_{k-2}, \bar{R}) \equiv \varphi(x, x_0, \cdots, x_{k-2}, \bar{R}^1)]$ Now we finish: $\varphi^*[a_\beta, P, R]$ iff β is divisible by 3 (for $\beta<\lambda_0(R)$.

Subcase III b: $|\bigcup_{c\in\mathcal{U}} (A_c - \{c\}| \geq \lambda_0(R)$ but not II a.

By case II we know A_c is definable (uniformly) from $\bar{b}{}^\frown\langle c\rangle$. Hence we can choose for $i < \lambda_0(R)$ c_i, e_i such that $e_i \in A_{c_i}$, $e_i \neq c_i$, and $c_i, e_i \notin \{c_j, e_j : j < i\}$. By Hajnal free subset theorem (See [H]) w.l.o.g.

$e_i \in A_{c_j}$ iff $i = j$, and e_i, c_i do not appear in $\bar{b}$.

Let g be the following permutation of $\mathcal{U}$:

$$g(e_{3i+1}) = e_{3i+2}$$

$$g(e_{3i+2}) = e_{3i+1}$$

$g(e) = e \qquad e \in \mathcal{U} - \{e_{3i+2}, e_{3i+2} : i < \lambda_0(R)\}$

Let $R_0 = R$, $R_1 = g(R)$, then

$$\psi(x, \bar{b}, R_0, R_1) \stackrel{def}{=} (\forall \bar{y})[\varphi(\bar{y}, x, b, R_1)]$$

is as required.

Case III c : not III a,b.

So for some $B \subseteq \mathcal{U}$, $|B| < \lambda_0(R)$, $[c \in \mathcal{U} - B \Longrightarrow A_c - \{c\} \subseteq B]$; w.l.o.g. $[c \in B \Longrightarrow A_c \subseteq B]$.

Let $\bar{d} \subseteq \mathcal{U} - B$, $l(\bar{d}) = l(\bar{z})$; now for every set $D \subseteq \mathcal{U}$, $|D| < \lambda_0(R)$ there are $\bar{d}_1 ^\frown \langle c_1 \rangle$ $\bar{d}_2 ^\frown, \langle c_2 \rangle$ disjoint to $D \cup B \cup \bar{d}$ without repetition, $\varphi(\bar{d}_1, c_1, \bar{b}, R) \equiv \neg\, \varphi(\bar{d}_2, c_2, \bar{b}, R)$ (by the choice of $\varphi, \bar{b}$). As $A_{c_1} - \{c_1\} \subseteq B$,

$$\varphi(\bar{d}_1, c_1, \bar{b}, R) \equiv \varphi(\bar{d}, c_1, \bar{b}, R).$$

and similarly

$$\varphi(\bar{d}_2, c_2, \bar{b}, R) \equiv \varphi(\bar{d}, c_2, \bar{b}, R)$$

hence

$$\varphi(\bar{d}, c_1, \bar{b}, R) \equiv \neg\, \varphi(\bar{d}, c_2, \bar{b}, R)$$

We can conclude that $\varphi(\bar{d}, x, \bar{b}, R)$ divide $\mathcal{U}$ to two subsets each of cardinality $\geq \lambda_0(R)$.

Remark: In case III the only use of "$\lambda_0(R)$ singular" is $[\sup_{c \in \mathcal{U}} |A_c| < \lambda_0(l) \Longrightarrow \sup_{c \in \mathcal{U}} |A_c|^+ < \lambda_0(R)]$, but with a little more work we can bound the numbers of copies of R used independently of R.

Proof of 2.2(2) :
If $\lambda_0(R) = |\mathcal{U}|$ we choose $R_1 = R$ and have nothing new to prove. If $\lambda_0(R) < |\mathcal{U}|$, let $\varphi_i(\bar{x}_i, \bar{y}_i, R)$ $(i < m)$ list all atomic formulas in $L(R)$, $l(\bar{x}_i) = k_i > 0$, $l(x_i) + l(y_i) \leq n(R)$, and w.l.o.g. $k_i = n(R) \Longrightarrow i = 0$. Let $d_i (0 < i < 2n(R))$ be distinct element of $\mathcal{U} - B$, B from case II above. Of course, we can concentrate on the case $n(R) > 1$. Let $R_1 = \{\langle a, \ldots, a \rangle : a \in B\} \cup \{\langle a_1, \ldots, a_{n(R)} \rangle : \models R[a_1, \ldots, a_{n(R)}], a_1, \ldots, a_{n(R)}$

are distinct members of $B\}\cup\{\langle d_i, a_1, \ldots, a_{k_i}, d_i, \cdots\rangle : 1 \le i < m,\ a_1, \ldots, a_{k_i}$ distinct members of B , and for all distinct b_l $(l < l(\bar{y}_i))$ from $\mathcal{U}-B$, $\models \varphi_i[\langle a_i, \ldots, a_{k_i}\rangle, \langle b_0, \ldots, \rangle, R]\}$.

Easily $\{\exists_R, Q_{\lambda_0(R)}\}^{mon}\} \le_{int} \exists_{R_1}$, and by case II above $\exists_{R_1} \le_{int} \exists_R$.

§3 The one-to-one function analysis

The aim of this section is similar to the previous one, going one step further, i.e. we want to analyse $\exists_R$, interpreting in it Q_λ^{1-1} for a maximal λ, hoping that " the remainder" has domain $\le\lambda$.

3.1 Definition : Let $\lambda_1 = \lambda_1(R)$ be Sup $\{ | \{tp_{bs}(a, A, R) : a \in \mathcal{U}-A\} | : A \subseteq \mathcal{U}\}$

3.2 Fact : $\lambda_1(R) \le \lambda_0(R)$

3.3 Claim : $Q_{\lambda_1(R)}^{1-1} \le_{int} \exists_R$ uniformly, if the sup is obtained.

Proof : Suppose h is a one-to-one, one place 'partial' function from $\mathcal{U}$ to $\mathcal{U}$, with $|\mathrm{Dom}\, h| \le \lambda_1(R)$. Let $A \subseteq \mathcal{U}$ be such that $\{tp_{bs}(a, A, R) : a \in \mathcal{U}-A\}$ has cardinality $\lambda \overset{def}{=} \lambda_1(R)$. So we can find $a_i \in \mathcal{U}-A\ (i<\lambda)$ such that $tp_{bs}(a_i, A, R)$ are pairwise distinct and w.l.o.g. $|\mathcal{U}-\{a_i : i<\lambda\}| = |\mathcal{U}|$. Let $h = \{\langle b_i, c_i\rangle : i<\lambda\}$, w.l.o.g. $b_i, c_i \notin A$ and we can find F_1, F_2 permutation of $\mathcal{U}$ which are the identity on A such that $F_1(a_i) = b_i$, $F_2(a_i) = c_i$. (they exist - see Def. 1.1(3).) Let $R_1 = F_1(R)$, and $R_2 = F_2(R)$ and define the monadic relations $P_0 = A$, $P_1 = \{b_i : i<\lambda\}$, $P_2 = \{c_i : i<\lambda\}$ (all of power $\le\lambda_0(R)$) . Let $\varphi(x, y, P_0, P_1, P_2, R_1, R_2)$ "say" that for every atomic $\varphi(x, \bar{z}, R) \in L(R)$ and $\bar{z} \in P_0$, $\varphi(x, \bar{z}, R_1) \equiv \varphi(y, \bar{z}, R_2)$ and $P_1(x), P_2(y)$.

3.4 Lemma : There is a set A such that

1) $|A| \le 5(n(R)+2)n(R)\lambda_1(R)$, furthermore, if the sup is not obtained in the definition of λ_1 then $|A| < \lambda_1$.

2) Let E_A be the equivalence relation:

$tp_{bs}(a, A, R) = tp_{bs}(b, A, R)$

If $\bar{b} \cong_\phi \bar{c}$ and $b_i E_A c_i$ for all $i < l(\bar{b})$ then $R(\bar{b}) \equiv R(\bar{c})$.

Proof : We define by induction on $l \le n(R)+2$ sets A_l such that $[m<l \to A_m \subseteq A_l]$, $|A_l| \le 5ln(R)\lambda_1(R)$, and if the sup is not obtained, $|A_l| < \lambda_1(R)$; we shall show that $A_{n(R)+2}$ satisfies the requirements of the lemma.

Let $A_0 = \phi$

If A_l is given, we define by induction on i a_i^l, A_i^l such that

1) $A_0^l = A^l$

2) A_i^l is increasing, continuous(in i).

3) $a_i^l \notin A_j^l$ for any i, j.

4) $|A_{i+1}^l - A_i^l| \le 2(n(R)-1)$

5) If $\alpha, \beta < i$, $\alpha \ne \beta$, then $tp_{bs}(a_\alpha^l, A_i^l, R) \ne tp_{bs}(a_\beta^l, A_i^l, R)$ hence $a_\alpha^l \ne a_\beta^l$.

For some $i = i(l)$ (which is necessarily $<\lambda_1(R)^+$) we cannot continue, i.e. A_i^l is defined but not a_i^l, A_{i+1}^l. Define $A'_{l+1} \overset{def}{=} A_l \cup \bigcup_{i<i(l)} A_i^l$,

$A_{l+1} \overset{def}{=} A'_{l+1} \cup \{a_i^l : i < i(l)\} \cup \{b$: the basic type realized over $A_l \cup \bigcup_{i<i(l)} A_i^l \cup \{a_i^l : i < i(l)\}$ by b is realized by $\le 3n(R)$ elements $\}$.

So $|A'_{l+1}| \le |A_l| + 2(n(R)-1)|i(l)| + |i(l)| = |A_l| + 2n(R)\lambda_1(R)$,
and $|A'_{l+1}| \le |A_l| + 2n(R)\lambda_1(R) + 3n(R)\lambda_1(R) \le 5n(R)(l+1)\lambda_1(R)$ if the sup in Definition 3.1 is not obtaine the inequality is strict. We prove $A = A_{n(R)+2}$ satisfies the requirements of the lemma. It is easy to see $|A|$ is as required (in demand (1) of 3.4).

Suppose $R(\bar b)$, $\neg R(\bar c)$ and $b_m E_A c_m$ for $m < n(R)$ and $\bar b \cong_A \bar c$. There are at most $n(R)$ l's such that $\bar b \cap A_{l+1} \ne \bar b \cap A_l$ so we choose l such that $\bar b \cap A_{l+1} \subseteq A_l$ and hence $\bar c \cap A_{l+1} \subseteq A_l$. Hence we may for simplicity assume:

$\bar c \cap A_{l+1} = \bar b \cap A_l = \phi$ and $\bar b$, $\bar c$ are without repetitions.

Let $B = A_l \cup \bigcup_{i<i(l)} A_i^l$ so $\bigwedge_{m<n(R)} b_m E_B c_m$ and $|b_m / E_B| \ge 3n(R)$. Now we can define $\bar d_k$ $(k=0, \ldots, n(R))$, each of length $n(R)$, $\bar b = \bar d_0$, $\bar c = \bar d_{n(R)}$, $\bar d_k$ with no repetitions, $\bigwedge_{m<n(R)} b_m E_B d_{k,m}$ and $|\{m : d_{k,m} \ne d_{k+1,m}\}| \le 1$. So, as in proof of the monadic case, we may assume $R(\bar b) \wedge \neg R(\bar c)$, $\bar b, \bar c$ without repetitions, $\bar b = \langle e \rangle {}^\frown \bar a$, $\bar c = \langle f \rangle {}^\frown \bar a$.

Notice there is $j < i(l)$ such that e, a_j^l realize the same basic type over $\bigcup_{i \le i(l)} A_i^l$ (as, if not, we could let $A_{i(l)+1}^l = A_{i(l)}^l$ and $a_{i(l)}^l = e$.) w.l.o.g. assume $R(a_j^l, \bar a)$. (otherwise use $\neg R$) and $R[e_j^l, \bar a]$, (otherwise interchange e and f).

3.4 A Claim : We can let $A_{i(l)+1}^l = A_{i(i)}^l \cup \bar a$, $a_{i(l)}^l = f$ and hence get a contradiction to the definition of $i(l)$.

Proof : Suppose $tp_{bs}(f, A_{i(l)+1}^l) = tp_{bs}(a_i^l, A_{i(l)+1}^l)$

if $i \neq j$ $tp_{bs}(f, A^l_{i(l)+1}) \supset tp_{bs}(f, A^l_{i(l)}) = tp_{bs}(e, A^l_{i(l)}) = tp_{bs}(a^l_j, A_{i(l)}) \neq tp_{bs}(a^l_i, A^l_{i(l)})$ contr.

If $i=j$ use $R(x,\bar{a})$.

So we have proved 3.4A, hence 3.4.

3.4B Claim : The sup is obtained in the definition of $\lambda_1(R)$

Proof : Suppose not, by the lemma, we can find A such that $|A|<\lambda_1(R)$ and $(\forall \bar{b},\bar{c})[\bigwedge_{i<n(R)} b_i E_A c_i \wedge \bar{b} \cong_\phi \bar{c} \rightarrow R(\bar{b}) \equiv R(\bar{c})]$. Clearly $\{a/E_A : a \in \mathcal{U}-A\}$ has power $<\lambda$, then for all B

$|\{tp_{bs}(a,B): a \in \mathcal{U}-B\}| \leq |A| + |\{tp_{bs}(a, B \cup A): a \in \mathcal{U}-B\}| < \lambda_1$, contradiction.

3.5 Conclusion : $\{\exists_R, Q^{mon}_{\lambda_0}\}$ is bi-interpretable with $\{Q^{mon}_{\lambda_0}, Q^{1-1}_{\lambda_1}, \exists_{R_1}, \exists_E\}$, where $|\text{Dom } R_1| \leq 5(n(R)+2)^2 \lambda_1(R)$, E an equivalence relation. This is done uniformly (i.e., the formulas depend on $n(R)$ only).

Proof : We've shown $Q^{1-1}_{\lambda_1(R)} \leq_{int} \{\exists_R, Q^{mon}_{\lambda_0(R)}\}$ (see 3.3). Let $A^1 \cap A = \phi$, $|A^1| = |A| = \lambda_1(R)$, A as in the lemma 3.4, $R_1 = R \restriction (A \cup A^1)$, $A \cup A^1$ includes $\geq Min\{3n(R)\}, |a/E_A|\}$ elements of each E_A equivalence class a/E_A. Now

$$R(x_1, \ldots, x_{n(R)}) \text{ iff}$$
$$(\exists y_1) \cdots (\exists_{n(R)})(\bigwedge_{1 \leq i \leq n(R)} x_i E_A y_i \wedge R_1(\bar{y}))$$

So $\exists_R \leq_{int} \{Q^{1-1}_{\lambda_1}, Q^{mon}_{\lambda_0}, \exists_{R_1}, \exists_{E_A}\}$

Now $\exists_{R_1} \leq_{int} \{\exists_R, Q^{mon}_{\lambda_0}\}$ by the definition of R_1, $\exists_{E_A} \leq_{int} \{\exists_R, Q^{mon}_{\lambda_0}\}$ directly, and $Q^{1-1}_{\lambda_1} \leq \{\exists_R, Q^{mon}_{\lambda_0}\}$ by 3.3, 3.4B. So $\{Q^{1-1}_{\lambda_1}, Q^{mon}_{\lambda_0}, \exists_{R_1}, \exists_{E_A}\} \leq_{int} \{\exists_R, Q^{mon}_{\lambda_0}\}$ and we finish.

3.5A Remark: Note the $Q^{1-1}_{|\text{Dom } R_1|}$ is uniformly interpretable (for fixed $n(R)$) in $Q^{1-1}_{\lambda_1}$ including the case λ_1 is finite, so 3.5 holds for it too.

§4 Above the local stability cardinal

We continue our analysis of $\exists_R$. For notational simplicity we make

4.1 Hypothesis : $|\text{Dom } R| = \lambda_1(R)$ (or, when $\lambda_1(R)$ is finite, $|\text{Dom } R| \leq 5(n(R)+2)\lambda_1(R)$. (and see 3.5A).

Also in this section (as well as in § 5 , § 6) we shall not prove the theorems "uniformly". This can be done, however we feel it will obscure the understanding by making us to deal with too many parameters. We also delay the treatment of the finite cases.

4.2 Definition : An m-type p is called $(\geq\lambda)$ -big if it is realized by λ pairwise disjoint sequences; let λ-big mean ($\geq\lambda^+$)-big. For this section big means $\lambda_2 = \lambda_2(R)$-big (λ_2 is defined below).

Let $(\exists^{\geq\lambda}\bar{x})\varphi(\bar{x},\bar{b})$ mean $\{\varphi(\bar{x},\bar{a})\}$ is $(\geq\lambda)$-big. We define $(\exists^{<\lambda}\bar{x})$, $(\exists^{\leq\lambda}\bar{x})$ similarly [as $\neg(\exists^{\geq\lambda}\bar{x})$, $\neg(\exists^{>\lambda}\bar{x})$, respectively.] Let "small" mean just the negation of big.

4.3 Remark : 1) Since we have monadic relations predicates and 1−1 permutations of power $|\mathrm{Dom}\ R|$ available, we can use one R (copies can be achieved easily).

2) Also, we can code any set of pairwise disjoint n-tuples, or any set of n -tuples forming a Δ-system (of power $\leq\lambda_1(R)$).

4.4 Definition : 1) M is an **admissible model** if it is an expansion of $(\mathcal{U},R)$ by countably many monadics relations and permutations of power $\leq\lambda_1(R)$.

4.5 Definition : $\lambda_2=\lambda_2(R)=$ least λ such that:

1) If M is admissible, Δ is a (finite) set of formulas, $A\subseteq M$, $|A|\leq\lambda$ and $m<\omega$. then $|S^m_\Delta(A,M)|\leq\lambda$.

2) $Q^{eq}_{\lambda,\lambda}\not\leq_{int}\exists_R$

4.6 Remark : the case $\lambda_2=\lambda_1$ is uninteresting as we want to prove now that it suffices to analyze R^*, $|\mathrm{Dom}\ R^*|=\lambda_2$, i.e. for some such R^* and some equivalence relation E

$\{Q^{mon}_{\lambda_0},Q^{1-1}_{\lambda_1},\exists_E,\exists_{R^*}\}\equiv_{int}\exists_R$ for some R^*, $|\mathrm{Dom}\ R^*|=\lambda_2(R)$ So we assume $\lambda_2<|\mathcal{U}|$. (but this is not essential).

4.7 Lemma : If M is admissible then there is A^*, $|A^*|\leq\lambda_2$ such that:

a) for any $\bar{a}$, $\bar{a}\cap A^*=\phi$ and finite Δ the type $q=tp_\Delta(\bar{a},A^*)$ is big.

b) For any such q, q is minimal; i.e., there is no $\varphi(\bar{x},\bar{y})\in\Delta$ and $\bar{b}$ such that both $q(\bar{x})\cup\{\pm\varphi(\bar{x},\bar{b})\}$ are big.

c) For any Δ,m, the number of such q's is $<\lambda_2$.

Remark : From c) we shall use only the "$\leq\lambda_2$".

Proof : We define, by induction on $i<\lambda_2$, $A_i\subseteq\mathcal{U}$, $|A_i|\leq\lambda_2$, A_i increasing, continuous, such that for all finite Δ :

0) for every $\bar{a}\in\mathcal{U}$ and $i<\lambda_2$, some $\bar{a}'\subseteq A_{i+1}$ realizes $tp_\Delta(\bar{a},A_i)$.

1) If $q_i=tp_\Delta(\bar{a},A_i)$ is not minimal, then some $\varphi_{q_i}(\bar{x},\bar{b}_{q_i})$ witnesses it for some $\varphi_{q_i}\in\Delta$, $\bar{b}_{q_i}\subseteq A_{i+1}$.

2) If $q_i = tp_\Delta(\bar{a},A_i)$ is not big, then for some $B_{\bar{a}} \subset A_{i+1}$, no sequence $\bar{b}$ realizing it is disjoint to B.

Let $A = A_{\lambda_2} \stackrel{def}{=} \bigcup\{A_i : i < \lambda_2\}$. Now 4.7 will follow from 4.8, 4.9.

4.8 Claim : If $A_{\lambda_2} \cap \bar{a} = \phi$ then for any (finite) Δ for some $i<\lambda_2$, $q_i = tp_\Delta(\bar{a},A_i)$ is minimal and big.

Proof : Clearly q_i is big (for every $i < \lambda_2$, by (2)). If q_i is not minimal, take $\varphi_{q_i} \in \Delta$, $\bar{b}_{q_i} \subseteq A_{i+1}$ witnessing this (by (1)). W.l.o.g., $\varphi_{q_i}(\bar{x},\bar{b}_{q_i}) \in q_{\lambda_2}$ and $q_i \cup \{\neg\varphi_{q_i}(\bar{x},\bar{b}_{q_i})\}$ is realized by the sequences $\langle \bar{a}_{i,\xi} : \xi < \lambda_2^+ \rangle$ which are pairwise disjoint.

w.l.o.g. $tp_\Delta(\bar{a}_{i,\xi},A_{\lambda_2})$ does not depend on ξ, and call it r_i; clearly r_i is λ_2-big.

Also, $r_i \neq r_j$ for $i<j$, since $\varphi_{q_i}(\bar{x},\bar{b}_{q_i}) \in q \restriction A_{i+1} \subseteq q \restriction A_j = q_j \subseteq r_j$ but $\neg\varphi_{q_i}(x,\bar{b}_{q_i})$ is satisfied by $\bar{a}_{i,\xi}$. W.l.o.g. $\bar{a}_{i,\xi}, \bar{a}_{j,\zeta}$ are disjoint when $\langle i,\xi \rangle \neq \langle j,\zeta \rangle$.

Now we can interpret $Q^{eq}_{\lambda_2,\lambda_2}$: we add a predicate A_{λ_2} and let xEy iff x codes $\bar{x}$, y codes $\bar{y}$ (remember 4.3(2)), and $\bar{x}$ and $\bar{y}$ realize the same Δ-type over A_{λ_2}. E has $\geq\lambda_2$ equivalence classes of power $\geq\lambda_2$, a contradiction.

4.9 claim : A_{λ_2} satisfies a), b) and c) of the lemma (4.7).

Proof : Let $q = tp_\Delta(\bar{a},A^*)$, $\bar{a} \cap A^* = \phi$. We know that for some $i<\lambda_2$ $q_i = q \restriction A_i$ is big and minimal, hence is realized by pairwise disjoint $\bar{c}_\xi$ $\xi<\lambda_2^+$.

For every $\varphi(\bar{x},\bar{b}) \in q$, $q_i \cup \{\varphi(\bar{x},\bar{b})\}$ is big [as $q_i \cup \{\varphi(\bar{x},\bar{b})\} \in q_j$, for some large enough $j<\lambda_2$], hence $q_i \cup \{\neg\varphi(\bar{x},\bar{b})\}$ is not big. There are $\leq\lambda_2$ such $\varphi(\bar{x},\bar{b})$, so omitting any tuple realizing any of them from our sequence $\langle \bar{c}_\xi : \xi<\lambda_2^+ \rangle$ still leaves λ_2^+ many, so each realizes q hence q is big.

If q is not minimal, then q_i is not minimal, contradiction. If 4.7(c) fails, we can interpret $Q^{eq}_{\lambda_2,\lambda_2}$ by taking Δ witnessing the fact that c) fails and defining E as before. This finishes lemma 4.7.

4.10 The Symmetry Lemma : There are no $\varphi(\bar{x},\bar{y},\bar{z})$, $\bar{d},\bar{a}_\alpha,\bar{b}_{\alpha,\beta}(\alpha,\beta<\lambda_2^+)$ such that:

1) for every $\alpha,\beta < \lambda_2^+$, $\models\varphi(\bar{b}_{\alpha,\beta},\bar{a}_\alpha,\bar{d})$ and these three sequences are disjoint.

2) For fixed α, $\bar{b}_{\alpha,\beta}(\beta<\lambda_2^+)$ are disjoint.

3) The $\bar{a}_\alpha$'s are disjoint.

4) $\varphi(\bar{b}_{\alpha,\beta},\bar{x},\bar{d})$ is not big.

Proof : We can throw away many $\bar{a}_\alpha$'s , $\bar{b}_{\alpha\beta}$'s, as long as their number remains so w.l.o.g. all the sequence $\bar{b}_{\alpha,\beta}(\alpha,\beta<\lambda_2^+)$ are pairwise disjoint. Let $\Delta=\{\varphi\}$.

We may assume that for each α, all $\bar{b}_{\alpha,\beta}$ realize the same Δ-type over $\{\bar{a}_\gamma:\gamma\le\alpha\}\cup\bar{d}$ (use part (1) of Def. 4.5 to thin the set $\{\bar{b}_{\alpha,\beta}:\beta<\lambda_2^+\}$). Similarly, we may assume that $\bar{a}_{\alpha_1}, \bar{a}_{\alpha_2}$ realize the same Δ-type over $\{\bar{a}_\gamma:\gamma<\alpha_1\cap\alpha_2\}\cup\{\bar{b}_{\gamma,i}:\gamma<\alpha_1\cap\alpha_2,\ i<\lambda_2\}\cup\bar{d}$ (use 4.7).

What is the truth value of $\varphi(\bar{b}_{\alpha,\beta},\bar{a}_\gamma,\bar{d})$ for $\alpha,\beta,\gamma<\lambda_2$?

True, if $\gamma=\alpha$.

False when $\gamma>\alpha$ (note that we have assumed $\alpha,\beta,\gamma<\lambda_2$ and not $\alpha,\beta,\gamma<\lambda_2^+$; note that $\varphi(\bar{b}_{\alpha,\beta},\bar{x},\bar{d})$ is not big, so only few $\bar{a}_i$ realize it, so no $\bar{a}_i$ realizes it for $i>\alpha$ (as then all such $\bar{a}_i$'s realize it).)

If $\gamma<\alpha$, the answer does not depend on β.

Let a_α code $\bar{a}_\alpha$, $b_{\alpha,\beta}$ code $\bar{b}_{\alpha,\beta}$. For notational simplicity we ignore the coding.

Let $P=\{a_\alpha:\alpha<\lambda_2\}$.

Let xEy iff $(\forall z\in P)(\varphi(x,z,\bar{d})\equiv\varphi(y,z,\bar{d}))$.

4.11 Fact : For $\alpha_1,\alpha_2,\beta_1,\beta_2<\lambda_2$, $b_{\alpha_1,\beta_1}E\ b_{\alpha_2,\beta_2}$ iff $\alpha_1=\alpha_2$.

Proof : We have just shown $(\Leftarrow)$.

Conversely, say $\alpha_1<\alpha_2$

$\varphi(b_{\alpha_1,\beta_1},a_{\alpha_2},\bar{d})$ is false but $\varphi(b_{\alpha_2,\beta_2},a_{\alpha_2},\bar{d})$ is true, so $(\Longrightarrow)$ is clear.

So we have interpreted $Q^{eq}_{\lambda_2,\lambda_2}$, contradiction, hence we have proven 4.10.

4.12 Lemma : For any admissible M , and any $\varphi(x,\bar{y})$, there is an admissible expansion M^* of M, and $\psi(\bar{y})$ such that $M^*\models(\exists^{\le\lambda_2}x)\varphi(x,\bar{y})\equiv\psi(\bar{y})$.

Proof : We define $A_i\subseteq M$ $i<\lambda_2^+$ increasing, continuous, $|A_i|\le\lambda_2$. Take A_0 to witness lemma 4.7.

A_{i+1} realizes all Δ-types over A_i for all finite Δ, and $\langle A_{i+1};A_0,\ldots,A_i\rangle$ is an elementary substructure of $\langle M;A_0,\ldots,A_i\rangle$ even allowing the quantifier $\exists^{\le\lambda_2}$.

Let E be the equivalence relation on $\mathcal{U}-A_0$: x_1Ex_2 iff $(\forall\bar{y}\subseteq A_0)(\varphi(x,\bar{y})\equiv\varphi(x_2,\bar{y}))$.

Clearly, every E- equivalence class is represented in each $A_{i+1}-A_i$.

We say that (i,j) is a **good pair** if $i<j$ and for any $\bar{a}$ such that $\bar{a}\cap(A_j-A_i)=\phi$, and $c\in A_j-A_i$, $\varphi(c,\bar{a})\equiv(\exists^{>\lambda_2}x)(xEc\wedge\varphi(x,\bar{a}))$.

4.13 Claim : If there are $i_0<j_0<i_1<j_1<\cdots<i_n<j_n$, $n>l(\bar{y})$, (i_l,j_l) good, then the lemma holds with $M^*=(M,A_{i_0},A_{j_0},\ldots,A_{i_n},A_{j_n})$ and $\psi(\bar{y})=\bigwedge_{l=0}^{n}$ [if $\bar{y}$ is disjoint from $A_{j_l}-A_{i_l}$ then there is no $c \in A_{j_l}-A_{i_l}$ such that $\varphi(c,\bar{y})$.]

Proof : Suppose $M^* \models \psi(\bar{y})$. Then for some l, $\bar{y}$ is disjoint to $A_{j_l}-A_{i_l}$, hence $M^* \models$ "there is no $c \in A_{j_l}-A_{i_l}$ such that $\varphi(c,\bar{y})$".

By the definition of a good pair, and as every E-equivalence class is represented in $A_{j_l}-A_{i_l}$, and there are $\le\lambda_2$ E-equivalence classes, clearly $M^* \models (\exists^{\le\lambda_2} x)\varphi(x,\bar{y})$.

For the converse, suppose $M^* \models (\exists^{\le\lambda_2} x)\varphi(x,\bar{y})$, and suppose $\bar{y}$ is disjoint to $A_{j_l}-A_{i_l}$ but $(\exists c \in A_{j_l}-A_{i_l})\varphi(c,\bar{y})$. This contradicts the definition of a good pair. So we have proved 4.13.

Now we assume there are few good pairs (i,j) i.e. there are no i_m, j_m as in 4.13 and get a contradiction, thus finishing the proof of 4.12.

For a club set $C \subset \lambda_2^+$, the following holds:

(*) $\delta \in C, i<\delta$ implies $\delta > sup\{j:(i,j)$ is a good pair $\}$ if the sup is $<\lambda$.

By the choice of the A_i's (and see [Sh4], beginning of §2 (or guarantee this in the A_i's definition) also e.g. 4.15 is a repetition of this):

(**) if $\langle c \rangle ^\frown \bar{b}_1 \subseteq A_\delta$, $\bar{b}_2 \cap A_\delta = \phi$, $M^* \models \varphi_1(c,\bar{b}_1{}^\frown\bar{b}_2)$ but $M^* \models (\exists^{\le\lambda_2} x)(xEc \wedge \varphi_1(x,\bar{b}_1{}^\frown\bar{b}_2))$, φ_1 is gotten from φ by permuting the variables **then** for every $\beta>\delta$ (but $\beta<\lambda_2^+$) there is such a $\bar{b}_2$ with $\bar{b}_2 \cap A_\beta = \phi$.

Let K be the set of $\langle c \rangle ^\frown \bar{b}_1$ such that $\varphi(c,\bar{b},\bar{z})$ is big (when $\langle c \rangle ^\frown \bar{b}_1 \in \bigcup A_i$ this is equivalent to: for arbitrarily large β there is $\bar{b}_2$ as in the antecedent (above), $\bar{b}_2 \cap A_\beta = \phi$.)

So again by the A_i's choice, if $\delta \in C$, $\bar{b}_1 \subseteq A_\delta$, $c \notin A_\delta$, $c \in \bigcup_{i<\lambda_2^+} A_i =_{df} A_{\lambda_2^+}$, $\langle c \rangle ^\frown \bar{b}_1 \in K$ then $(\forall \beta<\lambda_2^+)(\exists c^1 \in A_{\lambda_2^+})(\langle c^1 \rangle ^\frown \bar{b}_1 \in K \wedge c^1 \notin A_\beta)$. (This is by a similar hand-over-hand construction.)

Now if $\delta_1<\delta_2 \in C$, (δ_1,δ_2) not good, we can contradict lemma 4.7, 4.10.

4.14 Lemma : For M^* rich enough, for every $\varphi(x_1,\ldots,x_{n+1})$ there are $\vartheta_{i,j}$ $(i \le n,\ j<k)$ such that:

1) If $(\exists^{\le\lambda_2} y)\varphi(x_1,\ldots,x_n,y) \wedge \varphi(x_1,\ldots,x_n,x_{n+1})$, then $\bigvee_{i=1}^{n}\bigvee_{j=1}^{k}\vartheta_{i,j}(x_{n+1},x_i)$.

2) $\forall x \exists^{\le\lambda_2} y\, \vartheta_{i,j}(y,x)$.

Proof: : It suffices to find $\vartheta_{i,j}$ such that $\exists^{\le\lambda_2} y\, \varphi(x_1,\ldots,x_n,y) \wedge \varphi(x_1,\ldots,x_{n+1}) \rightarrow \bigvee_{i,j} [\vartheta_{i,j}(x_{n+1},x_i) \wedge \exists^{\le\lambda_2} y\, \vartheta_{i,j}(y,x_i)]$, as then the formulas $\vartheta_{i,j}(x_{n+1},x_i) \wedge \exists^{\le\lambda_2} y\, \vartheta_{i,j}(y,x_i)$ witness the lemma (using 4.12).

We prove by induction on n.

For $n=0,1$ trivial.

Assume for n, and we shall prove for $n+1$. We assume M^* is rich enough to contain the unary predicate A^* as in lemma 4.7 and the formulas ψ as in lemma 4.12. We shall define $n^{**}=4$ and (latter) a sequence of finite sets of formulas

Δ_i ($i \le n^{**}$), $\Delta_i \subseteq \Delta_{i+1}$, $\varphi \in \Delta_0$.

Remark : The $n^{**}=4$ is somewhat misleading: in a sense it is large compared to $n(R)$ but this is absorbed by some w.l.o.g. below. What is the point in having those Δ_i? Lemma 4.10 gives us a kind of symmetry (if a depends on b then b depends on a). But this is not true if we restrict ourselves to dependency witnessed by a formula from a finite Δ_i. but if we have long enough increasing sequence of Δ_i for some i, Δ_i-dependency is equivalent to Δ_{i+1}-dependency (for those sequences).

So suppose $\varphi(a_1, \ldots, a_{n+1}, c) \wedge \exists^{\le\lambda_2} x\, \varphi(a_1, \ldots, a_{n+1}, x)$. We want to prove that some $\vartheta \in \Delta_{n^{**}}$ satisifies $\vartheta(c, a_i) \wedge \exists^{\le\lambda_2} x\, \vartheta(x, a_i)$, for some i . W.l.o.g., there are no repetitions in $\langle a_1, \ldots, a_{n+1}, c \rangle$. (If $a_i = a_j$, use induction hypothesis on n; if $c = a_j$, we are done because we could have chosen to have $x=y \in \Delta_0$). W.l.o.g., no a_i satisfies any $\vartheta(x) \in \Delta_{n^{**}}$ such that $\exists^{\le\lambda_2} x\, \vartheta(x)$

Similarly for next observation, as then we use the induction hypothesis with the formula $\exists x [\varphi(a_1, \ldots, a_{i-1}, x, a_{i+1}, \ldots, a_{n+1}, c) \wedge$

$$(\exists^{\le\lambda_2} y) \qquad \varphi(a_1, \ldots, a_{i-1}, x, a_{i+1}, \ldots, a_{n+1}, y)$$
$$\wedge\, \vartheta(a_1, \ldots, a_{i-1}, x, a_{i+1}, \ldots, a_{n+1}) \wedge (\exists^{\le\lambda_1} z) \vartheta(a_1, \ldots, a_{i-1}, z, a_{i+1}, \ldots, a_{n+1})]:$$

W.l.o.g., for no $\vartheta \in \Delta_{n^{**}}$

$$\vartheta(a_1, \ldots, a_{i-1}, a_i, a_{i+1}, \ldots, a_{n+1}) \wedge \exists^{\le\lambda_2} x\, \vartheta(a_1, \ldots, a_{i-1}, x, a_{i+1}, \ldots, a_{n+1}).$$

Let $\bar a = \langle a_1, \ldots, a_n \rangle$ $\psi_i = \bigwedge \{\chi(\bar x, y, z) : \chi \in \Delta_i, \models \chi[\bar a, a_{n+1}, c]\}$.

We know that $\exists^{\le\lambda_2} x\, \psi_i(\bar a, a_{n+1}, x) \wedge \psi_i(\bar a, a_{n+1}, c)$ for $i \le n^{**}$; we say a_{n+1} i-depends on $\bar a$ if $(\exists^{\le\lambda_2} y)(\exists x)(\psi_i(\bar a, y, x)) \wedge \exists x\, \psi_i(\bar a, a_{n+1}, x))$.

We can assume that for $i < n^{**}$, a_{n+1} does not i-depend on $\bar a$ by putting $\psi_i(\bar a, y, x)$ into Δ_{i+1}. Similarly, $[i < n^{**} \Longrightarrow c$ does not i-depend on $\bar a]$.

Now by 4.10 (and the assumption, as Δ_1 is large enough and the uniformity of

4.10):

(*) for some $\vartheta^* \in \Delta_1$, $M^* \models \vartheta^*[a_1, \ldots, a_{n+1}, c]$ and $M^* \models (\exists^{\le\lambda_2}\langle x_1, \ldots, x_{n+1}\rangle)\vartheta^*(x_1, \ldots, x_{n+1}, c)$.

If $M^* \models (\exists^{>\lambda_2}\bar{x})\psi_1(\bar{x}, a_{n+1}, c)$ then by (*) the formula $\vartheta^a(y,z) = (\exists^{>\lambda_2}\bar{x})\psi_1(\bar{x}, y, z) \in \Delta_2$ necessarily satisfied $M^* \models \vartheta^a(a_{n+1}, c)$ and $M^* \models (\exists^{\le\lambda_2}y)\vartheta^a(y,c)$ hence for some $\vartheta \in \Delta_3, \vartheta(a_{n+1}, c) \wedge (\exists^{\le\lambda_2}z)\vartheta(a_{n+1}, z)$ contradiction. So assume $M^* \models \neg\vartheta^a(a_{n+1}, c)$. Also we can assume that $M^* \models (\exists^{>\lambda_2}\bar{x})(\exists y)\psi_2(\bar{x}, y, c)$ (otherwise use 4.10 and then the induction hypothesis on n) hence

$M^* \models (\exists^{>\lambda_2}\bar{x})(\exists y)[\psi_1(\bar{x}, y, c) \wedge \neg\vartheta^a(y,c)]$, hence there are pairwise disjoint $\bar{a}_\alpha(\alpha < \lambda_2^+)$ and elements b_α, such that $M^* \models \psi_1(\bar{a}_\alpha, b_\alpha, c) \wedge \neg\vartheta^a(b_\alpha, c)$. If there are λ_2^+ distinct b_α's, we easily contradict (*); so w.l.o.g. $b_\alpha = b$ for every α. But then $M^* \models \psi_1(\bar{a}_\alpha, b, c)$,$(\alpha < \lambda_2^+)$ implies $M^* \models (\exists^{>\lambda_2}\bar{x})\psi_1(\bar{x}, b, c)$ contradicting $M^* \models \neg\vartheta^a(b,c)$.

This proves lemma 4.14.

* * *

4.15 Lemma : For any $\varphi(\bar{x}, \bar{y})$ there are $\vartheta_i(z, \bar{x})$ such that:

1) If $\exists^{\le\lambda_2}\bar{y}\,\varphi(\bar{a}, \bar{y}) \wedge \varphi(\bar{a}, \bar{b})$ then $\bigvee_i \vartheta_j(b_j, \bar{a})$

2) $\exists^{\le\lambda_2}z\,\vartheta_j(z, \bar{a})$ for every $\bar{a}$

Proof : By induction on the length of $\bar{y}$ and of $\bar{x}$.

Instead of one ϑ we can produce a finite set. We shall define $\Delta_i(i < n^{**})$ be finite, increasing. $\varphi(\bar{x}, \bar{y}) \in \Delta_0$

Assume $\models \varphi[\bar{a}, \bar{b}]$, $\varphi(\bar{a}, \bar{y})$ small.

We can make similar assumptions as in the proof of the previous lemma and define ψ_i similarly.

Let $\bar{b} = \bar{c}^\frown\langle d\rangle$

By the induction hypothesis, for $i < n^{**}$ there are pairwise disjoint $\bar{c}^\alpha(\alpha < \lambda_2^+)$ such that $\exists z\,\psi_i(\bar{a}, \bar{c}^\alpha, z)$.

Say $\psi_i(\bar{a}, \bar{c}^\alpha, d^\alpha)$

if there are λ_2^+ distinct d^α's, we get a contradiction because $(\exists^{\le\lambda_2}\bar{y})\varphi(\bar{a}, \bar{y})$. So w.l.o.g. $d^\alpha = d^0$ all $\alpha < \lambda_2^+$

So $(\exists^{\ge\lambda_2^+}\bar{w})\psi_{i-1}(\bar{a}, \bar{w}, d)$ is one of the conjuncts of ψ_i

If $(\exists^{\geq\lambda_2^+} \bar{w})\psi_i(\bar{a},\bar{w},x)$ is not small we first define distinct d_i $(i<\lambda_2^+)$ such that $(\exists^{\geq\lambda_2^+}\bar{w})\psi_i(\bar{a},\bar{w},d_i)$, so $(\exists^{\geq\lambda_2^+}\bar{w})\psi_{i-1}(\bar{a},\bar{w},d_i)$ then define $\bar{c}^{i,\alpha},d^i$ pairwise disjoint for $\alpha<\lambda_2^+$ such that $\psi_{i-2}(\bar{a},\bar{c}^{i,\alpha},d_i)$. This shows $(\exists^{>\lambda_2}\bar{y})\psi_{i-1}(\bar{a},\bar{y})$. Contradiction.

If $(\exists^{\geq\lambda_2^+}\bar{w})\psi_i(\bar{a},\bar{w},x)$ is small, we get the desired conclusion.

4.16 Lemma : Every formula is equivalent to a Boolean combination of formulas of the form:

$\bigwedge_{i=1}^{n}\vartheta_i(y_i,y_0) \wedge \psi(y_0,\ldots,y_n,F_1(y_0,\ldots,y_n),\ldots,F_k(y_0,\ldots,y_n))$ such that :

$\forall y\exists^{\leq\lambda_2}z\,\vartheta_i(z,y) \wedge \forall z\exists^{\leq\lambda_2}y\,\vartheta_i(z,y)$ and for some ϑ^1 we have

$(\forall x_0,\cdots)\vartheta^1(F_j(x_0,\cdots)) \wedge (\exists^{\leq\lambda_2}x)\vartheta^1(x)$ and the F_i's are definable functions.

Proof : Let $\varphi(x_1,\ldots,x_n)$, $\langle a_1,\ldots,a_n\rangle$ be given. We define $n^{**}<\omega$, a sequence, increasing, of finite sets of formulas $\Delta_i(i\leq n^{**})$. Let

$I_i=\{l : a_l$ realizes a non-big formula in $\Delta_i\}$.

$T_i=\{\langle l,m\rangle:\vartheta(a_l,a_m)\frown\exists^{\leq\lambda_2}x\,\vartheta(x,a_m)$, for some $\vartheta\in\Delta_i\}$.

n^{**} is chosen big enough so that for some $i<j-8-2n(R)$, $j<n^{**}$, $I_i=I_j$, $T_i=T_j$. Note that T_i is an equivalence relation on $\{l:1\leq l\leq n, l\notin I_i\}$ when the appropriate ϑ's from the conclusion of lemma 4.10 and 4.12 are included in Δ_i $l+1$ for each l;

Since $T_i=T_j$, $j>i+8$, the necessary witnesses already appear in Δ_j. So T_j is an equivalence relation, as claimed.

Let $\bar{a}=\bar{a}_0\frown\bar{b}_1\frown\cdots\frown\bar{b}_m$ where $\bar{a}_0=\langle a_l:l\in I_i\rangle$ and such that a_l and a_{l^1} appear in the same $\bar{b}_j$ iff $a_l,a_{l^1}\in\bar{a}$ and $\langle l,l^1\rangle\in T_i$

We may assume that for each Δ_k there is a predicate A_k^* as in lemma 4.7and $A_k^*(x)\in\Delta_{k+1}$, so $|A_k^*|\leq\lambda_2$ and every complete Δ_k-type over A_k^* is minimal.

Also, using a few permutations, we have in some admissible expansion of M the predicates R_l^k such that R_l^k codes $\{\bar{b}_{l,\xi}^k:\xi<\xi_l<\lambda_2\}$, pairwise disjoint sequences of length $l=l(\bar{b}_l^k)$ such that R_l^k contains exactly one code for a sequence from each complete big type in $S_{\Delta_k}^l(A_k^*)$, and $\bar{b}_{l,\xi}^k\in A_{k+1}^*$ and it realizes a big Δ_k-type over $A_k^*\cup\bigcup_{\zeta<\xi}\bar{b}_{l,\zeta}^k$. So we may assume $R_l^k(x)\in\Delta_{k+1}$ $(l=1,\ldots,m)$. Similarly, we may assume for the functions F_l mapping sequences of the appropriate length which realize some big type in $S_{\Delta_k}(A_k^*)$ satisfying $\bigwedge_{i,j=1}^{n} \vartheta^*(b_i,b_j)$ [for $\vartheta^*(x,y)=\bigvee \{\vartheta(x,y)\in\Delta_i: \forall x \exists^{\leq\lambda_2} y\, \vartheta(x,y)$ and $\forall y \exists^{\leq\lambda_2} y\, \vartheta(x,y)\}$ which is in Δ_{k+1}] to the unique sequence in R_l^k realizing the same Δ_k- type over $A_k^*\}$, (i.e. $(F(\bar{x}_1) = y) \in \Delta_{k+1}$)

This proves lemma 4.16.

4.17 Theorem : Q_R is bi-interpretable with $\{Q_{\lambda_0}^{mon}, Q_{\lambda_1}^{1-1}, Q_E, Q_{R^*}\}$ with E an equivalence relation, and $|\mathrm{Dom}\ R^*|\leq\lambda_2$.

Proof : By what we already know, we may assume $|\mathrm{Dom}\ R|\leq\lambda_1$. We know $R(x_1, \ldots, x_{n(R)})$ is equivalent to some Boolean combination of formulas as in the statement of the previous lemma. There appear there formulas $\psi_l, \vartheta_l(x), \vartheta_l(x,y)$. Without loss of generality, $\forall x \exists^{\leq\lambda_2} y\, \vartheta_l(y,x) \wedge \forall y \exists^{\leq\lambda_2} x\, \vartheta_l(y,x)$. Let $B^0=\{x; \bigvee_l \vartheta_l(x)\}$. Let $\vartheta^1(y,x)=\bigvee_l [\vartheta_l(y,x) \vee \vartheta_l(x,y)] \vee x=y$ (so ϑ^1 is symmetric but not necessarily transitive.)

On $\mathcal{U}-B^0$ we have the equivalence relation E^0= the transitive closure of $\vartheta^1(y,x)$.

By our assumption, each equivalence class of E^0 has power $\leq\lambda_2$.

Let $B^1=B^0 \cup \{x:(\exists^{\leq\lambda_2} y)(|y/E^0|=|x/E^0|)\}$ and let $B^2=\{x\in\mathcal{U}-B^1$: $(\exists y)(yE^0x \wedge (\exists^{\geq\aleph_0} z)[\vartheta^1(z,x) \vee \vartheta^1(x,z)])\}$

$E^1=E^0 \restriction B^2$

We want to interpret E^1 and analyze $E^0 \restriction (\mathcal{U}-B^0 \cup B^1)$. Note that if we want to "express" our life will be much easier. For each equivalence class C of E^1 we do the following:

Case I : There is $b_C \in C$ such that $|\{x\in\mathcal{U}-B^0: \vartheta^1(x,b_C)\}|=|C|$.

Let $D_C=\{x\in C: x\neq b_C, \vartheta^1(x,b_C)\}$.

Case II : Not I , so $|C|$ is singular.

Choose a regular $\lambda_C<|C|$ in such a way that ($\forall\mu\leq\lambda_2, \mu$ singular) ($\forall\lambda<\mu, \lambda$ regular) [$|\{C:|C|=\mu, C$ an E^1-equivalence class $\}|$ $=|\{C:|C|=\mu$, $\lambda_C=\lambda$, C an E^1-equivalence class $\}|$.]

This is possible as

$\lambda_2<|\{C:|C|=\mu\}|$ (else $C\subseteq B^1$); and choose $b_C\in C$, $D_C=\{x\in C: x\neq b_C, \vartheta^1(x,b_C)\}$ such that $|D_C|\geq\lambda_C$.

Let $P=\{b_C: C$ an E^1 equivalence class $\}$

$Q = \bigcup \{D_C : C$ an E^1 equivalence class $\}$.
W.l.o.g., P and Q are predicates of M, as $|P| \le \lambda_0$, $|Q| \le \lambda_0$.
Let $yE^* z$ iff
$Q(y) \wedge Q(z) \wedge (\forall x)(P(x) \rightarrow \vartheta^1(y,x) \equiv \vartheta^1(z,x))$. The E^* equivalence classes are the sets D_C.

E^* will serve as the E mentioned in 4.17, so we have proved $Q_E \le_{int} Q_R$. Now we shall start to prove the other direction (we still have to define R^*).
We shall now interpret E^1.
Take some isomorphic copies of E^*, say E_0^*, E_1^*, such that for each E^1-equivalence class C satisfying $|C| = \mu$ is singular, E_0^* decomposes C into cf μ equivalence classes, each of power $< \mu$; and some E_1^* equivalence class includes exactly one element from each and is disjoint from all other C's, and E_0^*, E_1^* refine E_1.
For $|C|$ regular, C is an E_0^* equivalence class and an E_1^* equivalence class. So we have interpreted E^1.
(If $\lambda_1 = |\mathcal{U}|$, it may happen that such a choice of E_0^* and E_1^* is not possible, but then split $\mathcal{U}$ into two parts closed under E^1 and do this on each part.)
Let $\Delta = \{\psi_l, \vartheta_l(x), \vartheta_l(x,y) : l\}$.
Let $S = \bigcup_{k=1}^{n(R)} S_\Delta^k(B^1, M)$
For each $p \in S$, choose $\bar{x}_p$ to realize p.
Let $B_2 = B^1 \cup \bigcup_{p \in S} \bar{x}_p$.
Let $R^* = R \restriction B_2$.
Suppose $|C| = \mu$, C an E^1-equivalence class. Then E^1 has $\ge \lambda_2^+$ equivalence classes of power μ, else C would be contained in B^1. So we can use several copies of E^1 to code whatever we want on C(for all C's simultaneously). In particular, we can have elements of C code sequences from C. We can also interpret the equivalence relation $xEy \overset{def}{=}$"x *and* y code sequences realizing the same Δ-type over B^1."
Use another few copies of E^1 together with $1-1$ functions of power $\lambda_1 \ge |\text{Dom } R|$ to interpret the functions F_l (for coding F_l it is enough to have Rang (F_l) and $E_{F_l} : xE_{F_l}y$ iff $x = y$ $\vee (\exists z)[(y$ *code* $z) \wedge y \in B_2 \wedge (x = F_l(y) \vee y = F_l(x) \vee F_l(x) = F_l(y))]$.)

Def. 5.1. So λ_2^+ satisfies 5.3A below. So from 5.10, 5.12 it follows that $|A| \le \lambda_2^+ \Rightarrow |S_\Delta^m(A)| \le \lambda_2^+$. So λ_3^+ satisfies the demands of λ_2, hence $\lambda_2 \le \lambda_3^+$.

So there are only two possibilities: $\lambda_2=\lambda_3$ or $\lambda_2=\lambda_3^+$. For this section:

5.3 Hypothesis : $\lambda_2=\lambda_3^+$. Let $\lambda=\lambda_2$.

We shall eventually prove that $\{\exists_R, Q_\lambda^{mon}, Q_\lambda^{1-1}\}$ ($|\mathrm{Dom}\, R|=\lambda$) is bi-interpretable with Q_λ^{word} = { well orderings of A of order type λ: $|A|=\lambda\le|\mathcal{U}-A|$ }. Together with the preceding theorems, this completely analyzes the case $\lambda_2\ne\lambda_3$.

However we want to do this in a somewhat more general case, so for the rest of this section:

5.3A Hypothesis : λ is regular , $\lambda\le\lambda_2$ and for finite Δ, m and admissible M, if $A\subseteq\mathcal{U}, |A|<\lambda$, then $|S_\Delta^m(A,M)|<\lambda$ (hence as in 5.2's proof, $Q_{\lambda,\lambda}^1 \not\le_{int} Q_R$).

5.4 Definition : We say q is a pure extension of p (both are m-types) if $x_l=c\in q \Rightarrow x_l=c\in p$; we write $p\subseteq_{pr} q$. We call p pure if $\phi\subseteq_{pr} p$.

5.5 Definition : For every admissible M, $|A|<\lambda$, $p\in S_\Delta^m(A,M)$ we define rank $Rk(p)=\langle\alpha,\beta\rangle$ (α,β *may be* ∞) (really we should write Rk_Δ^m):

$Rk(p)\ge\langle 0,0\rangle$ if p is realized by some $\bar{a}$.

$Rk(p)\ge\langle\alpha,\gamma\rangle$ $(0<\gamma<\infty)$ if for every $\beta<\gamma$, and $A^1\supseteq A$ such that $|A^1|<\lambda$, p has an extension $q\in S_\Delta^m(A^1,M)$ such that $Rk(q)\ge\langle\alpha,\beta\rangle$ and if $\alpha>0$, q is a pure extension of p.

$Rk(p)\ge\langle\alpha,\infty\rangle$ if $Rk(p)\ge\langle\alpha,\gamma\rangle$ for every γ.

$Rk(p)\ge\langle\alpha,0\rangle$ when $\alpha>0$, if for every $\beta<\alpha$ there are $A^1\supseteq A$, $|A^1|<\lambda$, and $[q_1\supseteq_{pr} p$ and $q_2\supseteq_{pr} p]$ or $[\alpha=1, q_1\supseteq p, q_2\supseteq p]$ such that $q_1, q_2\in S_\Delta^m(A^1,M)$, $q_1\ne q_2$, $Rk(q_1)\ge\langle\beta,\infty\rangle$, $Rk(q_2)\ge\langle\beta,\infty\rangle$.

Now $Rk(p)=\langle\alpha,\beta\rangle$ iff $Rk(p)\ge\langle\alpha,\beta\rangle$ and $Rk(p)\not\ge\langle\alpha,\beta+1\rangle$.

$Rk(p)=\langle\alpha,\infty\rangle$ iff $Rk(p)\ge\langle\alpha,\beta\rangle$ all β, $Rk(p)\not\ge\langle\alpha+1,0\rangle$.

$Rk(p)=\langle\infty,\infty\rangle$ if $Rk(p)\ge\langle\alpha,\beta\rangle$ for all α,β.

5.6 Remark : We can show that if $Rk(p)=\langle\alpha,\beta\rangle$ then $\beta\in\{0,\infty\}$. Note that there is no connection between ranks for different Δ's.

5.7 Claim: : Fix Δ, m, p, q will be complete Δ–m–types.

0) $p\subseteq_{pr} q \Rightarrow Rk(p)\ge Rk(q)$

It is easy to interpret R.

The analysis is complete, getting the biinterpretability except that we have forgotten B^3=Dom $R-B^1 \cup B^2$. On B^3, E^0 may have countable equivalence classes but $(\forall x \in B^3)(\exists^{<\aleph_0} y)\vartheta^1(y,x)$. We shall deal with the new points only.

First we can define a partition of B^3 to $B_l^3(l=0,1,2,3)$ such that $\vartheta^1(x,y)$, $x \in B_l^3$ implies $y \in B_{l-1}^3 \cup B_l^3 \cup B_{l+1}$ (where $l-1, l+1$ is computed mod 4) [e.g. choose $x_C \in C$ from each E^0-equivalence class C ($\subseteq B^3$) and let $y \in C$ be in B_l^3 if $d(y,x) \equiv l$ mod 4 where $d(y,x) = Min\{k:$ there are $z_1, \ldots, z_k, y = z_1, x = z_k, \vartheta^1(z_i, z_{i+1})$ for each $i\}$].

Next for $x \in B^3$ let $\lambda_x = |\{y : \vartheta^1(y,x)\}|$ and $\mu(\lambda) = |\{x \in B^3 : \lambda_x \geq \lambda\}$ (for $\lambda < \aleph_0$.) We can assume each $\mu(\lambda)$ is 0 or $\geq \aleph_0$, (and even $\geq \lambda_2^+$) and note $\mu(\lambda)$ is decreasing in λ, hence eventually constant, say for $k \leq \lambda < \aleph_0$.

Now we can interprate $\exists_E$, E an equivalence relation which for $\chi \leq k$ has exactly $\mu(\lambda)$ classes.

For the converse, let us e.g. interprate $\vartheta_i(x,y)$. It suffices to code for $l<4$, $S = \{\langle x,y \rangle : \vartheta_1(x,y) \wedge (x \in B_l^3)\}$ Note that $|B_l^3| = |B^3| > \lambda_2$ (by the definition of B^1).

Let F be a one-to-one function from S into B_{l+2}^3, and let E_1, E_2 be equivalence relations. The E_1-equivalence classes are $\{x\} \cup \{F(\langle x,y \rangle) : \langle x,y \rangle \in S\}$, for $x \in B_l^3$, and the E_2 equivalence classes are $\{y\} \cup \{F(\langle x,y \rangle) : \langle x,y \rangle \in S\}$. (we can assume B_{l+2}^3 has the right cardinality as we are dealing with $\geq \lambda_2^+$ equivalence classes hence could have chosen it suitably). Together with monadic predicates the reconstruction is easy; as well as dealing with the ψ's.

§5 In the first stability cardinal

5.1 Definition :

Let $\lambda_3 = \lambda_3(R)$ be the least λ such that

$(\forall A \subseteq \mathfrak{U})(|A| \leq \lambda \rightarrow |S_\Delta^{\mathfrak{T}}(A,M)| \leq \lambda$

Δ finite, M admissible.

5.2 Fact : λ_2 is λ_3 or λ_3^+

Proof : Clearly $\lambda_3 \leq \lambda_2$.

Suppose $\lambda_3 \neq \lambda_2$. We cannot interpret $Q^{eq}_{\lambda_3^+, \lambda_3^+}$ because otherwise for some admissible M, finite Δ, $A \subseteq M$, $|A| = \lambda_3$ we would have $|S_\Delta(A,M)| = \lambda_3^+$, contradiction to

1) $Rk(p) \geq \langle 0,0 \rangle$ iff $Rk(p) \geq \langle 0,\infty \rangle$.

2) If p is realized by no λ pairwise disjoint m-tuples outside Dom p then $Rk(p) \leq \langle 1,0 \rangle < \langle 1,\infty \rangle$.

3) If p is realized by $\geq\lambda$ pairwise disjoint outside Dom p m-tuples then $Rk(p) \geq \langle 1,\infty \rangle$.

Proof : 0) is obvious.

1) Let $\bar{a}$ realize p. Suppose $Rk(p) \geq \langle 0,\beta \rangle$ all $\beta<\alpha$. Suppose $A^1 \supseteq A$ is given, $|A^1|<\lambda$. then $q=tp_\Delta(\bar{a},A^1)$ extends p, so $Rk(p) \geq \langle 0,\alpha \rangle$.

2) If p is realized by no λ pairwise disjoint m-tuples, let A^1 be such any no sequence disjoint to A^1-Dom p realize p, Dom $p \subseteq A^1$, and $|A^1|<\lambda$. There is no $q \supseteq_{pr} p$, $q \in S_\Delta^m(A^1,M)$, hence $Rk(q) \not\geq \langle 1,1 \rangle$. So $Rk(p) \leq Rk(q) \leq \langle 1,0 \rangle$.

3) Suppose p is realized by $\geq\lambda$ pairwise disjoint outside Dom p sequences, pairwise disjoint outside Dom p. We prove $Rk(p) \geq \langle 1,\gamma \rangle$ by induction on γ.

$\gamma=0$: Let $\bar{a} \neq \bar{b}$ realize p. Let $A^1=A \cup \bar{a} \cup \bar{b}$ and let $q_1=tp_\Delta(\bar{a},A^1)$ $q_2=tp_\Delta(\bar{b},A^1)$. Easily $\{q_1,q_2\}$ witnesses $Rk(p) \geq \langle 1,0 \rangle$.

$\gamma>0$: Let $A^1 \supseteq A$, $|A^1|<\lambda$, $\beta<\gamma$. We know $|S_\Delta^m(A^1,M)|<\lambda$.

So by Hypothesis 5.3A p has $<\lambda$ extensions in $S_\Delta^m(A^1,M)$. Since p is realized by λ pairwise disjoint outside Dom p sequences, some extension q of p in $S_\Delta^m(A^1,M)$ is realized by λ pairwise disjoint sequences, by regularity of λ.

By the induction hypothesis, $Rk(q) \geq \langle 1,\beta \rangle$, as required. This proves the claim.

5.8 Claim : Assume $p \in S_\Delta^m(A,M)$, $Rk(p)=\langle \alpha,\infty \rangle$, $0<\alpha<\infty$, $A \subseteq B$, $|B|<\lambda$. Then p has one and only one pure extension $q \in S_\Delta^m(B,M)$ of the same rank.

: **Proof** : Take γ^* so large that $Rk(p^1) \geq \langle \alpha,\gamma^* \rangle \Rightarrow Rk(p^1) \geq \langle \alpha,\infty \rangle$ (possible, as there are only set-many types). We know $Rk(p) \geq \langle \alpha,\gamma^*+1 \rangle$ so p has a pure extension $q \in S_\Delta^m(B,M)$ with $Rk(q) \geq \langle \alpha,\gamma^* \rangle$. Hence $Rk(q) \geq \langle \alpha,\infty \rangle$. If there are two such q, then $Rk(p) \geq \langle \alpha+1,0 \rangle$, contradiction.

5.9 Claim : 1) If $\lambda>\aleph_0$, then for any A of cardinality less than λ, and finite Δ,m, there is $B \supseteq A$, $|B|<\lambda$, such that

$$p \in S_\Delta^m(B,M) \Rightarrow [Rk(p)=\langle \alpha,\infty \rangle \text{ for some } \alpha<\infty \text{ or } Rk(p \restriction A)=\langle \infty,\infty \rangle]$$

2) We can do (1) simultaneously for all Δ.

Proof : 1) Define A_n $(n \in \omega)$ by induction:

$A_0 = A$.

Suppose A_n has been defined. For each $p \in S^m_\Delta(A_n, M)$ such that $Rk(p) = \langle \alpha, \gamma \rangle$, $\gamma < \infty$, take $B_p \supseteq A_n$ $|B_p| < \lambda$ such that p has no extension in $S^m_\Delta(B_p, M)$ of rank $\langle \alpha, \gamma \rangle$. (i.e. B_p witnesses $Rk(p) \not\geq \langle \alpha, \gamma+1 \rangle$).

Let $A_{n+1} = A_n \cup \bigcup \{B_p : p \in S^m_\Delta(B_p, M)\}$.

Let $p \in S^m_\Delta(\bigcup_n A_n, M)$. Since $Rk(p \restriction A_0) \geq Rk(p \restriction A_1) \geq Rk(p \restriction A_2) \geq \cdots$, we can find N such that $Rk(p \restriction A_N) = Rk(p \restriction A_{N+1}) = \cdots$. Suppose $n > N$ and $Rk(p \restriction A_n) = \langle \alpha, \gamma \rangle$, $\gamma \neq \infty$. $Rk(p \restriction A_n) = Rk(p \restriction A_{n-1})$, $A_{n-1} \subseteq A_{n-1} \cup B_{p \restriction A_{n-1}} = B_{p \restriction A_{n-1}} \subseteq A_n$ so $p \restriction B_{p \restriction A_{n-1}}$ is an extension of $p \restriction A_{n-1}$ of the same rank, contradicting the definition of B_p. So $n > N \rightarrow [Rk(p \restriction A_n) = \langle \alpha, \infty \rangle$ *for some* $\alpha]$.

Let $A^* = \bigcup_n A_n$.

If $Rk(p \restriction A_N) \neq \langle \infty, \infty \rangle$, take $\alpha < \infty$ such that $Rk(p \restriction A_n) = \langle \alpha, \infty \rangle$ for every large enough n. $p \restriction A_N$ has a unique extension $q \in S^m_\Delta(A^*, M)$ such that $Rk(q) = \langle \alpha, \infty \rangle$. Also $p \restriction A_N$ has a unique extension in $S^m_\Delta(A_n, M)$ of rank $\langle \alpha, \infty \rangle$, but $q \restriction A_n$, $p \restriction A_n$ are such extension for large n.

So $p \restriction A_n = q \restriction A_n$ for large n, so $p = q$, so $Rk(p) = \langle \alpha, \infty \rangle$.

2) Same proof.

5.10 Fact : If $Rk(p) < \langle \infty, \infty \rangle$ for every $p \in S^m_\Delta(A, M)$, $|A| < \lambda$, then $\lambda_2 \leq \lambda$ (hence w.l.o.g. $|\mathrm{Dom}\ R| \leq \lambda$.

Proof : Easy noting $A \subseteq B \Longrightarrow |S^m_\Delta(A)| \leq |S^m_\Delta(B)|$.

5.11 Lemma : Suppose for some large enough finite Δ, for each Δ-type p in m variables $Rk(p) < \langle 2, \infty \rangle$. Then,

1) $\{\exists_R, Q^{mon}_{\lambda_0}, Q^{1-1}_{\lambda_1}\}$ can be analyzed as before (in §4, with λ for λ_2^+), and $\{\exists_R, Q^{mon}_{\lambda_0}, Q^{1-1}_{\lambda_1}\} \equiv_{int} \{Q^{mon}_{\lambda_0}, Q^{1-1}_{\lambda_1}, \exists_E, \exists_{R^*}\}$ for some equivalence relation E, and relation R^*, $|\mathrm{Dom}\ R^*| < \lambda$.

or 2) $\{\exists_R, Q^{1-1}_\lambda\}$ is bi-interpretable with $\{Q^{word}_\lambda, Q^{eq}_{\kappa,\lambda}\}$, some κ.

or 3) $\{\exists_R, Q^{1-1}_{\lambda_1}\}$ is bi-interpretable with $\{\exists_{R^*}, \exists_E\}$, E an equivalence relation $|\mathrm{Dom}\ R^*| < \lambda = \aleph_0$.

Proof : Notice that if $p \in S^1_\Delta(A, M)$ has rank $\langle 1, \infty \rangle$, then p is minimal big.

We shall determine Δ later.

Let A^* be as in the previous (of 5.9), so the rank of any Δ-type in one variable over A^* is either $\langle 0,\infty\rangle$ or $\langle 1,\infty\rangle$. (If $\lambda=\aleph_0$ we can still get this by Konig Lemma and 5.7(1).

Let $P_m=\{p\in S^m_\Delta(A^*,M):Rk(p)=\langle 1,\infty\rangle\}$, $\kappa_m=|P_m|$ and $\kappa=\kappa_{m(*)}$ where $m(*)$ is large enough with respect to Δ let $P=\bigcup_{m\leq m(*)} P_m$. We can interpret $Q^{eq}_{\kappa_m,\lambda}$ - in fact, for $m=1$, the equivalence relation of realizing the same Δ-type over A^* with domain $\{a:a$ realizes some $p\in P_m$, $a\in\mathrm{Dom}(R)\}$, is an equivalence relation of this form. For $m>1$, remember we can code sets of $\leq\lambda$ pairwise disjoint sequences so we can interpret $Q^{eq}_{\kappa,\lambda}$.

Define $A_i(i<\lambda)$ continuous, increasing, such that:

1) $A_0=A^*$.

2) $A_{i+1}\supseteq A_i\cup\bigcup_p B_p$ where B_p is defined as before.

3) $|A_i|<\lambda$ for $i<\lambda$.

4) Dom $(R)=\bigcup_{i<\lambda} A_i$ (see 5.10).

We know that every $p\in P_m$ has a unique pure extension $p^{[i]}\in S^m_\Delta(A_i,M)$ of the same rank. We shall show that every pure $p\in S^m_\Delta(A_i,M)$ is of this form, provided that Rank $(p)=\langle 1,\infty\rangle$.

If $p\restriction A_0\notin P_m$, then it has rank $\langle 0,\infty\rangle$, so $Rk(p)\leq Rk(p\restriction A_0)=\langle 0,\infty\rangle<\langle 1,\infty\rangle=Rk(p)$, contradiction.

If $p\restriction A_0=q\in P_m$ but $p\neq q^{[i]}$, then for some $\varphi\in q^{[i]}$, $\neg\varphi\in p$. But by Def. 5.5, $p,q^{[i]}$ exemplify $Rk(q)\geq(2,0)$, contradiction.

This proves every p of rank $\langle 1,\infty\rangle$ in $S^m_\Delta(A_i,M)$ is $q^{[i]}$ for some $q\in P_m$.

We assume for a while:

Hypothesis A $:(\forall i)(\exists j>i)(\exists m\leq m(*))(\exists p_i\in S^m_\Delta(A_j,M))$ [$Rk(p_i)\geq\langle 1,\infty\rangle$ and p_i Δ_1 -splits over A_i] ($\Delta_1\supseteq\Delta$ to be determined.)

where we define: $p\in S^m_\Delta(A,M)$ Δ_1-splits over $B\subseteq A$ if there are $\bar b,\bar c\subseteq A$ realizing the same Δ_1-type over B and there is $\varphi\in\Delta$ such that $\varphi(\bar x,\bar b)\in p$, $\neg\varphi(\bar x,\bar c)\in p$.

Clearly for all i, $p_i=q_i^{[j]}$ for some $j=j_i$, and some $q_i\in P$ (when we restrict ourselves to Δ-types in one variable.) As $|P_{m(*)}|<\lambda$, we may assume all q_i are the same, $q=q_i$, and q is pure. For notational simplicity, let $j_i=i+1$.

For each i, let $\bar a_i\subseteq A_{i+1}-A_i$ realize $q^{[i]}$ and $\bar b_i,\bar c_i\subseteq A_{i+1}$ be such that

$\varphi_i(x,\bar{b}_i)$, $\neg\varphi_i(x,\bar{c}_i)\in q^{[i+1]}$. We may assume all the $\varphi_i(\bar{x},\bar{y})$ are the same, $\varphi_i=\varphi$. Now $\varphi(\bar{a}_\alpha,\bar{b}_\beta)\wedge\neg\varphi(\bar{a}_\alpha,\bar{b}_\beta)$ holds whenever $\alpha>\beta$ (as $\bar{a}_\alpha$ realizes $q^{[\alpha]}\supseteq q^{[\beta+1]}\supseteq\{\varphi(\bar{x},\bar{b}_\beta),\neg\varphi(\bar{x},\bar{c}_\beta)\}$) and $\varphi(\bar{a}_\alpha,\bar{b}_\beta)\wedge\neg\varphi(\bar{a}_\alpha,\bar{c}_\beta)$ fails if $\alpha<\beta$ when we choose Δ_1 appropriately, namely, when we ensure $\bar{b}_\beta$ and $\bar{c}_\beta$ realize the same $\{\psi(\bar{y};\bar{x})\}$-type over A_β where $\psi(\bar{y},\bar{x})=\varphi(\bar{x},\bar{y})$..
So some formula well-orders $\{\bar{a}_\alpha{}^\frown\bar{b}_\alpha{}^\frown\bar{c}_\alpha:\alpha<\lambda\}$. There is a subset of power λ which is a Δ-system, (as λ is regular $\geq\aleph_0$) so we can code the elements of that subset (with a few permutations) by elements of M and thus interpret Q^{word}_λ so $\{Q^{eq}_{\kappa,\lambda},Q^{word}_\lambda\}\leq_{int}\{Q_R,Q^{1-1}_\lambda\}$.
To see $Q_R\leq_{int}\{Q^{eq}_{\kappa,\lambda},Q^{word}_\lambda\}$, for simplicity we show that this holds when R is binary. ($|\mathrm{Dom}\,R|=\lambda$, of course). With a well-order and a set we code an equivalence relation E whose equivalence classes are $A_{i+1}-A_i$. Recall $\kappa=|\mathcal{P}_{m(*)}|$. On each E-equivalence class C, we can code (by more well orderings) $R\restriction C$ and for every $q\in\mathcal{P}$ and $a\in C$ we have to say whether a realizes q and whether $R(x,a)\in q^{[i+1]}$. We can do this with $Q^{eq}_{\kappa,\lambda}$ and Q^{word}_λ. So we have proved the desired conclusion (5.11(2)).

So we finish the case Hypothesis A holds, so assume

Hypothesis B : Hypothesis A is false.
By relabelling and taking A_i for some large i as our A_0, we can assume no $q^{[t]}$ Δ_1-splits over A_0. (for every $q\in\mathcal{P}$).
Now we ask:

If $A_0\subseteq A_1\subseteq A_2$ are as above, $\bar{a}_1\subseteq A_2$, $\bar{b}_1\subseteq A_2$, $\bar{a}_2\subseteq\mathcal{U}-A^*_2$, $\bar{b}_2\subseteq\mathcal{U}-A^*_2$,

$p_2=tp_{\Delta_1}(\bar{a}_2,A_2)\supseteq p_1=tp_{\Delta_1}(\bar{a}_1,A_1)$

$q_2=tp_{\Delta_1}(\bar{b}_2,A_2)\supseteq q_1=tp_{\Delta_1}(\bar{b}_1,A_1)$.

$Rk_{\Delta_1}(p_1)=Rk_{\Delta_1}(p_2)$, $Rk_{\Delta_1}(q_1)=Rk_{\Delta_1}(q_2)$.

Must $tp_\Delta(\bar{a}_1{}^\frown\bar{b}_2,A_1)=tp_\Delta(\bar{a}_2{}^\frown\bar{b}_1,A_1)$?

(Caution: Unlike first order types, the answer may depend on the specific $\bar{a}_l,\bar{b}_l$ used and not just the types they realize.)
If the answer is yes, (for every Δ, for some A_0 for every A_1,A_2), then we can essentially copy the analysis (in § 4) of reducing from $|\mathrm{Dom}\,R|=\lambda_1$ to $|\mathrm{Dom}\,R|=\lambda_2$ and get the desired conclusion (5.11(1) if $\lambda>\aleph_0$, or 5.11(3) if $\lambda=\aleph_0$),

If the answer is no (for some Δ, for every A_0), then by inductively choosing counter- examples, thinning to a Δ-system, and coding via permutations, we can interpret Q_λ^{word} and, as before, we get $\{Q_R, Q_\lambda^{1-1}\} \equiv_{int} \{Q_{\lambda,k}^{eq}, Q_\lambda^{word}\}$. This proves lemma 5.11.

Now we are reduced to the case that $Rk(p)=\langle \infty,\infty \rangle$ for some p or $Rk(p) \geq \langle 2,0 \rangle$ for some Δ-type p in one variable.

5.12 Lemma : For no $p \in S_\Delta^m(A,M)$ is $Rk(p)=\langle \infty,\infty \rangle$,$(|A|<\lambda)$.

Proof : We assume $Rk(p)=\langle \infty,\infty \rangle$ and reach a contradiction by interpreting $Q_{\lambda,\lambda}^{eq}$.

5.13 Definition : Suppose (by adding dummy variable) that Δ is a (finite) set of formulas of the form $\varphi(\bar{x},\bar{y})$ (with a fixed $\bar{y}$) and p is a Δ-type in the sequence of variables $\bar{x}$. Let Δ^c be the set of formulas obtained by reversing the role of $\bar{x}$ and $\bar{y}$; i.e. a Δ^c-type would consist of formulas $\varphi(\bar{a},\bar{y})$.

5.13A Fact : If $p=tp_\Delta(\bar{a},A)$, $Rk(p)=\langle \infty,\infty \rangle$, then for some $B \supseteq A$, $|B|<\lambda$, $q=tp_\Delta(\bar{a}^1,B) \supseteq_{pr} p$, $Rk(q)=\langle \infty,\infty \rangle$ and q Δ^c-splits over A.

Proof : Choose $B_0 \supseteq A, |B_0|<\lambda$ such that every Δ^c-type over A realized in M is realized in B_0.

We take $p_0 \in S_\Delta^m(B_0,M)$, $p_0 \supseteq_{pr} p$, $Rk(p_0)=\langle \infty,\infty \rangle$. So there exists $\varphi(\bar{x},\bar{b})$ such that both $p_0 \cup \{\varphi(\bar{x},\bar{b})\}$ and $p_0 \cup \{\neg\varphi(\bar{x},\bar{b})\}$ can be completed to Δ-types of rank $\langle \infty,\infty \rangle$.

So there is $\bar{c} \subseteq B_0$, $tp_{\Delta^c}(\bar{c},A)=tp_{\Delta^c}(\bar{b},A)$.

Without loss of generality, $\varphi(\bar{x},\bar{c}) \in p_0$.

So $p_0 \cup \{\varphi(x,\bar{c}) \neg\varphi(x,\bar{b})\}$ can be completed to a Δ -type rank $\langle \infty,\infty \rangle$ which Δ^c-splits over A (and is a pure extension of p_0).

5.13B Fact : We can interpret $Q_{\lambda,<\lambda}^{eq}$.

Proof : Take $A_i,(i<\lambda)$ as in lemma the proof of 5.11.

For each i, take $p_i=tp_\Delta(\bar{a}_i,A_i)$ to have rank $\langle \infty,\infty \rangle$ and w.l.o.g. is pure. By fact 5.13A we can take $\bar{b}_i$, $\bar{c}_i$ such that $tp_{\Delta^c}(\bar{b}_i,A_i)=tp_{\Delta^c}(\bar{c}_i,A_i)$ and $p_i \cup \{\varphi_i(\bar{x},\bar{b}_i), \neg\varphi_i(\bar{x},\bar{c}_i)\}$ has a pure completion of rank $\langle \infty,\infty \rangle$.

So for some $\psi_i(\bar{x},\bar{d}_i)$ both $p_i \cup \{\varphi_i(\bar{x},\bar{b}_i), \neg\varphi_i(\bar{x},\bar{c}_i), \pm\psi_i(\bar{x},\bar{d}_i)\}$ have completions of rank $\langle \infty,\infty \rangle$.

Let $\bar{a}_{i,\alpha}^l$ be pairwise disjoint, $\bar{a}_{i,\alpha}^l \cap A_i=\phi$ $(\alpha<i$, $l=0,1)$, $\bar{a}_{i,\alpha}^l$ realize $p_i \cup \{\varphi_i(\bar{x},\bar{b}_i), \neg\varphi_i(\bar{x},\bar{c}_i)\}$ and $\psi_i(\bar{a}_{i,\alpha}^l,\bar{d}_i)$ iff $l=0$.

Without loss of generality, $\bar{d}_i\frown\bar{b}_i\frown\bar{c}_i\frown\bar{a}^l_{i,\alpha}\subseteq A_{i+1}$. for $\alpha<|A_{i+1}|$; w.l.o.g., $\varphi_i=\varphi$, $\psi_i=\psi$ do not depend on i. W.l.o.g., $\bar{b}_i\frown\bar{c}_i\frown\bar{d}_i\cap A_i$ is constant, (if $\lambda>\aleph_0$, by applying Fodor's theorem to $F(i)=$ at least j such that $A_i\cap(\bar{b}_i\frown\bar{c}_i\frown\bar{d}_i)\subseteq A_j$; and then using that there are λ-many i but less than λ-many finite sequences from A_j; if $\lambda=\aleph_0$, by the Δ-system lemma and renaming.) W.l.o.g. , $\bar{a}^l_{i,\alpha}$ is disjoint to $\bar{b}_i\frown\bar{c}_i\frown\bar{d}_i$.

We can interpret $P=\{\bar{b}_i\frown\bar{c}_i\frown\bar{d}_i;\ i<\lambda\}$ since we have arranged that they form a Δ-system.

Let f be the permutation

$f(\bar{a}^l_{i,\alpha})=\bar{a}^{1-l}_{i,\alpha}$, f is the identity elsewhere.

When does $\varphi(\bar{a}^l_{i,\alpha},\bar{b}_j)\wedge\neg\varphi(\bar{a}^l_{i,\alpha},\bar{c}_j)\wedge[\psi(\bar{a}^l_{i,\alpha},\bar{d}_j)\equiv\neg\psi(f(\bar{a}^l_{i,\alpha}),\bar{d}_j)]$ hold? For $i=j$, the formula is true by inspection.

For $i<j$, the answer is no, as $\bar{b}_j,\bar{c}_j$ realize the same Δ^c-type over $A_j\supseteq\bar{a}^l_{i,\alpha}$.

For $i>j$, the answer is no; since $\bar{d}_j\subseteq A_i$, and $\bar{a}^l_{i,\alpha},f(\bar{a}^l_{i,\alpha})$ realize p_i which is a complete Δ-type over A_i, contrary to the third conjunct.

So we can interpret E with domain $\{\bar{a}^0_{i,\alpha}:\ \alpha<i<\lambda\}$, $\bar{a}^0_{i_1,\alpha_1}E\,\bar{a}^0_{i_2,\alpha_2}\equiv_{df}i_1=i_2$. (using P and f to do so, remember 4.3(2)).

But E is in $Q^{eq}_{\lambda,<\lambda}$.

5.13C Fact : We can interpret Q^{word}_λ.

Proof : By Fact 5.13B we can interprete an equivalence relation E with equivalence classes $A_{i+1}-A_i$.

Let E_i be the equivalence relation on finite sequences of suitable length m from $A_{i+1}-A_i$:

$\bar{a}_1E_i\bar{a}_2$ iff $\bar{a}_1,\bar{a}_2$ realize the same Δ^c-type over A_i.

We can code $\bigcup_{i<\lambda}E_i=E^1$ by fact 5.13B, since $|A_{i+1}-A_i|<\lambda$.

Let $x\in A_{i+1}-A_i$ $y\in A_{j+1}-A_j$.

If $j<i$,

$\vartheta(x,y)\stackrel{def}{=}$" If $\{b_1,\ldots,b_m,c_1,\ldots,c_m,x\}$ are in the same E-equivalence class and $\bar{b}E^1\bar{c}$ then $(\forall\bar{z}$ *such that* $z_lEy)\bigwedge_{\varphi\in\Delta}(\varphi(\bar{z},b_1,\ldots,b_m)\equiv\varphi(\bar{z},c_1,\ldots,c_m))$ holds."

Obviously if $j<i$, $\vartheta(x,y)$ holds.

If $j>i$, $p_i\cup\{\varphi_i(\bar{x},\bar{b}_i),\neg\varphi_i(\bar{x},\bar{c}_i)\}$ has a completion of rank $\langle\infty,\infty\rangle$, so w.l.o.g. it is realized in $A_{j+1}-A_j$, so $\vartheta(x,y)$ fails.

This proves Fact 5.13C.

5.13D Fact : We can interpret $Q^{eq}_{\lambda\lambda}$.

Proof : We can find $\bar{x}_{i,j}, \bar{y}_{i,j} \subseteq A_{j+1} - A_j$, $(i<j)$ such that $tp_\Delta(\bar{x}_{i,j}, A_i) = tp_\Delta(\bar{y}_{i,j}, A_i)$, $tp_\Delta(\bar{x}_{i,j}, A_{i+1}) \neq tp_\Delta(\bar{y}_{i,j}, A_{i+1})$ for all $i<j<\lambda$.
(we can take them to realize $p_i \cup \{\pm\psi_i(\bar{x}, \bar{d}_i)\}$ from the proof of 5.13A).
In fact there are $|A_i|$ such pairwise disjoint pairs.
So, w.l.o.g., $\bar{x}_{i_1,j_1}, \bar{x}_{i_2,j_2}, \bar{y}_{i_1,j_1}, \bar{y}_{i_2,j_2}$ are all disjoint for $(i_1,j_1) \neq (i_2,j_2)$. Since we can interpret Q^{word}_λ, we can interpret the equivalence relation $\bar{x}_{i_1,j_1} E \bar{x}_{i_2,j_2}$ iff $i_1 = i_2$.
So we have proved 5.13D, hence 5.12.

5.14 Lemma : For no p (and Δ) $Rk(p) > \langle 2, \infty \rangle$).

Proof : We know $(\forall p)(Rk(p) < \langle \infty, \infty \rangle)$. If the lemma fails we shall interpret $Q^{eq}_{\lambda\lambda}$ getting a contradiction. By Def. 5.5 we can find p $Rk(p) = <2,\infty>$.
We can define A_i $(i<\lambda)$, A_i increasing continuous as in 5.11's proof, $|A_i| < \lambda$, $p_0 \in S^m_\Delta(A_0)$, $Rk(p_0) = \langle 2, \infty \rangle$; $p_i = p_0^{[i]} \in S^m_\Delta(A_i)$ $p_0 \subseteq p_i$, $Rk(p_i) = <2,\infty>$, $p_i \subseteq_{pr} q_i \in S^m_\Delta(A_{i+1})$, $Rk(q_i) = \langle 1, \infty \rangle$, $\bar{a}_{i,j} \subseteq A_{j+1} - A_j$, $tp_\Delta(\bar{a}_{i,j}, A_j) \supseteq_{pr} q_i$ has rank $\langle 1, \infty \rangle$, and $\varphi_i(\bar{x}, \bar{b}_i) \in q_i$, $\neg\varphi_i(\bar{x}, \bar{b}_i) \in p_{i+1}$. W.l.o.g. the $\langle \bar{b}_i : i<\lambda \rangle$ form a Δ-system. And even $\bar{a}_{i,j}$, $\bar{b}_i$ are pairwise disjoint outside some $\bar{b}^*$.
If for every i for some j, $p_0^{[j]}$ Δ^c-split over A_i, we can easily interpret Q^{word}_λ. Otherwise we can easily interpret first $Q^{eq}_{\lambda,<\lambda}$, with which we can code $\{A_{i+1} - A_i : i<\lambda\}$ and relation over the $A_{i+1} - A_i$; so we can again code Q^{word}_λ. (really the first case occurs as for every $i \leq \lambda$, there are $i < j_1 < j_2 < \lambda$, $tp_{\Delta^c}(\bar{b}_{j_1}, A_i) = tp_{\Delta^c}(\bar{b}_{j_2}, A_i)$ In both cases we finish as in 5.13D.

Now 5.11, 5.12, 5.14 give a complete analysis of the case $\lambda_2 \neq \lambda_3$

* * *

During our investigations, we came across the following quantifier:

5.14A Definition : Let $K^{word}_\alpha = \{R : R$ **a two place relation, (** Dom R, R **) is a well ordering of order-type** $\alpha\}$.

5.15 Claim : 1) If $\alpha \leq \beta$ **then** $Q^{word}_\alpha \leq Q^{word}_\beta$.

2) $Q^{eq}_{|\alpha|,<|\alpha|} \leq_{int} Q^{word}_\alpha$ **for infinite** α **(hence** $Q^{1-1}_\lambda \leq Q^{word}_\lambda$).

3) $Q^{eq}_{\kappa,\lambda} \leq_{int} Q^{word}_{\lambda\kappa}$, **(** $\lambda > \kappa \geq \aleph_0$ **are cardinals).**

4) For λ singular $Q^{word}_\lambda \equiv_{int} Q^{eq}_{\lambda,\lambda}$.

5) If $\alpha = \lambda^2$ **then** $Q^{eq}_{\lambda,\lambda} \leq Q^{word}_\alpha$.

Proof : Easy (for (4) use 6.4).

5.16 Lemma : 1) If λ is regular, $\kappa<\mu\le\lambda$, $\mu\le 2^\kappa$, then $Q^{eq}_{\mu,\lambda} \le_{int} \{Q^{word}_\lambda, Q^{eq}_{\kappa,\lambda}\}$

2) If λ is regular $\kappa\le\mu\le\lambda$, $\mu\le\beth_n(\kappa)$ then $Q^{eq}_{\mu,\lambda} \le_{int}\{Q^{word}_\lambda, Q^{eq}_{\kappa,\lambda}\}$

3) Assume $\alpha<\lambda$, $\lambda\alpha\le\beta<\lambda(\alpha+1)$, λ regular, $|\alpha|=\kappa$. Then $Q^{word}_\beta \equiv_{int} \{Q^{word}_\lambda, Q_{\kappa,\lambda}\}$.

Proof : 1) Let $S=\{\delta<\lambda: \delta$ divisible by $\kappa\}$, and let E be an equivalence relation on S with $\le\mu$ equivalence classes each of power λ (we shall define and interpret him). As the number of models $(\kappa,<,P)$ is 2^κ, we can find $P\subseteq\lambda$ such that:

(*) for $\delta_1,\delta_2\in S, \delta_1 E\delta_2$ iff for every $i<\kappa$, $\delta_1+i\in P \Longleftrightarrow \delta_2+i\in P$.

Now let E_0 be $\{\langle \delta_1+i,\delta_2+i\rangle : \delta_1\in S,\ \delta_2\in S,\ i<\kappa\}$. Easily we can interpret E by $<,P$ and E_0, all interpretable by $\{Q^{word}_\lambda, Q^{eq}_{\kappa,\lambda}\}$.

2) By induction on n.

3) Easy.

5.17 Lemma : 1) $Q_{\aleph_0,\lambda} \not\le_{int} Q^{word}_\lambda$ for λ regular.

2) $Q^{eq}_{\kappa,\lambda} \not\le_{int}\{Q^{word}_\lambda, Q^{eq}_{\mu,\lambda}\}$ for λ,κ regular, $\lambda\ge\kappa\ge\aleph_0$ and $\mu\ge\beth_\omega(\kappa)$.

Proof : 1) We can prove that if $\bigwedge_{l=1}^{n} (Q^{word}_\lambda R)(\forall \bar{x})[R(\bar{x})\equiv R_l(\bar{x})]$, then the model $M=(\bigcup_{l=1}^{n}\mathrm{Dom}\, R_l, R_1, \dots, R_n)$ can be represented as $\sum_{i<\lambda} M_i$ where:

(A) each M_i is a model of power $<\lambda$.

(B) the $|M_i|$ are pairwise disjoint

(C) the meaning of $M=\sum_{i<\lambda} M_i$ is that if $\bar{a}_l\subseteq M_{i(l)}, i(1)<\cdots<i(k)$, we can compute the basic type of $\bar{a}_1{}^\frown\cdots{}^\frown\bar{a}_n$ in M from the basic types of $\bar{a}_l$ in $M_{i(l)}$ (not depending on the particular $i(l)$'s.)

Now by Feferman-Vaught theorem the conclusion follows.

2) Like (1); but for formulas to depth n, we use the F.V. theorem for formulas of $L_{\infty,\kappa}$ of (quantifier) depth $\le n$.

5.18 Lemma : $Q^{word}_\lambda \equiv_{\exp} Q^{eq}_{\lambda,\lambda}$.

Proof : Clearly $Q^{word}_\lambda \le_{\exp} Q^{eq}_{\lambda,\lambda}$ (in fact $Q^{word}_\lambda \le_{int} Q^{eq}_{\lambda,\lambda}$). Ordinal addition on λ gives a pairing function, and on a subset of cardinality λ, and we can define addition as we can quantify over one-to-one functions.

§6 Below the first Stability cardinal

Hypothesis : We now assume $\lambda=\lambda_2=\lambda_3$.
We try to approach λ from below. W.l.o.g. $|\mathrm{Dom}\, R|\le\lambda$, and we analyze $\{\exists_R, Q_\lambda^{1-1}\}$.

6.1 Construction : Let M be admissible, rich enough, Δ finite large enough. We define by induction on i:

$\bar{d}^l_{i,\alpha}, \bar{a}^l_{i,\alpha}$ $(l=0,1\ \alpha<i)$, $\varphi_i(\bar{x},\bar{y},\bar{z},\bar{b}_i)$, $\psi_i(\bar{x},\bar{c}_i)$ such that , for
$A_i=\{\bar{d}^l_{i,\alpha}, \bar{a}^l_{j,\alpha}, \bar{b}_j, \bar{c}_j : \alpha<j<i, l<2\}$:

a) $\varphi_i, \psi_i \in \Delta$.
b) $\varphi_i(\bar{x},\bar{y},\bar{z},\bar{b}_i)$ is not realized in A_i or at least by no $\bar{a}^0_{j,\alpha}{}^\frown\bar{a}^1_{j,\alpha}{}^\frown\bar{d}_{j,\alpha}$ $(i>j>\alpha)$.
c) $\varphi_i(\bar{x},\bar{y},\bar{z},\bar{b}_i)\rightarrow(\psi_i(\bar{x},\bar{c}_i)\wedge\neg\psi_i(\bar{y},\bar{c}_i))$.
d) $\bar{a}^0_{i,\alpha}$, $\bar{a}^1_{i,\alpha}$ realize the same Δ-type over A_i.
e) $\varphi_i(\bar{a}^0_{i,\alpha}, \bar{a}^1_{i,\alpha}, \bar{d}_{i,\alpha}, \bar{b}_i)$ for all $\alpha<i$.
f) all the sequences $\{\bar{d}_{j,\alpha}{}^\frown\bar{a}^0_{j,\alpha}{}^\frown a^1_{j,\alpha} : \alpha<j\le i\}$ are pairwise disjoint.

Continue until i^*, when the process breaks down.
Let $\chi=$ card (i^*), $A^0=A_{i^*}$ so $|A^0|=\chi$ if χ is infinite $|A^0|<2^{2^\chi}$ if χ is finite.

6.2 Claim 1) We can interpret $Q^{eq}_{\chi,<\chi}$ if χ is infinite.
2) We can interpret $Q^{eq}_{\chi_1,<\chi_1}$ if χ is finite, $\chi_1=\chi^{1/2^{m(\Delta)}}$.

Proof : 1) If χ is regular, we can make the parameters $(b_i{}^\frown\bar{c}_i)$ into a Δ-system and proceed as in fact 5.13B previously.
So suppose $\kappa=cf\,\chi<\chi=\sum_{i<\kappa}\chi_i$, $\kappa<\chi_i$ all;
We can find a subsequence of $\langle \bar{b}_i{}^\frown\bar{c}_i : i<i^*\rangle$ of length κ which is a Δ-system, and with it interpret an equivalence relation E with κ equivalence classes of arbitrarily large powers less than χ.
For each i, there is a set $S_i\subseteq\chi_i^+$, of cardinality χ_i^+ such that $\langle \bar{b}_j{}^\frown\bar{c}_j : j\in S_i\rangle$ is a Δ-system. Let $\bar{e}_i$ be the heart of this Δ-system.
There is $T\subseteq\kappa$, $|T|=\kappa$, such that $\langle \bar{e}_i : i\in T\rangle$ is a Δ-system with heart $\bar{e}$.
Let $\gamma_i\in S_i$ for each i.
By hand over hand thinning of each S_i we may assume $\bar{b}_\alpha{}^\frown\bar{c}_\alpha\cap\bar{b}_{\alpha^1}{}^\frown\bar{c}_{\alpha^1}\subseteq\bar{e}$ if $\alpha\in S_i$, $\alpha^1\in S_j$, $i\neq j$.
We may assume $S_i\cap S_j=\phi$ for $i\neq j$, $i,j\in T$. Let $i(\alpha)$ be the unique i, such that

$\alpha \in S_i$. By permutations, we can code $\{\bar{e}_i : i \in T\}$ and $\{\bar{b}_\alpha{}^\frown \bar{c}_\alpha - \bar{e}_{i(\alpha)} : \alpha \in S_i\}$ and $\{\bar{b}_{\gamma_i}{}^\frown \bar{c}_{\gamma_i} : i \in T\}$.

We need to code the equivalence relation E^1: $\bar{b}_\alpha{}^\frown \bar{c}_\alpha{}^\frown \bar{e}_{i(\alpha)} E^1 \ \bar{b}_\beta{}^\frown \bar{c}_\beta{}^\frown \bar{e}_{i(\beta)}$ iff $i(\alpha)=i(\beta)$.

By our reduction, this can be accomplished if we can do it for singletons rather than sequences. This we can do, with the equivalence relation E.

2) Left to the reader (really we need just that χ_1 as a function of χ diverge to $\aleph_0$)

6.3 Conclusion : $\chi \le \lambda$.

6.4 Claim : 1) If χ is singular, $Q^{eq}_{\chi,\chi} \equiv_{int} Q^{eq}_{\chi,<\chi}$.

2) $Q^{eq}_{\lambda,\chi} \equiv_{int} Q^{eq}_{\lambda,<\chi}$ if $\lambda > \chi$, χ singular.

3) If χ is finite then $Q^{eq}_{\chi/2,\chi/2} \le_{int} Q^{eq}_{\chi,<\chi} \le_{int} Q^{eq}_{\chi,\chi}$.

Proof : 1) Now we know $Q^{eq}_{\chi,<\chi} \le_{int} Q^{eq}_{\chi,\chi}$. Let us do the other inequality. Say E is the following equivalence relation on $\{\langle i,j \rangle : i<j<\chi\}$:

$\langle i,j \rangle E \langle k,l \rangle$ iff $j=l$; clearly $E \in Q^{eq}_{\chi,<\chi}$.

Let $\langle \chi_i : i<\kappa \rangle$ be as before, and let $E^1 \in Q^{eq}_{\chi,<\chi}$ be an equivalence relation on $\{\langle 0,j \rangle : 0<j<\chi\}$ with χ equivalence classes each equivalence class unbounded in χ of power less than χ.

$xE^* y \equiv \exists x^1 \exists y^1 (x^1 Ex \wedge y^1 Ey \wedge x^1 E_1 y^1)$ is an equivalence relation with χ classes of power χ.

2) Similarly.

3) Easy.

6.5 Claim : At least one of the following occurs (if χ finite, we should use $3n(R)\ \chi$) (A^0 etc are from 6.1):

(1) For no $m<m(\Delta), l(\bar{x})=m, \varphi(\bar{x},\bar{y}) \in \Delta, \bar{a}$ (finite) and pure $p \in S^m_\Delta(A^0,M)$, are both $p \cup \{\pm\varphi(\bar{x},\bar{a})\}$ realized by χ pairwise disjoint sequences.

(2) For no $\varphi(\bar{x},\bar{y}) \in \Delta$ and $\bar{a}$ is $\varphi(\bar{x},\bar{a})$ realized by no $\bar{b} \subseteq A^0$, but $(\exists^{\ge\chi}\bar{x})[\varphi(\bar{x},\bar{c}) \wedge \bar{x} \cap A^0 = \phi]$.

Proof : Suppose $\varphi(\bar{x},\bar{b}), p \in S^m_\Delta(A^0,M)$ exemplify (1) fail, i.e. there are $\bar{a}^l_\alpha (l<2, \alpha<i^*)$ realizing p, pairwise disjoint and disjoint to A^0 (as p is pure) such that $\models \varphi(\bar{a}^l_\alpha, \bar{b})$ iff $l=0$.

Suppose further that $\psi(\bar{x},\bar{c})$ exemplify (2) fail i.e. there are $\bar{d}_\alpha (\alpha<3n(R)i^*)$, pairwise disjoint and disjoint to A^0 such that $\models \psi[\bar{d}_\alpha, \bar{c}]$ and $\neg(\exists \bar{x} \subseteq A)\psi(\bar{x},\bar{c})$.

We now can, by thinning, have $\bar{a}^l_\alpha, \bar{d}_\alpha (\alpha < i^*)$ which are pairwise disjoint. However we could have chosen $a^l_{i^*,\alpha} = a^l_\alpha$ ($l<2, \alpha<i^*$), $\bar{b}_{i^*} = \bar{b}$, $\varphi_{i^*} = \varphi(\bar{x}, \bar{b})$, $\bar{d}_{i,\alpha} = \bar{d}_\alpha$, $\psi_{i^*} = \psi(\bar{z}, \bar{c}_i)$, $\bar{c}_{i^*} = \bar{c}$, contradicting the choice of i^*.

6.6 Lemma : Suppose that 6.5(1) holds, and let $A^1 = \{a \notin A^i : tp_\Delta(a, A_0)$ is realized by exactly one element $\}$.

$A^2 = \mathcal{U} - A^0 \cup A^1$.

Then

1) If $A^0 \subseteq B$, $p = tp_\Delta(\bar{a}, A^0, M)$ then the number of $q \in S^{m}_\Delta(B, M)$ extending p is at most $|B|$ (or $\leq \chi\, |B|^{m(\Delta)} |\Delta|$ when χ is finite).

2) $|S^{m(\Delta)}_\Delta(B, M)| \leq |S^1_\Delta(A^0, M)| + |B|$ (or $\leq S^1_\Delta(A^0, M)\, \chi |\Delta|\, |B|^{m(\Delta)}$ if χ is finite).

3) $\lambda \leq 2^\chi$ and $\lambda = |S^{\leq m(\Delta)}_\Delta(A^0, |M|)|$ for χ infinite.

Proof : 1) Immediate from 6.5(1): let B be infinite ($\supseteq A^0$) and suppose, $tp_\Delta(\bar{a}_\alpha, B)$ $(\alpha < |B|^+)$ are distinct and

(*) $\bar{a}_\alpha (\alpha < |B|^+)$ realizes p.

W.l.o.g. $\bar{a}_\alpha = \bar{a}^+_\alpha \frown \bar{a}^*$, where $\bar{a}^* \subseteq B$, $\bar{a}^+_\alpha \cap B = \phi$, and the $\bar{a}^+_\alpha (\alpha < |B|^+)$ are pairwise disjoint. As 6.5(1) holds, for every $\bar{c}, \psi$ one of the sets $\{\alpha < |B|^+ : \models \psi(a, \bar{c})\}$, $\{\alpha < |B|^+ : \models \neg \psi(\bar{a}, \bar{c})\}$ has cardinality $< \chi$. Now we get contradiction to (*).

2) Follows from (1), as w.l.o.g. we can count pure types only.

3) Clearly $|S^{\leq m(\Delta)}_\Delta(B, M)| \leq 2^\chi$ for B of power $\leq 2^\chi$. This is closely related to the definition of $\lambda_3 = \lambda$ but there is a difference: M and Δ are here fixed. But we could have repeat §4 , §5 for a fix larged enough Δ, M (with Δ depending on $n(R)$ and not on R). If λ is regular use §5 with hypothesis 5.3A, for λ singular 6.2, 6.4. (alternatively repeat this section for any Δ).

6.7 Lemma : Suppose 6.5(2), and that Δ is closed under permuting the variables.

(1) There is $A^1, A^0 \subseteq A^1, |A^1| \leq |S^{\leq m(\Delta)}_\Delta(A^0, M)|\ m(\Delta)$, such that for every B extending A^1, and $\bar{b}$ disjoint to B, (of length $\leq m(\Delta)$) $tp(\bar{b}, B)$ does not Δ-split over A^0.

2) $\lambda \leq |S^{\leq m(\Delta)}_\Delta(A^1, M)|$

3) $\lambda \leq 2^\chi$ when χ is infinite.

Proof : 1) Immediate.

2) Follows from (1), as in 6.6 (as by (1) every pure $p \in S^{\leq m(\Delta)}_\Delta(B)$ is

determined by $p \restriction A^l$)

3) By (1), (2) it suffices to prove that if $A^0 \subseteq B$, $|B| \le 2^\chi$, $m \le m(\Delta)$ then $|S_\Delta^m(B,M)| \le 2^\chi$. Suppose B, m form a counterexample. Then for some $\bar{e} \subseteq B, \varphi(\bar{x},\bar{y},\bar{e}) \in \Delta$, $|S_{\{\varphi(\bar{x};\bar{y},\bar{e})\}}(B,M)| > 2^\chi$, and we choose an example with minimal $l(\bar{x}\frown\bar{y})$ and so there are $\bar{b}_i$ disjoint to B, for $i<(2^\chi)^+$, with $tp_{\{\varphi(\bar{x};\bar{y},\bar{e})\}}(\bar{b}_i,B,M)$ distinct, $l(\bar{b}_i)=l(\bar{x})$, and w.l.o.g. $tp(\bar{b}_i,A^0)=p$ for a fixed p. If for some $i \ne j$, there are $\bar{a}_\alpha(\alpha<\chi)$ pairwise disjoint, disjoint to A^0 such that $\varphi(\bar{b}_i,\bar{a}_\alpha) \wedge \neg\varphi(\bar{b}_j,\bar{a}_\alpha)$, we get contradiction to 6.5(2).

So for every $j>0$ there is $B_j \subseteq B, |B_j| < \chi$ such that if $\bar{a} \subseteq B-B_j \cup A^0$ then $\varphi(\bar{x},\bar{a}) \in p_0 \Leftrightarrow \varphi(x,\bar{a}) \in p_j$. The number of possible B_j is $\le |B|^{<\chi}$ so w.l.o.g. $B_j=B_i$ for $j>0$. But now let $\{\varphi_\alpha(\bar{x},\bar{y}_\alpha,\bar{e}_\alpha):\alpha<\chi\}$ be the formulas we get from $\varphi(\bar{x},\bar{y},\bar{e})$ by substituting one member of $\bar{y}$ by a member of $A^0 \cup B_1$. Clearly $0<i<j$ implies $\bigvee_{\alpha<\chi} tp_{\varphi_\alpha(\bar{x};\bar{y}_\alpha,\bar{e}_\alpha)},(\bar{a}_i,B,M) \ne tp_{\varphi_\alpha(\bar{x};\bar{y},\bar{e}_\alpha)}(\bar{a}_j,B,M)$. Hence for some α $|S_{\varphi_\alpha(\bar{x};\bar{y}_\alpha,\bar{e}_\alpha)}(B,M)| > 2^\chi$, contradicting the minimality of $l(\bar{x}\frown\bar{y})$.

6.8 Theorem : There is a function $f:\omega\to\omega$, diverging to infinity such that: if $\chi=\chi(R)$ is finite, then for some R^* and equivalence relation E, $\{\exists_R, Q_\chi^{1-1}\} \equiv_{int} \{\exists_{R^*}, \exists_E\}$, $n=|\mathrm{Dom}\ R^*|$ finite, and $Q^{eq}_{f(n),f(n)} \le_{int} \exists_R$.

Proof : Combine the previous lemmas.

By 6.6 or 6.7 there is A^1, with $|A^1|$ not too large than χ, such that every pure $p \in S_\Delta^{\le m(\Delta)}(A^1,M)$ has no two explictely contradicting χ-big extensions. Now as in §5, we can apply §4 to get $\exists_R \equiv_{int} \{\exists_{R^*}, \exists_E\}$ with $|\mathrm{Dom}\ R^*|$ not too large than $|A^2|$.

As for $Q^{eq}_{f(n),f(n)} \le_{int} \exists_R$, use 6.2(2).

6.9 Claim : We can interpret $Q^{eq}_{\chi,\chi}$ if χ is infinite.

Proof : We are done if χ is singular by 6.2, 6.4. So we assume χ is regular. If $|S_\Delta^{\le m(\Delta)}(A,M)| < \chi$ whenever $|A| < \chi$, we repeat the case $\lambda_2=\lambda_3^+$ with λ_2 replaced by χ everywhere.

So we assume $\kappa=|A^1|<\chi$, $|S_\Delta^1(A^m,M)| \ge \chi$. Now $|S_\Delta^1(A^1,M^1)| \ge \chi$, for some A^1 of cardinality $<\chi$, and some admissible M^1 [as if $tp_\Delta(\bar{a}_i,A) \in S_\Delta^m(A,M)$ are distinct $(i<\chi)$, then w.l.o.g. $\{\bar{a}_i\}$ form a Δ-system. We can expand A to include its heart and use permutations to get distinct elements of $S_\Delta^1(A^1,M^1)$]

Since Δ is finite, there is $\varphi(x,\bar{y})$ such that $|S^1_{\{\varphi\}}(A^1,M)| \ge \chi$.

Let $m=$ length $(\bar{y})$.

Let $I=\{\bar{a}\subseteq A^1:\varphi(x,\bar{a})$ belongs to at least χ types $p\in S^1_{\{\varphi\}}(A^1,M)$, and $\neg\varphi(x,\bar{a})$ belongs to at least χ types $p\in S^1_{\{\varphi\}}(A^1,M)\}$.
Let $S_\varphi(I)=\{p\cap\{\pm\varphi(x,\bar{a}) : \bar{a}\in I\}:p\in S_{\{\varphi\}}(A^1,M)\}$.
Note that $|S_\varphi(I)|\geq\chi$, as follows:
Let $F:S^1_\varphi(A^1,M^1)\to S_\varphi(I)$ be the obvious projection.
If $|S^1_\varphi(I)|<\chi$, we take $q\in S_\varphi(I)$ with $\geq\chi$ pre-images p_i $(i<\chi)$, $p_j\neq p_j$, so each, except possibility one contains a formula which belongs to fewer than χ of the p_i. But there are fewer that χ-many formulas in all, contradiction.
Also note that for every $\bar{a}\in I$: $|\{p\in S_\varphi(I):\varphi(x,\bar{a})\in p\}|\geq\chi$ and
$|\{p\in S_\varphi(I):\neg\varphi(x,\bar{a})\in p\}|\geq\chi$.
(otherwise the pre-image under F of a set of size $<\chi$ would have cardinality $\geq\chi$, but we just showed any $q\in S_\varphi(I)$ can have only $<\chi$ elements in its pre-image).
On I, define the following equivalence relation E:
$\bar{a}E\bar{b}$ iff $|\{p\in S_\varphi(I):\varphi(x,\bar{a})\in p\equiv\varphi(x,\bar{b})\notin p\}|<\chi$.
Let $J\subseteq I$ be a set of representatives and $G:S_\varphi(I)\to S_\varphi(J)$ the natural map.

6.10 Fact : $|G^{-1}(q)|<\chi$ for any q.

Proof : Suppose $G(p_i)=q$ and $p_i\neq p_j$ for $i<j<\chi$. For each i, take $\bar{b}_i\in I$, such that $(\varphi(x,\bar{b}_i)\in p_i)\Longleftrightarrow(\varphi(x,\bar{b}_i)\notin p_{i+1})$.
Let $\bar{a}_i\in J$, $\bar{a}_iE\bar{b}_i$.
Since $|J|\leq|I|\leq|A^1|<\chi$, there are $\bar{a}^*$, and $\bar{b}^*$, such that $S=\{i:\bar{a}_i=\bar{a}^*$ and $\bar{b}_i=\bar{b}^*\}$ has cardinality $=\chi$, so $\bar{a}^*E\bar{b}^*$. W.l.o.g., $\varphi(x,\bar{a}^*)\in q$.
For all $i<\chi$, $\varphi(x,\bar{a}^*)\in p_i$ and $\bar{a}^*E\bar{b}^*$ so for all but fewer than χ ordinals $i\in S$, $\varphi(x,\bar{b}^*)\in p_i$. Similarly, for all but $<\chi$ ordinals $i\in S$, $\varphi(\bar{x},\bar{b}^*)\in p_{i+1}$. So for some $i\in S$, $\varphi(x,\bar{b}^*)\in p_i$ and $\varphi(\bar{x},\bar{b}^*)\in p_{i+1}$, but $\bar{b}^*=\bar{b}_i$ contradiction.
So 6.10 holds.

Thus $|S_{\{\varphi\}}(J)|\geq\chi$.
We define B_i $(i<\chi)$ by induction; such that
1) B_i is disjoint from $A^1\cup\bigcup_{j<i}B_j$.
2) $|B_i|=\kappa<\chi$ (remember $|J|\leq|I|\leq|A^1|^m=\kappa$).
3) No two elements of $\bigcup_{j<i}B_j$ realize the same $S^1_\varphi(J,M^1)$ type. (Possible, as $|\bigcup_{j<i}B_j|<\chi\leq|S^1_\varphi(J)|$).

4) If $\bar{a} \neq \bar{b}$ and $\bar{a}, \bar{b} \in J$, then for some admissible $c \in B_i$, $\varphi(c, \bar{a}) \equiv \neg\varphi(c, \bar{b})$ (possible, as $|J| \leq \kappa = |B_i|$ and the choice of J).

Since $Q^{eq}_{\chi,\kappa} \leq_{int} Q^{eq}_{\chi,<\chi}$, we can interpret an equivalence relation E^1 relation with equivalence classes the B_i's. Also, since $|J| \leq \kappa$, we can code sequences from J by single elements.

Let $\langle c_{i,\alpha}: \alpha < \kappa \rangle$ enumerate B_i, and $\{\bar{a}_\alpha : \alpha < \kappa\}$ enumerate J.

With equivalence relations from $Q^{eq}_{\chi,<\chi}$ we can code pairs from B_i by elements of B_i, so with a monadic predicate we can interpret

$Q = \{\langle c_{i,\alpha}, c_{i,\beta} \rangle : \varphi(c_{i,\alpha}, \bar{a}_\beta)\}$. Now we can interpret $S = \{\langle c_{i,\alpha}, \bar{a}_\alpha \rangle : i < \chi, \alpha < \kappa\}$ by the formula $\Theta(x, \bar{y}) \equiv x \in \bigcup_{i<\chi} B_i \wedge \bar{y} \in J \wedge (\forall z)(zE^1x \rightarrow (\varphi(z, \bar{y}) \equiv Q(z, x)))$.

Suppose $R = \{\langle d^1_\xi, d^2_\xi \rangle : \xi < \chi\}$ is a binary relation on $\bigcup_{i<\chi} B_i$. Let $P^l = \{c_{\xi,\alpha}: \varphi(d^l_\xi, \bar{a}_\alpha), \xi < \chi, \alpha < \kappa\}$ for $l = 1,2$.

Now $\langle b^1, b^2 \rangle \in R$ iff $(\exists x \in \bigcup B_i)(\forall y \in x/E^1) \bigwedge_{l=1}^{2} (y \in P^l \text{ iff } (\exists z \in J)(\langle y, z \rangle \in S \wedge \varphi(b^l, z)))$.

(We are coding b^l by the $\{\varphi\}$-type it realizes over J. Even though b^l might be in some other B_n, the code is on level B_ξ for b^l in the ξth pair of R).

So $Q^{eq}_{\chi,\chi} \leq_{int} Q_R$.

6.10 Remark : So we have proved that if in $(\mathcal{U}, R)$ for some $A, \mu = |S^1_\Delta(A, M)| > |A| = \kappa \geq \aleph_0$ (κ minimal) then $Q^{eq}_{\mu,\mu} \geq_{int} \{\exists_R, Q^{eq}_{\mu,\kappa}\}$.

6.11 Theorem : Suppose χ is infinite. Then

1) $\exists_R \equiv_{int} \{Q^{eq}_{\chi,\chi}, \exists_E, \exists_{R_1}\}$

Where E is an equivalence relation, $|\text{Dom } R_1| \leq 2^\chi$.

2) Also for some M and (finite) Δ, there are A^0, A^1, $|A^0| = \chi$, $|S^{m(0)}_\Delta(A^0, M)| \geq |A^1|$. $|S^{m(1)}_\Delta(A^1, M)| = |\text{Dom } R_1|$.

Proof : Combine the previous proofs.

§7 Summing Positive Results.

7.1 Theorem : If $V = L$, then any R is uniformly invariantly bi-expressible with Q_E, where E is some equivalence relation, or with $\{Q^{mon}_{\lambda_0}, \exists_{R_1}\}$, Dom R_1 finite.

Proof : Clearly $\lambda_3 = \lambda_2 \leq 2^\chi$ is the only remaining case. We can find A such that $|A| = \chi$, $|S^1_\Delta(A, M)| = \chi^+$, $Q^{eq}_{\chi,\chi} \leq_{int} \exists_R$. ($\Delta = \{$ atomic and negated atomic formulas in the language of $M\}$ (and is finite).

We shall show that we can express $Q^{word}_{\chi^+}$.

On A we can interpret the structure $\langle L_\chi, \in \rangle$. For every a, $tp_\Delta(a, A, M)$ can be viewed as a subset of L_χ.

We express an ordering

$a \le b$ *iff* ($\exists$ well-founded $\mathcal{L}$) [$\mathcal{L} \models$ "I am an L_α for some α" and $\mathcal{L}$ extends (as $\langle L_\chi, \in \rangle$ already interpreted) and the subsets of L_χ which a and b represent appear in $|\mathcal{L}|$ and the subset representing a occurs earlier].

$\le$ is a well-founded linear quasi-ordering.

Use a monadic predicate to pick out one element from each of the induced equivalence classes. This gives us a well-ordering of order-type χ^+. By 5.18 we finish. Q.E.D.

7.2 Conclusion : ($V=L$) for every K either for some family $\mathbf{E}$ of equivalence relation, $\exists_K, \exists_{\mathbf{E}}$ are uniformly invariantly bi-expressible or for some finite family K_f of finite relations and λ, $\exists_K, \{Q^{mon}_{<\lambda}, \exists_{K_f}\}$ are uniformly invariantly bi-expressible (if we omit uniformly we can omit the second case.)

Remark : On analysing $\mathbf{E}$ see 1.5.

* * *

For some χ we can close the gap (χ, λ) more easily, so such χ are impossible.

7.3 Lemma : Suppose M is admissible. And for some finite Δ, m and $A, |A| = \chi$, $|S^m_\Delta(A)| = \mu > \chi$ and $B \subseteq A$, $[|B| < \chi \Longrightarrow |S^m_\Delta(B)| \le \kappa]$, $\chi \le \kappa < \mu$ and χ is singular,

Then 1) $\{\exists_R, Q^{eq}_{\kappa,\kappa}, Q^{eq}_{\mu, cf\,\chi}\} \le_{int} Q^{eq}_{\mu,\mu}$

2) If $\aleph_0 < cf\,\chi$, $2^{cf\,\chi} < \chi$, μ regular, then $\{\exists_R, Q^{eq}_{\kappa,\kappa}\} \le_{int} Q^{word}_\mu$.

Remark : For (1) note that if $cf\,\chi = \aleph_0$, then $Q^{eq}_{\mu,\aleph_0} \le_{inex} Q^{1-1}_\mu$.

Proof : We can interpret in (an admissible expansion of) M, a tree T of power κ, with $cf\,\chi$ levels, and μ branches $\{B_i : i < \mu\}$ (of order type $cf\,\chi$).

If μ is regular, we can assume that $x \in B_i \Longrightarrow |\{j : x \in B_j\}| = \mu$ so each B_i can be coded by a set W_i of length $cf\,\chi$ branches, as its limit, with $[i \ne j \Longrightarrow B_i \cap B_j = \phi]$ and (1) follows.

If μ is singular, we can similarly code $Q_{\mu,<\mu}$ and finish (1) by 6.4.

For (2) we can consider $\{B_i : i<\mu\}$ as a set of function in $\kappa^{cf\,\chi}$, which are pairwise eventually distinct. By [Sh 7] for some ultrafilter D over $cf\,\chi$ and $I\subseteq\mu$, $\{B_i : i\in I\}$ is well ordered by $<_D$.

§8 Complementary Independence Results

8.1 Lemma : Suppose $\lambda=\lambda^{<\lambda}>\aleph_0$, $\mu>\lambda$, and P is the forcing for adding μ functions $F_i:\lambda\to 2$ $(i<\mu)$ (equivalently a function $F:\mu\times\lambda\to 2$, $F(i,\alpha)=F_i(\alpha)$ by conditions of power $<\lambda$) . Let for $S\subseteq\mu$, $\underset{\sim}{R}_s$ be the following partial order $<_S$ on ${}^{\lambda>}2\cup S$: $x\leq y$ iff $(\exists\alpha<\lambda)(y\in{}^{\alpha}2\wedge y=x\restriction\alpha)$ v $(\exists\alpha<\lambda)(x\in{}^{\alpha}2\wedge y\in S\wedge x\subseteq F_y)$ (so $\underset{\sim}{R}_S$ is defined in V^P). We let for $x\in S$ and $\alpha<\lambda$, $x\restriction\alpha=^{def}F_i\restriction\alpha$; for x an ordinal $\notin S$, let $x\restriction\alpha=\langle -1\rangle$.

Then $Q_{\lambda}^{word}\not\leq_{int}Q_{R_S}$ (we assume $\{\alpha\cup{}^{\lambda>}2=\mathcal{U}$ for some ordinal α such that $S\subseteq\alpha$).

Proof : Suppose not, then for some $p\in P$ and first order $\varphi(x,y,\bar{c},\bar{H},R_S)$,

$p\Vdash_P$ " $\bar{c}$ a finite sequence of elements of $\mathcal{U}$,$\underset{\sim}{\bar{H}}=\langle \underset{\sim}{H}_l : l<n^*\rangle$ a finite sequence of permutations of $\mathcal{U}$, and $\{\langle x,y\rangle : \varphi(x,y,\underset{\sim}{c},\underset{\sim}{\bar{H}},\underset{\sim}{R}_S)\}$ is a well ordering of order-type λ^+ and w.l.o.g. $\{\underset{\sim}{H}_l^{-1} : l<n^*\}=\{\underset{\sim}{H}_l : l<n^*\}$."

As P satisfies the λ^+-chain condition, there is $S_0\in V$, $S_0\subseteq\mathcal{U}$, $|S_0|\leq\lambda$ such that

$p\Vdash_P$ " ${}^{\lambda>}2\cup S_0$ is closed under $\underset{\sim}{H}_l$ for $l<n^*$ ".

Let $\underset{\sim}{M}=(\mathcal{U},\underset{\sim}{R}_s,\underset{\sim}{H}_0,\ldots,\underset{\sim}{H}_{n^*-1})$. For notational simplicity let $\mathcal{U}={}^{\lambda>}2\cup S$. Let

$K=\{I : I$ a model of the form $(|I|,f_0^I,\ldots,f_{n^*-1}^I)$, each f_l a permutation of $|I|$, and I has no proper (non empty) submodel $\}$.

Clearly, $K\in V$, and each $I\in K$ as cardinality $\leq\aleph_0$.

We let $K^1\subseteq K$ be a set of representatives of the isomorphism types, and $I\in K^1\Longrightarrow|I|\leq\aleph_0$ hence $K^1\in V$. In V^P we define, for each $I\in K$:

(a) we call $\langle x_t : t\in I\rangle$ a component if $H_l(x_t)=x_{f_l^I(t)}$.

(b) $A_I=\{\langle\eta_t : t\in I\rangle$: $\eta_t\in{}^{\lambda>}2$ for every $t\in I$, and there are function $G_i:I\to\mathcal{U}(i<\lambda^+)$ with pairwise disjoint ranges, such that $\eta_t=G_i(t)\restriction l(\eta_t)$ and $\bigwedge_{i<\lambda^+}(\forall t\in I)[\bigwedge_{l<n^*}G_i(f_l^I(t))=H_l(G_i(t))]\}$.

Note that $\mathcal{U}(={}^{\lambda>}2\cup S)$ is partitioned into components. (x,y are in the same components iff $x=z_0$, $y=z_k$ and $\bigwedge_{m<k}\bigvee_{l<n}, z_{n+1}=H_l(z_m)$ for some k and $\langle z_m : 0<m<k\rangle$).
So in V^P there is S_1, $|S_1|\le\lambda$, ${}^{\lambda>}2\subseteq S_1$ and for every $x\in\mathcal{U}-S_1$, its component $\langle x_t : t\in I\rangle$ is disjoint to $\mathcal{U}-S_1$, and for every $\alpha<\lambda$, $\langle x_t\restriction\alpha : t\in I\rangle\in A_I$.
So again as P satisfies the λ^+- chain condition, we can assume $S_1\in V$, and that forcing of $\langle F_i{:}i\in S_1\rangle$ also determines $\langle A_I{:}I\in K^1\rangle$ So let $V^1=V[\langle F_i{:}i\in S_1\rangle]$, P^2 the quotient forcing (which is just forcing $\langle \underset{\sim}{F}_i : i\in\mu-S_1\rangle$ by approximations of power $<\lambda$.)
Notice that in order to know in $V^1[G^2]$ that $\models\varphi[x,y]$ (which holds), $(x,y\in S)$ it is not enough to know $\langle x\restriction\alpha,y\restriction\alpha\rangle$ for large enough α, though it is enough to know $p_1\in G$ for some large enough $p_1\in P^2$, which force it! $V^1[G^2]$). A simple example is $\psi(x,y)=[H_0(x)=y]$. But something similar and more general holds.

Fact : 1) If ψ is a formula from L_{ω_1,ω_1}, $\langle x_t^\xi{:}t\in I_\xi\rangle$ for $\xi<\xi_0$ distinct components disjoint to S_0, and $\bar{y}$ is a countable sequence from S_1,$(I_\xi\in K^1)$ $\models\psi[\cdots x_t^\xi,\ldots,\bar{y}]_{\xi<\xi_0,t\in I_\xi}$ for $t\in I_\xi$, $\xi<\xi_0$
then for some $\alpha<\lambda$:

(*) if $z_t^\xi\in\mathcal{U}-S_1$, $\langle z_t^\xi{:}t\in I_\xi\rangle$ distinct components and $z_t^\xi\restriction\alpha=x_t^\xi\restriction\alpha$ **then** $\models\psi[\cdots z_t^\xi,\ldots,\bar{y}]_{\xi<\xi_0,t\in I_\xi}$.
2) We could also have assumed that for any such ψ, $I_\xi(\xi<\xi_0)$ and $\bar{y}$, the P-name

$T_{\psi,<I_\xi:\xi<\xi_0>}=$ $\{\langle\ldots,\eta_t^\xi,\ldots\rangle : t\in I_\xi,\ \xi<\xi_0$ for λ^+ pairwise distinct components, $\langle\cdots x_t^{\xi,i}\cdots : t\in I_\xi\rangle$,

$\models\psi[\cdots x_t^{\xi,i},\ldots,\bar{y}]_{\xi<\xi_0,t\in I_\xi}$

and for some γ, $x_t^{\xi,i}\restriction\gamma=\eta_t^\xi$ for every $t\in I_\xi,\xi<\xi_0,i<\lambda^+\}$
depend only on $\langle \underset{\sim}{F}_i{:}i\in S_1\rangle$

Proof: For 2) close S_1 λ or just $\aleph_1$ times, and if $\langle\eta_t^\xi : t\in I_\xi,\ \xi<\xi_0\rangle$ is not in $T_{\psi,\langle I_\xi:\xi<\xi_0\rangle}$, but $\models\psi[\cdots x_t^\xi,\ldots,\bar{y}]$, $x_t^\xi\restriction\gamma=\eta_t^\xi$ for every $t\in I_\xi,\xi<\xi_0$ then $\{x_t^\xi : t\in I_\xi,\ \xi<\xi_0\}\cap S_1\neq\phi$.

Now we can prove by induction on the depth of ψ (in V^1) that the fact is forced (i.e. $\Vdash_{P^s}$)

From the fact, and the Tarski-Vaught criterion we can conclude (in V^P) that if $S_1 \subseteq \mathcal{U}_1 \subseteq \mathcal{U}$, $\mathcal{U}_1$ closed under H_l $(l<n^*)$, then $M \upharpoonright \mathcal{U}_1$ is an L_{ω_1,ω_1}- elementary submodel of M. By increasing S_1 further we get that this holds for any $\mathcal{U}_1 \subseteq \mathcal{U}$, extending S_1 and closed under H_l $(l<n^*)$.

Now if $Q_{\lambda^+}^{word} \leq_{int} Q_{R_S}$ then we can find $I \in K$ $\psi \in L_{\omega_1,\omega}, \bar{y} \subseteq S_1$, and distinct components $\langle X_t^\xi : t \subseteq I \rangle$ disjoint to S_1 such that for $\xi,\zeta<\lambda^+$, $M \models \psi(\cdots x_t^\xi, \ldots; \ldots, x_t^\zeta, \ldots, \bar{y})$ iff $\xi<\zeta$.

Now, for some $\xi_0<\lambda^+$, for every $\xi \in (\xi_0,\lambda^+)$ and $\alpha<\lambda$ for arbitrarily large $\zeta<\lambda^+$, $\bigwedge_{t\in I} x_t^\xi \upharpoonright \alpha = x_t^\zeta \upharpoonright \alpha$. Using the fact, the contradiction is easy.

8.2 Lemma : 1) In 8.1 if $S, S^1 \subseteq M$, $|S-S^1|>\lambda$, then $\exists_{R_S} \not\leq_{int} \exists_{R_{S^1}}$.

2) If $\kappa<\lambda$, in 8.1 we can get $Q_{\lambda^+}^{word} \not\leq_{int} \{\exists_{R_S}, Q_{|\mathcal{U}|,\kappa}^{eq}\}$ and even $\exists_{R_S} \leq_{int} \{\exists_{R_{S^1}}, Q_{|\mathcal{U}|,\kappa}^{eq}\}$.

3) If $\kappa<\lambda$, in 8.1 we get $Q_{\lambda^+,\kappa^+}^{eq} \not\leq_{int} \{\exists_{R_S}, Q_{|\mathcal{U}|,\kappa}^{eq}\}$.

Proof : 1) Similar.

2) We use L_{κ^+,κ^+} instead L_{ω_1,ω_1}, and κ permutations $H_i(i<\kappa)$ (instead n^*), and repeat the previous proofs - but any $E \in K_{|\mathcal{U}|,\kappa}^{eq}$ can be defined by an L_{κ^+,κ^+}-formula using suitable κ permutations.

3) Similar proof.

8.3 Lemma : (G.C.H.) If $\lambda>\aleph_1$ is regular $\mu=\lambda^+$, we can build $\langle F_i : i<\mu \rangle$ as required in 8.1, 8.2 without forcing.

Proof : See [Sh 6], 2.1.

The following lemma shows that we cannot prove 8.2 without some set-theoretic hypothesis.

8.4 Lemma : Suppose $V \models \chi=\chi^{<\chi} \wedge \chi<\lambda \wedge cf\,\lambda=\lambda$, then for some χ-complete forcing notion P of power $\lambda^{<\chi}$, satisfying the χ^+– c.c., $\Vdash_P$ "$2^\chi \geq \lambda$", and

1) $\Vdash_P$ "if $S \subseteq {}^\chi 2$, $cf\,|S| \neq \lambda$, R_S as in 8.1, then $Q_{|S|,|S|}^{eq} \leq_{int} \{\exists_R, Q_{\chi,\chi}^{eq}\}$".

2) in V^P, $\chi(R)=\chi_1<\lambda_3(R)$ implies $\lambda_3(R)=\chi^+=\lambda$.

Proof : We let P be the limit of the χ-support iteration $\langle P_i, \underset{\sim}{Q}_i : i<\lambda \rangle$ where $\underset{\sim}{Q}_i \in V^{P_i}$ is defined as follows:

let $({}^{\lambda}2)^{V^{P_i}}=\{f^i_\alpha:\alpha<(2^\chi)^{V^{P_i}}\}$ and

$Q_i=\{(F,A)$: F a function from a subset of ${}^{\chi>}2$ of power $<\chi$, into χ, $A\subseteq(\chi)^{V^{P_i}}$, $|A|<\chi$, and for $\beta,\gamma\in A, \alpha<\chi$, if $f^i_\beta\restriction\alpha=f^i_\gamma\restriction\alpha$, then $f^i_\beta\restriction\alpha\in\mathrm{Dom}\ F\}$,

$(F_1,A_1)\le(F_2,A_2)$ *iff* $F_1\subseteq F_2, A_1\subseteq A_2$ and if $\eta\ne\nu$ are in A_2 and

$[\beta\in A_1\wedge\gamma\in A_1\wedge\alpha<\chi\wedge f^i_\beta\restriction\alpha\ne f^i_\gamma\restriction\alpha\wedge\beta<\gamma$
$\wedge f^i_\gamma\restriction\alpha\notin\mathrm{Dom}\ F_1\Longrightarrow F_2(f^i_\beta\restriction\alpha)<F_2(f^i_\gamma\restriction\alpha)]$.

8.6 Conjecture : It is consistent with ZFC that every $\exists_K$ is biinterpratable with some $\exists_{\mathbf{E}}$, $\mathbf{E}$ a family of equivalence classes.

8.7 Question: Prove it is consistent with ZFC that some $\exists_K$ is not bi-expressible with any $\exists_{\mathbf{E}}$,$\mathbf{E}$ a family of equivalence classes.

References.

[BS] J. Baldwin and S. Shelah, Classification of theories by second order quantifiers, Proc. of the 1980/1 Jerusalem model theory year, **Notre Dame J. of Formal logic**. to appear.

[H] A. Hajnal, Proof of a conjecture of S. Ruziewicz, *Fund. Math*. 50 (1961/62) 123-128.

[Sh1]
S. Shelah, First order theory of permutation groups, **Israel J. Math** 14 (1973), 149-162.

[Sh2]
S. Shelah, Errata to: First order theory of permutation groups, **Israel J. Math** 14 (1973) 262-279.

[Sh3]
S. Shelah, There are just for second order quantifiers. **Israel J. Math** 15 (1973) 282-300.

[Sh4]
S. Shelah, Models with second order properties I, **Annals of Math Logic**, 14 (1978) 52-72.

[Sh5]
S. Shelah, Models with second order properties III, omitting types in λ^+ for $L(Q)$ **Archiv. Math Logic**, 21 (1981) 1-11.

[Sh6]
S. Shelah, Models with second order properties IV **Annals of Math Logic**. 25 (1983), 183-212.

[Sh7]

S. Shelah, Powers of singular cardinals, **Notre Dame J. of Formal logic.**

[Sh8]

S. Shelah, The monadic theory of order, **Annals of Math**, 102 (1975) 379-419.

Classification over a Predicate II

§1 Introduction and preliminary facts.

Let T be a fixed complete first-order theory, and $P \in L(T)$ be a fixed monadic predicate.

Question: Describe the structure of $M \models T$ knowing $M \restriction P$.

When $(\forall x) \neg P(x) \in T$, this is a problem addressed [Sh 1], [Sh 4].

If $\forall x \; P(x) \in T$, there is an extremely strong structure theory. Gaifman dealt with the case "M has few ($\leq |P^M|$) automorphisms over P^M " and gets a representation theorem.

But for us the maximal structureness will be

"M is prime and even primary over P^M".

This is parallel to the case "T categorical in λ"; but this is stronger: remember that by Loewenheim Skolem Theorem T (if non- trivial) has models in all $\lambda \geq |T| + \aleph_0$. So the exact parallel will be "$\|M\|$, $M \restriction P$ determine M", or at least "dim (M,P), $M \restriction P$ determine M." If we are interested in the "categoricity theorem" (= uniqueness) we can restrict oneself to the case:

1.0 Hypothesis : $(\forall M \models T)(|P^M| = \|M\|)$ and even $(\exists \psi \in T)(\forall M \models \psi)$ $(|P^M| = \|M\|)$ (to avoid having to deal with the possibility that T is uncountable, and $(\forall M) \models T)[|P^M| = \|M\|]$ because of Chang's two cardinal theorem failing for all $\lambda \geq |T|$). The last condition is equivalent to: $[N \prec M \models T,\; P^M \subseteq N \Longrightarrow N = M]$.

We add * to the theorems assumings Hypothesis 1.0 (in our main conclusion here we shall do.)

This means that generally from P^M we cannot reconstruct M, not even its power.

We have start to deal with the problem in [Sh 2], but reading of it is not required (see there on other works on the subject of Gaifman Hodges and Pillay).

Section 3-4 are given almost as they were lectured in the seminar, hence are less formal but are more detailed and repetitious then usual. We do not try to save on set theoretic assumptions. In [Sh 1] the following classification is discussed.

unstable	*stable* - *superstable*	*superstable* - $\aleph_0$*-unstable* (*only* for *categories of models of countable theories*)	$\aleph_0$*-stable*

This corresponds to, roughly:

for every $p \in S(A)$:
stability $\Longrightarrow$ each $p \restriction \varphi$ is definable
superstability $\Longrightarrow p$ is almost definable over some finite $B \subset A$
$\aleph_0$-stability $\Longrightarrow p$ definable over some finite $B \subset A$.

We expect that the classification will be (this) $\times\ \omega$ with ω levels of complexity. Each time, for the unstable case, a non-structure theorem for $|T|^+$-saturated models, and for the unsuperstable T a non-structure theorem for $\aleph_\varepsilon$-saturated models. Only in the stable case we can continue to the next level. In fact it seemed that in order to get non-structure from unsuperstability we need first stability for all levels. We expect that the solution will be long, involving many branches. We concentrate on the stable/unstable dichotomy and quite saturated models. We shall use in "non-structure" proofs hypothesis like $G.C.H$, $V = L$ freely. If we do not do this we maybe forced to look at diagrams we get at approximation of less comfortable cofinalities;

if the properties are distinct the picture will be even more elaboborate. Let us explain more the expected classification.

n = - 1. Is every relation on P^M definable in M, also definable in $M \restriction P^M$?

1.1 Hypothesis: We assume, yes and even:

"every formula is equivalent (by T) to an atomic relation." (see [Sh 2])

n = 0. If M is saturated, $\|M\| = \lambda > |T|$, is M determined by $M \restriction P$? Its isomorphism type, yes but its isomorphism type over $M \restriction P$ not necessarily.

1.2 Hypothesis : For every $\bar{a} \in M \models T$ and φ, $p = tp_\varphi(\bar{a}, P^M)$ is definable (i.e. for some ψ_φ, and $\bar{c} \in P^M$; $\forall \bar{b} \in P^M[\varphi(\bar{x}, \bar{b}) \in p \Leftrightarrow \psi_\varphi(\bar{b}, c)]$. (see [Sh 2])

1.3 Theorem : If M is saturated, $\|M\| = \lambda > |T|$, then M is λ-prime over P^M among the λ saturated models, and is even λ-primary over it (i.e. $|M| = \{a_i : i < \alpha\}$, $tp(a_j, P^M \cup \{a_i, i < j\})$ is λ-isolated for λ regular; this proves uniqueness over P^M).

This is a weak structure theorem.

Proof: Note:

1.4 Fact: For every $\bar{c} \in M \models T$, $tp(\bar{c}, P^M)$ is $|T|^+$-isolated, in fact if $M \prec N$, then $tp(\bar{c}, P^M) \vdash tp(\bar{c}, P^N)$.

This follows from Hypothesis 1.2: for every φ there are $\psi_\varphi, \bar{c}_\varphi$ (ψ_φ does not depend on $\bar{c}$, only on $\ell(\bar{c}), \bar{c}_\varphi \subset P$) such that:

$$(\forall \bar{y} \subset P)[\varphi(\bar{c}, \bar{y}) \equiv \psi_\varphi(\bar{y}, \bar{c}_\varphi)].$$

So the formula $\Theta_\varphi(\bar{x}, \bar{c}_\varphi) = (\forall \bar{y} \subset P)[\varphi(\bar{x}, \bar{y}) \equiv \psi_\varphi(\bar{y}, \bar{c}_\varphi)]$ is satisfied by $\bar{c}$, its parameters are from P^M, so $\Theta_\varphi(\bar{x}, \bar{c}_\varphi) \in tp(\bar{c}, P^M)$ and easily $\Theta_\varphi(\bar{x}, \bar{c}_\varphi) \vdash tp_\varphi(\bar{c}, P^M)$. Hence,

$$\{\Theta_\varphi(x, \bar{c}_\nu) : \varphi \in L\} \begin{matrix} \subseteq tp(\bar{c}, P^M) \\ \vdash tp(\bar{c}, P^M) \end{matrix}$$

So $tp(\bar{c}, P^M)$ is $|T|^+$-isolated.

If $M \prec N$, then $N \models \Theta_\varphi(\bar{c}, \bar{c}_\varphi)$ hence

$$\{\Theta_\varphi(x,\bar{c}_\varphi):\varphi \in L\} \begin{matrix} \subseteq tp(\bar{c},N) \\ \vdash tp(\bar{c},N) \end{matrix}$$

but $\{\Theta_\varphi(x,\bar{c}_\varphi):\varphi \in L\} \subseteq tp(\bar{c},P^M)$.

Proof of the Theorem 1.3: Let $|M| = \{a_i : i < \lambda\}$. As $\lambda > |T|$, by the fact for $j < \lambda$ $tp(\langle a_i, i \le j\rangle, P^M)$ is isolated by a subset of power $\le |T| + |j| < \lambda$ (taking union on all finite subsequences). Hence $tp(a_j, P^M \cup \{a_i : i < j\})$ is λ-isolated. So M is λ-primary over P^M, etc. (see [Sh 1], Ch. IV).

1.5 Notation: Let $\mathfrak{C}$ be a very saturated model on T; we restrict ourselves to "small" elementary submodels of it. (see [Sh 1], Ch. I, §1).

1.6 Definition : $A \subseteq \mathfrak{C}$ is complete if $\mathfrak{C} \restriction (A \cap P) \prec \mathfrak{C} \restriction P^{\mathfrak{C}}$ and for every $\bar{a} \in A$ and φ there is $\bar{c}_{\varphi,\bar{a}} \subseteq A \cap P$ such that $\models \Theta_\varphi(\bar{a},\bar{c}_{\varphi,\bar{a}})$ (Θ_φ as previously). An equivalent formulation is: for every formula $\varphi(\bar{x},\bar{y})$ and $\bar{b} \in A$, if $\models (\exists \bar{x} \subseteq P)\, \varphi(\bar{x},\bar{b})$ then for some $\bar{a} \subseteq A \cap P$, $\models \varphi[\bar{a},\bar{b}]$.

Hence if $M \cap P^{\mathfrak{C}} \subseteq A \subseteq M$, then A is complete.

1.7 Remark: 1) If A,T are countable, this means (by the omitting type theorem):

$$\exists M (A \subseteq M \wedge M \cap P = A \cap P)$$

2) if $A \cap P$ is λ-saturated, $\lambda = |A|$ this means the same.

1.8 Definition : $S_*(A) = \{tp(\bar{a},A) : A \cup \bar{a}$ complete, $\bar{a} \cap P = \phi\}$. Of course A complete, and let $S_*^m(A) = \{p \in S_*(A) : p = tp(\bar{a},A), \ell(\bar{a}) = m\}$.

1.9 Explanation: We are reconstructing M from P^M. It is reasonable to try to do this using intermediate A, $P^M \subseteq A \subseteq M$ but then the types in which we may be interested in realizing are only those from $S_*(A)$.

1.10 Explanation: From where comes the ω levels of the classification? We try to reconstruct M from P^M (e.g. in the case of categoricity). We let $\|M\| = \lambda_0$, let $M = \bigcup_{i<\lambda_0} M_i$, M_i increase continuously, $\|M_i\| = |T| + |i|$. This can be decomposed to λ_0 problems of:

"reconstruct M_{i+1} over $P^M \cup M_i$"

By Hypothesis 1.2 (see Fact) $tp_*(M_{i+1}, P^{M_{i+1}} \cup M_i) \vdash tp_*(M_{i+1}, P^M \cup M_i)$, so we have the reconstruction problem of M_{i+1} over $P^{M_{i+1}} \cup M_i$. We can decompose the diagram again, decreasing the power while increasing the diagram to 2^n sets. This is similar to [Sh 3] (and [Sh 4] XII §5, but there only the good cases occur). Note that if we allow $|P^M| < \|M\|$ an extra complication arises.

If we have "good" behaviour for one power, every n, we can prove it for all larger powers. For each n we look at n-dimensional diagram $A = \bigcup_{w \subset n} A_w$ ($A_w \prec P^M$ if $0 \notin w$, $A_w \prec \mathfrak{C}$ if $0 \in w$), and ask about $|S_*^m(A)$. If we get stability (i.e. $|S_*^m(A)| \le |A|^{|T|}$), we can define $(n+1)$-diagrams [as we like to have that $tp_*(A_u, A_v)$ is determined by $tp_*(A_u, A_{v \cap u})$, and get some uniqueness we deal with them mainly when stability for n-diagrams was already proved, and 1.0 help simplifying]. If e.g. for every n the parallel of $\aleph_0$-stability holds, we would be able to prove "M is prime over P^M". From instability we will try to get non-structure theorems. We shall deal with ranks corresponding to stability (unstability .)

1.11 Definition : For every complete set A, for Δ_1, Δ_2 (sets of formulas $\varphi(\bar{x})$) we define $R = R_A^m(p, \Delta_1, \Delta_2, \lambda)$ (we sometimes omit A).
[the rank measure how close we are to:
p has a perfect set $\neq \phi$ of extensions in $S_*^m(A)$
Δ_1 is for "many extension"
Δ_2 is for "$A \cup \bar{x}$ is complete".]

We now define by induction when $R \ge \alpha$.

1) $R = -1 \iff p \vdash \bigvee_{\ell < \ell(\bar{x})} P(x_\ell)$

2) $R \ge 0 \iff R \ne -1 \iff p \cup \{\neg P(x_\ell) : \ell < \ell(\bar{x})\}$ is finitely satisfiable.

3) $R \ge \delta$ (δ limit) $\iff R \ge i$ for every $i < \delta$.

4) $R \ge \alpha + 1 \iff$ for every finite $q \subseteq p$ and cardinals μ, κ where [α odd $\Rightarrow \mu = 0$] , [α even $\Rightarrow \kappa = 0$] and $\mu + \kappa < \lambda$, and for every formula $\varphi_i(\bar{x}, \bar{y}_i \bar{z}_i) \in \Delta_2$ and $\bar{b}_i \in A$ $(i < \kappa)$ there are $\Delta_1 - m -$types $r_j (j \le \mu)$ pairwise

explicitly contradictories and $\bar{d}_i \in P^{\mathfrak{C}} \cap A(i \le \kappa)$, such that:

$$R^m(\ q \cup r_j \cup \{(\forall \bar{z}_i \subset P)[\varphi_i(\bar{x},\bar{b}_i,\bar{z}_i) \equiv \psi_{\varphi_i}(\bar{z}_i,\bar{d}_i):i < \kappa\}],\quad \Delta_1,\Delta_2) \ge \alpha$$

1.12 Remark: It does not matter whether we fix $\langle \psi_\varphi : \varphi \in L \rangle$ or just asked for "some suitable ψ_φ".

1.13 Definition : If K is a category of complete $A \subset \mathfrak{C}$ and some embeddings $f:A \to \mathfrak{C}$, then we can define R for K rather than for A, allowing in the definition to replace A by K-extension (i.e. the r_j but not $\bar{d}_i$ can be found there).

1.14 Claim: If $\mathfrak{C} \restriction A$ can be expanded to an $\aleph_0$-saturated model, and Δ_1,Δ_2 are finite, *then* $R_A^m(p,\Delta_1,\Delta_2,\aleph_0)$ is finite or ∞. (We make explicit the dependency on A).

Proof: By compactness. (similarly to [Sh 1], ch. II §2)

§2 Ranks and non-structure for n=1,2.

2.1 Remark: We concentrate on the case $\Delta_1,\Delta_2,\lambda$ finite, this lead to the "stable/unstable" dichotomy.
Of course the rank has obvious monotonicity and the finite character properties.

2.2 Claim: For every finite m $\Delta_1,\Delta_2,\lambda,n$ and $\varphi(\bar{x},\bar{y})$ there is a formula $\Theta(\bar{y})$ such that for any complete A and $\bar{a} \in A$

$$R_A^m(\varphi(\bar{x},\bar{a}),\Delta_1,\Delta_2,\lambda) \ge n \quad iff \quad \mathfrak{C} \restriction A \models \Theta[\bar{a}]$$

Proof: By induction on n.

2.3 Definition : 1) We say p is Δ_1-big (for A) *if* A is complete and $R_A^m(p,\Delta_1,\Delta_2,2) \ge \omega$ for ever finite Δ_2

2) A is unstable if for some finite Δ_1, $\{\bar{x} = \bar{x}\}$ is Δ_1-big for A.

2.4 Lemma : Suppose A is complete and stable. Then $|S_*^m(A)| \le |A|^{|T|}$.

Proof : For every $p \in S_*^m(A)$ we can find a complete $q_p \subset p$, of cardinality $\le |T|$ such that for every finite Δ_1,Δ_2 : $R_A(p,\Delta_1,\Delta_2,2) = R_A^m(q_p,\Delta_1,\Delta_2,2)$. If

$|S^m_*(A)| > |A|^{|T|}$,

then for some finite Δ_1, $\{p \restriction \Delta_1 : p \in S^m_*(A)\}$ has power $> |A|^{|T|}$, there are $B \subseteq A, |B| \leq |T|$, q and $p, p_i \in S^m_*(A)$ for $i < (|A|^{|T|})^+$ such that $q_{p_i} = q_p \in S^m(B)$ hence $p_i \restriction B = p \restriction B$, and the $p_i \restriction \Delta_1$ are pairwise distinct. The rest is easy noting:

2.5 Fact: If A is complete, $p \in S_*(A)$, then $R_A(p, \Delta_1, \Delta_2, 2)$ is ∞ or is even.

2.6 Lemma : If $|A| = \lambda$, A complete, $\mathfrak{C} \restriction A$ saturated, A unstable, then $|S^m_*(A)| = 2^\lambda$.
In fact: there is a finite Δ_1 such that $|\{p \restriction \Delta_1 : p \in S^m_*(A)\}| = 2^\lambda$.

Proof: There is Δ_1 such that $R^m_A(\bar{x} = \bar{x}, \Delta_1, \Delta_2, 2) > n$ for every finite Δ_2 and n. We define by induction on $\alpha < \lambda$ for every $\eta \in {}^\alpha 2$ an m-type p_η over A such that:

(1) $|p_\eta| < \aleph_0 + |\ell(\eta)|^+$

(2) for every finite Δ_2 $R^m_A(p_\eta, \Delta_1, \Delta_2, 2) \geq \omega$

(4) If $\alpha = \beta+1$, $\nu \in {}^\beta 2$ then for some $\varphi \in \Delta_1$, $\bar{c} \in A, \varphi(\bar{x}, \bar{c}) \in p_{\eta\frown\langle 0\rangle}$ and $\neg\varphi(\bar{x}, \bar{c}) \in p_{\eta\frown\langle 1\rangle}$.

(5) For every formula $\varphi(\bar{x}, \bar{a}, \bar{z})$, $\bar{a} \in A$, for some α, for every $\eta \in {}^\alpha 2$, for some $\bar{c} \in A \cap P$ $(\forall \bar{z} \subseteq P)\,(\varphi(\bar{x}, \bar{a}, \bar{z}) \equiv \psi_\varphi(\bar{z}, \bar{c})) \in p_\eta$.

For $\alpha = 0$, α limit no problem.

How to satisfy (4)?:

As Δ_1 is finite we can code it by one formula (see [Sh 1] II 2.1); so let $\Delta_1 = \{\varphi(\bar{x}, \bar{y})\}$. What are the demands on $\bar{c}$? Write $\bar{z}$ for $\bar{c}$: $\{R^m_A(q \cup \{\varphi(\bar{x}, \bar{z})^{\mathbf{t}}\}, \Delta_1, \Delta_2, 2) \geq n$: for finite $q \subseteq p$ any $\mathbf{t}$ and any finite $\Delta_2, n\}$

(where $\mathbf{t}$ is false or truth, $\varphi^{truth} = \varphi$, $\varphi^{false} = \neg\varphi$)

By claim 2.2 each demand is first order in $\mathfrak{C} \restriction A$. As $\mathfrak{C} \restriction A$ is saturated, $|p_\nu| < \lambda = |A|$, it is enough to show any finitely many demands are satisfiable. By monotonicity in rank just *one* is enough; say $R^m_A(q \cup \{\varphi(\bar{x}, \bar{z})^{\mathbf{t}}\}, \Delta_1, \Delta_2, 2) \geq n$. But $R^m(q, \ldots) \geq n+2$ and use this.

A similar proof works for (5).

2.7 Remark: Now there are theorems which give us for unstable A and $\mu \geq |T|$ an $A' \equiv A$, $|S^m_*(A')| > \mu \geq |A'|$.
But we shall be "easy" on the non-structure side, as this is not our main concern in these notes.

2.8 Question: Is some (= every) model stable?
Meanwhile we assume *no* and get some non-structure theorems, then we will assume *yes* and continue.

2.9 Note: We shall observe that: no $\Longrightarrow$ $(\exists M \models T)(|M| > |P^M|)$

2.10 Theorem : Suppose that for some models $M \subset N$, cardinal μ, and finite Δ_1, $P^N \subset M$, $||M|| \leq \mu$, $|\{tp_{\Delta_1}(\bar{a}, M) : \bar{a} \in N\}| \geq \mu^+$.
If $|T| < \lambda = \lambda^{<\lambda}, \diamondsuit_\lambda$, $2^\lambda < 2^{\lambda^+}$ and *then* there are 2^{λ^+} non-isomorphic models of T of power λ^+, with the same restriction to P.

Proof : Expand N to have enough set theory and get N^+, let $Q^{N^+} = M$. Let $N_{<>}$ be a saturated model of $Th(N^+)$ of power λ.
We define by induction on $\alpha < \lambda^+$ N_η, Γ_η (for $\eta \in {}^\alpha 2$) such that :

(1) N_η is saturated of power λ, elementarily equivalent to N^+, Γ_η a family of $\leq \lambda$ types omitted by N_η, moreover no one has a support over N_η in the sense of [Sh 5] (for carrying this we need $\diamondsuit_\lambda$).

(2) For $\beta < \ell(\eta)$; $N_{\eta \restriction \beta} \prec N_\eta$, $\Gamma_{\eta\restriction\beta} \subset \Gamma_\eta$, $P^{N_\eta} = P^{N_{<>}}$, and even $Q^{N_\eta} = Q^{N_{<>}}$.

(3) For $\alpha = \beta+1$, $\nu \in {}^\beta 2$, there is a Δ_1—m-type over $P^{N_{<>}}$ realized by $N_{\nu\frown<0>}$ and belonging to $\Gamma_{\nu\frown<1>}$.

For the continuation of the process in the limit we have to have more induction hypothesis as in the paper above; in the case $\alpha = \beta+1$, $\nu \in {}^\beta 2$ N_ν has a λ-saturated extension in which λ^+ Δ_m—m-types complete over $Q^{N_{<>}}$ are realized. So there is one p_ν with no support $<\lambda$ over N_η. So let $\Gamma_{\nu\frown<1>} = \Gamma_\nu \cup \{p_\nu\}$, $N_{\nu\frown<0>}$ realizes p_ν, (we can get also the dual demand).

So, let for $\eta \in {}^{\lambda^+}2$: $N_\eta = \bigcup_\alpha N_{\eta \restriction \alpha}$. Over $Q^{N_{<>}}$ they are pairwise non-isomorphic; as $2^\lambda < 2^{\lambda^+}$, 2^{λ^+} of them are not isomorphic (even over ϕ) (easily, by [Sh 1, 1.2] and $N_\beta \restriction P^{N_\eta} = N_{<>} \restriction P^{N_{<>}}$ is the same.

Remark: We can eliminate the use of $\diamondsuit_\lambda$ by forgetting Γ_η by demanding that for $\alpha = \beta + 1$, $\nu \in {}^\beta 2$ there is a $\Delta_1 - m -$type p over $P^{N_{<>}}$ which $N_{\eta\hat{}<0>}$ realize it whereas $N_{\eta\hat{}<1>}$ "says" it is omitted (and you can demand that you can interchange them.)

2.11 Remark: We can replace λ-saturated by λ-compact.

2.12 Theorem: Suppose that some model is unstable, but the hypothesis of the last theorem fails.
If $|T| \leq \lambda = \lambda^{<\lambda}, \diamondsuit_{\{\delta<\lambda^+ : cf(\delta)=\lambda\}}$, then the conclusion of the last theorem holds.

Remark: We can replace diamond by weak diamonds.

Proof : We define by induction on α for every $\eta \in {}^\alpha 2$ a model N_η such that:

(1) N_η is λ-saturated when $\ell(\eta)$ is a successor or $cf(\ell(\eta)) = \lambda$

(2) $|N_\eta| = \lambda(1+\ell(\eta))$

(3) $N_{\eta \restriction \beta} \prec N_\eta$, $P^{N_\eta} = P^{N_{<>}}$

Let $\langle\langle \eta_\delta, \nu_\delta, F_\delta \rangle : \delta < \lambda^+, cf\ \delta = \lambda, \lambda^\omega$ divides δ(λ^ω is ordinal exponeniation)$\rangle$ be a $\diamondsuit$-sequence i.e. $F_\delta : \delta \to \delta$, $\eta_\delta \neq \nu_\delta \in {}^\delta 2$ and for every $\eta \neq \nu \in {}^{\lambda^+}2$, and function $F: \lambda^+ \to \lambda^+$ for some (in fact a stationary set of) $\delta : \langle \eta_\delta, \nu_\delta, F_\delta \rangle = \langle n \restriction \delta, \nu \restriction \delta, F \restriction \delta \rangle$; so F maps δ to δ.

(4) For each δ, there is a type q over N_{η_δ} which is realized in $N_{\eta_\delta\hat{}<0>}$ and also in $N_{\eta_\delta\hat{}<1>}$ but $F_\delta(q)$ is not realized in any λ- saturated extension N^+ of $N_{\nu_\delta\hat{}<0>}$ or $N_{\nu_\delta\hat{}<1>}$ with $P^{N^+} = P^{N_{<>}}$.
If we succeed; there will be no problem.

For $\alpha = 0$, α **limit** : no problem.

$\alpha = \beta+1$ β **successor** : Over N_ν there is a Δ_1-big $p \in S_*^m(N_\nu)$. Let it be realized by $\bar{c}$, $N_\nu \cup \bar{c}$ is complete, hence (as $\mathfrak{C} \restriction P^{N_\nu}$ is λ-saturated of power λ) there are (for $e = 0,1$) λ-saturated $N_{\nu^\frown<e>}$ for power λ, such that $N_\nu \cup \bar{c} \subseteq N_{\nu^\frown<e>}$, $P^{N_{\nu^\frown<e>}} = P^{N_\nu}$.

$\alpha=\beta+1$, $cf\ \beta < \lambda$: $N_\eta = \bigcup_{\gamma<\beta} N_{\eta\restriction\gamma}$ is a complete set with P^{N_η} saturated (see below); hence we can find $N_{\eta^\frown<e>} \supseteq N_\eta$ saturated with the same P. *We use freely:*

2.13 Claim: If A is complete, $\mathfrak{C} \restriction (A \cap P)$ λ-saturated, $|A| = \lambda$[and $|T| < \lambda = \lambda^{<\lambda}$], then we can find $N, P^N \subseteq A \subseteq N$ [and N is λ- saturated]. (like the proof of the unstability Lemma 2.6, but simpler).

The next case is:

$\alpha=\beta+1$, $cf\ \beta = \lambda$ and w.l.o.g. $\langle N_\delta, \nu_\delta, F_\delta \rangle$ is defined. We define by induction on i a model N^i of power λ, $N^0 = N_{\nu_\delta}$, $N^j \prec N^i$ for $j < i$, $P^{N_i} = P^{N_0}$ and there is $\bar{c}_i \in N_{i+1}$ such that $tp_{\Delta_1}(\bar{c}_i, N^0)$ is not realized in N^i. We define as long as we can for $i < \lambda^+$.
If we can continue for $i < \lambda^+$ we get the hypothesis of the previous theorem. As for limits we have no problem, there is a last N^{i^*}, w.l.o.g. (by 2.13) it is λ-saturated. Let $N_{\nu_\delta^\frown<e>} = N^{i^*}$ for $e = 0,1$. Now $|S_*^m(N_{\eta_\delta})| > \lambda$, $|N^{i^*}| \le \lambda$, so for some $q_\delta \in S_*^m(N_{\eta_\delta})$, $F_\delta(q_\delta)$ is not realized in N^{i^*}. Choose $N_{\eta_\delta^\frown<e>}$ to realize q_δ (possible as $q_\delta \in S_*^m(N_{\eta_\delta})$ not just $\in S^m(N_{\eta_\delta})$). For $\rho \in 2^\beta \setminus \{\nu_\delta, \eta_\delta\}$ you have more freedom. (We could have made the situation symmetric).

* * *

So we have shown non-structure when some M is unstable. Let us relist our hypothesis:

T complete, P one place predicate

n = - 1 Hypothesis A=1.1: every formula is equivalent to a relation

n = 0 Hypothesis B=1.2: For every $\bar{a} \in \mathfrak{C}$, $tp(\bar{a}, P^{\mathfrak{C}})$ is definable

n = 1 **Hypothesis C**: For every $M, |S_*^m(M)| \le ||M||^{|T|}$.

Note: For every M by $B, tp_*(M,P^M) \vdash tp_*(M,P^{\mathfrak{C}})$. The next stage is:

n = 2 **Question D**: Is every $M_0 \cup P^{M_1}$ stable, where $M_0 \prec M_1 \prec \mathfrak{C}$?

2.14 Theorem : Suppose the answer to question D is yes, $\lambda^{<\lambda} = \lambda, \lambda > |T| \ge \aleph_0$. If M is λ-saturated of power λ^+, then over P^M there is a λ-prime model
(So if $(\forall N \models T)((|N| = ||P||)$ then M is λ -prime over P^M)

2.15 Remark: Really: $\lambda^{<\lambda} \le \lambda^+, \lambda > |T|$ is enough.

Proof: If $|P^M| \le \lambda$ use the previous theorem 1.3.

Let $P^M = \bigcup_{i<\lambda^+} A_i$ increasing continuous, $\mathfrak{C} \restriction A_i \prec \mathfrak{C} \restriction P^M \prec \mathfrak{C} \restriction P$ and for $i = 0$, and i successor $\Longrightarrow \mathfrak{C} \restriction A_i$ is λ-saturated.

We define by inducton on i models M_i, increasing continuous, $M_i \cap P^{\mathfrak{C}} = A_i$, such that

(*) for every $\bar{c} \in M_{i+1}$ $tp(\bar{c}, M_i \cup A_{i+1})$ is λ-isolated

(**) M_0, M_{i+1} are λ-saturated, $||M_i|| = \lambda$.

(***) $tp(\bar{c}, A_0)$ is λ-isolated for $\bar{c} \in M_0$.

Why is this enough?
Let $M_0 = \{c_\alpha : \alpha < \lambda\}$ $M_{i+1} \setminus M_i = \{c_\alpha : \lambda(1+i) \le \alpha < \lambda(1+i+1)\}$, maybe with repeatitions.

Now $tp_*(\{c_\beta : \beta \le \alpha\}, A_0)$ is λ-isolated (as union of $< |\alpha|^+ + \aleph_0$ such types) but $tp_*(\{c_\beta, \beta \le \alpha\}, A_0) \vdash tp_*(\{c_\beta, \beta \le \alpha\}, P^M)$ so the latter is λ-isolated too; hence $tp_*(c_\alpha, P^M \cup \{c_\beta : \beta < \alpha\})$ is λ-isolated. Also for $(1+i)\lambda \le \alpha < (1+i+1)\lambda$ $tp(c_\alpha, P^M \cup \{c_\beta, \beta < \alpha\})$ is λ- isolated by:

2.16 Fact: If $A \cup \bar{a}$ is complete, then

$$tp(\bar{a}, A) \vdash tp(\bar{a}, A \cup P^{\mathfrak{C}})$$

Proof of Fact 2.16: For every $\bar{b} \in A$, $tp(\bar{a}^\frown \bar{b}, A \cap P) \vdash tp(\bar{a}^\frown \bar{b}, P^{\mathfrak{C}})$, hence $tp(\bar{a}, (A \cap P^{\mathfrak{C}}) \cup \bar{b}) \vdash tp(\bar{a}, P^{\mathfrak{C}} \cup \bar{b})$, taking unions over all $\bar{b} \in A$ we get the fact.

∎

We know $tp_*(\{c_\beta : (1+i)\lambda \le \beta \le \alpha\}, M_i \cup A_{i+1})$ is λ-isolated and

$$tp_*(\{c_\beta : (1+i)\lambda \le \beta \le \alpha\}, M_i \cup A_{i+1}) \vdash tp_*(\{c_\beta : (1+i)\lambda < \beta \le \alpha\}, M_i \cup P^M)$$

by the Fact.

Hence the latter is λ-isolated, hence $tp(c_\alpha, \{c_\beta : (1+i)\lambda \le \beta < \alpha\} \cup M_i \cup P^M)$ is λ-isolated, but this is $tp(c_\alpha, \{c_\beta : \beta < \alpha\} \cup P^M)$. So $tp(c_\alpha, \{c_\beta : \beta < \alpha\} \cup P^M)$ is λ-isolated for every $\alpha < \lambda^+$, and this is enough.

We still have to define M_i

For $i = 0$, as A_0 is complete λ-saturated of power λ, there is M_0, $P^{M_0} = A_0$, M_0 λ- saturated and we know M_0 satisfies (***) necessarily.

Note:

2.17 Fact: If B is complete, $\lambda = \lambda^{<\lambda} > \|T\|$, $\mathfrak{C} \restriction (B \cap P^{\mathfrak{C}})$ is λ- saturated, $|B| = \lambda$, then there is a λ-saturated $N \supset B, N \cap P^{\mathfrak{C}} = B \cap P^{\mathfrak{C}}$.

For $i+1$: As $M_i \cup A_{i+1}$ is complete, and its intersection with $P^{\mathfrak{C}}$ is (A_{i+1}, which is) λ-saturated, clearly by 2.17 there is $N_i \supseteq M_i \cup A_{i+1}, N_i$ λ-saturated $P \cap N_i = A_{i+1}$. We define by induction on $\alpha < \lambda, c_\alpha \in N$ such that $tp(c_\alpha, M_i \cup A_{i+1} \cup \{c_\beta : \beta < \alpha\})$ is λ-isolated. By standard bookkeeping it is enough to prove that if $p(x_\alpha)$ is a type over $M_i \cup A_{i+1} \cup \{c_\beta : \beta < \alpha\}$ of power $< \lambda$ then it has a λ-isolated extension (over this set).

By the induction hypothesis there is a type

$$q(x_\beta : \beta < \alpha) \subseteq tp_*(\langle c_\beta, \beta < \alpha \rangle, M_i \cup A_{i+1})$$

of power $< \lambda$ such that $q(x_\beta : \beta < \alpha) \vdash tp_*(\langle c_\beta, \beta < \alpha \rangle, M_i \cup A_{i+1})$. Replace in $p(x_\alpha)$ the c_β's by x_β and get $p'(x_\beta : \beta \le \alpha)$. So $p' \cup q$ is finitely satisfiable (in N_i) and of power $< \lambda$ and is over $M_i \cup A_{i+1}$. Let $\{(\bar{y}_\gamma, \Delta_1^\gamma, \Delta_2^\gamma) : \gamma < |\alpha| + |T|\}$ be the list of all triples $(\bar{y}, \Delta_1, \Delta_2)$; $\bar{y} \subseteq \{x_\beta : \beta \le \alpha\}$ is finite and $\Delta_1, \Delta_2 \subseteq L(T)$ are finite.

We define by induction on γ a type r_γ in N_i over $M_i \cup A_{i+1}$ where r_γ is increasing of cardinality $< \lambda$, $r_{\gamma+1} = r_\gamma \cup r^\gamma(\bar{y}_\gamma)$, r^γ finite over $M_i \cup A_{i+1}$, the union consistent and $R(r^\gamma(\bar{y}_\gamma),\Delta_1,\Delta_2,2)$ is minimal where the rank is for $M_i \cup A_{i+1}$ (minimality: under the constraints required). As $M_i \cup A_{i+1}$ is stable and as A_{i+1} is λ-saturated, $N_i \cap P = A_{i+1}$, we can extend $r_{|\alpha|+|T|}$ to r' so that its domain is a set $C \subseteq A_{i+1} \cup M_i$ and $r' \restriction \bar{y} \in S^m_*(C)$ for any finite $\bar{y} \subseteq \{x_\beta : \beta \leq \alpha\}$ of length m. Simply let $\langle c'_\beta : \beta \leq \alpha \rangle$ be a sequence in N_i realizing $r_{|\alpha|+|T|}$; now choose $C_0 \subseteq A_{i+1}$ so that $\forall \bar{d} \subseteq \{c'_\beta : \beta \leq \alpha\} \cup \mathrm{Dom}\ r_{|\alpha|+|T|}$, $tp(\bar{d}, P^{N_{i+1}}) = tp(\bar{d}, A_{i+1})$ is definable over C_0 and let $C = C_0 \cup \mathrm{Dom}\ r_{|\alpha|+|T|}$, $r' = tp(\langle c'_\beta : \beta \leq \alpha \rangle, C)$.

By the definition of $R(\cdots)$, 2.5, and as for no Δ_1 $(\forall n)$ ($\forall$ finite Δ_2) $R_{A_{i+1} \cup M_i}(\bar{x} = x\ ,\ \Delta_1,\Delta_2,2) \geq n$, clearly r' has a unique complete extension over $A_{i+1} \cup M_i$ (using the construction of r^1).

So we have finished proving 2.14.

2.18 Theorem : Suppose the answer (to Question D) is no, $\lambda = \lambda^{<\lambda} > |T|$. Let Q be the forcing of adding λ^+ Cohen subsets to λ. Then for some $A \prec P^{\mathfrak{C}}, |A| = \lambda^+$:

$\Vdash_Q$ "there are 2^{λ^+} λ-saturated models $M, P^M = A$, $\|M\| = \lambda^+$, pairwise non-isomorphic *over* A."

2.19 Remark: We can replace forcing by appropriate diamonds and get such models. Note that the answers to all our questions so far are absolute.

Proof : By assumption:

There is a triple: $P^M \overset{\subset M^*}{\underset{\subset P^{N^*}}{}}$, $M^* \prec N^*$ whose union, $P^{N^*} \cup M^*$, is unstable. We can prove that there are many such triples. But for us it is enough to do the following. We define (in V) by induction on $i < \lambda^+, A_i$ such that A_i is strictly increasing, continuous, $|A_i| = \lambda$, $\mathfrak{C} \restriction A_i \prec \mathfrak{C} \restriction P$, A_0, A_{i+1} are λ-saturated and when $cf\ i = \lambda$ $(\mathfrak{C} \restriction A_{i+1}, A_i) \equiv (\mathfrak{C} \restriction P^{N_*}, P^{M^*})$ and when $cf\ i \in \{0,1,\lambda\}$ $(\mathfrak{C} \restriction A_{i+1}, A_i)$ is λ-saturated.

For $i = 0$, i **limit** : no problem.

$1+i$, i *successor* or $cf\ i < \lambda$: easy.

$cf\ i = \lambda$: $\mathfrak{C} \restriction A_i$ is λ a saturated of power λ by the induction-assumption. $Th(\mathfrak{C} \restriction P^{N^*}, P^{M^*})$ has the λ- saturated model of power λ say (A, A^0), the A^0- part is saturated of power λ and has the theory of $\mathfrak{C} \restriction P$, hence is isomorphic to A_i. We can identify them and choose A_{i+1} as A^1.

Now for any sequence $\langle r_i : i < \lambda^+, cf\ i = \lambda \rangle = \bar{r}$ of Cohen subsets of λ we describe how to build a λ-saturated model $M_{\bar{r}}$ of T with $P^{M_{\bar{r}}} = \bigcup A_i$.

Before this:

2.20 Fact: If M is a λ-saturated model of T, $\|M\| = \lambda$, $M \cap P^{\mathfrak{C}} = A_i$, $cf\ i = \lambda$; *then* $M \cup A_{i+1}$ is a λ-saturated model of $Th(M^* \cup P^{N^*})$, and even $(M \cup A_{i+1}, A_i, M, A_{i+1})$ is a λ-saturated model of $Th(M^* \cup P^{N^*}, P^{M^*}, M^*, P^{N^*})$ (same argument as before plus use of 1.3).

We shall define $M_{\bar{r}} = \bigcup_{i<\lambda^+} M_{\bar{r},i}$, $M_{\bar{r},i}$ depends on $\bar{r} \restriction i$ only, $M_{\bar{r},i} \cap P^{\mathfrak{C}} = A_i$, $M_{\bar{r},i+1}$ is λ-saturated. So in $S^m(M_{\bar{r},i} \cup A_{i+1})$ there is a perfect set homeomorphic to ${}^{\lambda}2$; we can (see 2.6) choose a tree $\{p_\eta : \eta \in {}^{\lambda>}2\}$ of types $p_\eta \in S^m(C^\eta_{\bar{r},i})$ $C^\eta_{\bar{r},i}$ increasing with η, $p_{\eta^\frown\langle 0\rangle}$, $p_{\eta^\frown\langle 1\rangle}$ explicitely contradictory, $C^\eta_{\bar{r},i} \subseteq M_{\bar{r},i} \cup A_{i+1}$ has power $\leq \ell(\eta) + \aleph_0$ and $(\forall c \in M_{\bar{r},i} \cup A_{i+1})$ $(\exists \alpha)$ $(\forall \eta \in 2^\alpha)$ $[c \in C^\eta_{\bar{r},i}]$.

Now r_i define a branch $\eta_i \in {}^{\lambda}2$ and we demand that M_{i+1} realizes $\bigcup_{\xi<\lambda} p_{\eta_i \restriction \xi}$. We can carry this as under our hypothesis since:

Fact A: $tp_*(M, P^M) \vdash tp_*(M, P^{\mathfrak{C}})$

Fact B: If A is complete, $|A| \leq \lambda$, $A \cap P$ saturated of power λ then $(\exists N \supseteq A)[N$ is λ-saturated and $N \cap P = A \cap P]$.

Now, if we add to λ λ^+ Cohen subsets, there is no problem to define $\bar{r}_E$ (for

$E \subset \lambda^+, E \in V$ and $\langle\langle (S(E), g_E\rangle : E \subset \lambda^+, E \in V\rangle$ from V such that:

$\bar{r}_E \in V[r_j : j \in S(E) \subset \lambda^+]$,

$\bar{r}_E(i) = r_{g_E(i)}$, where $g_E : \lambda^+ \to S(E)$ is one to one and $g_E \in V$.

$E_1 \neq E_2 \Longrightarrow \{i < \lambda^+ : cf\ i = \lambda,\ r_{E_1}(i)$ does not appear as $\bar{r}_{E_2}(j)\}$ is stationary

[Easy, as there are $S_\xi \subseteq \{i < \lambda^+ : cf\ i = \lambda\}$ for $\xi < \lambda^+$ stationary pairwise disjoint]

Suppose $\underset{\sim}{f}$, a Q-name, is forced to be an isomorphism. As the forcing satisfies λ^+-cc there is a club $D \subset \lambda^+$, $D \in V$ such that :

f maps $M_{\bar{r}_{E_1},i}$ onto $M_{\bar{r}_{E_2},i}$ for $i \in D$ and $f \restriction M_{\bar{r}_{E_2},i}$ does not depend on $r_{E_1}(i)$ (in fact depend only on the generic sets $\{\bar{r}_E(j) : j<i\} \cup \{r : r$ does not appear in $\bar{r}_{E_1}, \bar{r}_{E_2}\}$). Choose $i \in D$, $cf\ i = \lambda$, $\bar{r}_{E_1}(i)$ does not appear in this $\bar{r}_{E_2}$. Let $V^+ = V[r_j : r_j \neq r_{E_1}(i)]$. Now $f \restriction M_{\bar{r}_E,i}$ is in the universe V^+, as well as the tree of types we have for $M_{\bar{r}_{E_1},i}$ after Fact 2.20. But in $M_{\bar{r}_{E_1}}(i+1)$ there is a type realized which $\notin V^+$, a contradiction.

§3 Introducing n-dimensional diagrams and on uniform local atomicity

3.1 Remark: In our non-structure theorems we prove something like: If ....., and λ is special e.g. $\lambda = \mu^+ = 2^\mu$, $\diamondsuit_\mu$ and $\diamondsuit_{\{\delta<\lambda : cf\ \delta = \mu\}}$ then over some $A \subset P^{\mathfrak{C}}$, $|A| = \lambda$, there are 2^λ models M with $P^M = A$ pairwise non-isomorphic over it. This excludes e.g. singular cardinals even if $V = L$. However in the cases we have dealt with we can really get 2^{λ^+} non-isomorphic models M_i, $P^{M_i} = A$ (non-isomorphic over it) with $|A| = \chi$ for any $\chi > \lambda$. Just iterate taking ultraproduct for D an ultrafilter over ω. So when our proof rests on omitting types of power μ, $\mu > \aleph_0$ this does not change much. For e.g. $\mu = \aleph_0$, we have to use indiscernibles instead; we shall return to this.

3.2. Let $\mathcal{P}(n) = \{w; w \subseteq \{0,1,\ldots,n-1\}\}$

$$\mathcal{P}^-(n) = \mathcal{P}(n) - \{n\} = \{w : w \subset \{0,1,\ldots,n-1\}\}$$

We shall deal with $I \subseteq \mathcal{P}(n)$ closed under subsets, mainly with $\mathcal{P}(n)$, $\mathcal{P}^-(n)$ and with (λ, I)-system $\langle A_s : s \in I \rangle$ $\lambda = \Sigma |A_s|$ such that

$$0 \notin s \Longrightarrow A_s \prec P^{\mathfrak{C}}$$

$$0 \in s \Longrightarrow A_s \prec \mathfrak{C}$$

$$A_s \cap P = A_{s \setminus \{0\}},\ A_s \cap A_t = A_{s \cap t}$$

and more.

We first deal with small n; for such systems we may ask about stability (of $\bigcup_{s \in I} A_s$), and existence (of M, $P^M \subseteq \bigcup_{s \in I} A_s \subseteq M$)

* * *

Note that:

for $\mathcal{P}^-(0)$ we get nothing

for $\mathcal{P}(0)$ we have just A_ϕ which is $\prec P^{\mathfrak{C}}$ (i.e. $A \subseteq P^{\mathfrak{C}}$ and $\mathfrak{C} \restriction A_\phi \prec \mathfrak{C} \restriction P^{\mathfrak{C}}$).

$\mathcal{P}^-(1) = \{\phi\}$

.sp $\mathcal{P}(1) = \{\phi, \{0\}\}$

So a $\mathcal{P}(1)$-system is $\langle A_{\{0\}}, A_\phi \rangle$

$A_{\{0\}}$ a model

A_ϕ its P-part

a $\mathcal{P}^-(1)$-system is just $A_\phi \prec P^{\mathfrak{C}}$ and the existence-problem is $\exists M (P^M = A_\phi)$. The stability just asks on $S_*(A)$ when $A \prec P^{\mathfrak{C}}$.

n=2: A $\mathcal{P}(2)$-system is $\langle A_\phi, A_{\{0\}}), A_{\{1\}}), A_{\{0,1\}} \rangle$.

For $\mathcal{P}^-(2)$ we have dealt with stability and existence. In this case automatically $tp_*(A_{\{0\}}, A_\phi) \vdash tp_*(A_{\{0\}}, A_{\{1\}})$.

n = 3: We have a cube, we add the demand

$$(A_{\{1\}},A_{\phi}) \prec (A_{\{1,2\}},A_{\{2\}}).$$

We shall assume that T absolutely has no two cardinal model (i.e. 1.0) (not always we shall use it).

3.3 Claim *: If $P^M \subseteq A \subseteq M$, A stable (and complete), then M is locally atomic over A[that is $\forall \bar{b} \in M, tp(\bar{b},A)$ is locally isolated which means that for every $\varphi = \varphi(\bar{x},\bar{z})$, there is $\psi(\bar{x},\bar{a}) \in tp(\bar{b},A)$, $\psi(\bar{x},\bar{a}) \vdash tp_\varphi(\bar{b},A)$] and even uniformly so (i.e. ψ depends on φ only and not on $\bar{b}$, though $\bar{a}$ may still depend on $\bar{b}$).

Proof : First assume (M,A) is saturated of power λ. Then (see 3.4(2)) we can find $N, P^N \subseteq A \subseteq N$, $|N| = \{a_i : i < \lambda\}$ $tp(a_i, A \cup \{a_j, j < i\})$ is λ-isolated, hence we can embed N into M over A, by 1.6 the embedding is onto A, hence w.l.o.g. $N = M$. So for every $\bar{b} \in M, tp(\bar{b},A)$ is λ-isolated. For some $q \subseteq tp(\bar{b},A)$, $|q| < \lambda, q \vdash tp(\bar{b},A)$. For every $\varphi = \varphi(\bar{x},\bar{y})$ let

$$\Gamma = q(\bar{x}_1) \cup q(\bar{x}_2) \cup \{\varphi(\bar{x}_2,\bar{y}), \neg\varphi(\bar{x}_1,\bar{y}), \bigwedge_{\ell < \ell(\bar{y})} y_\ell \in A\}$$

(we have a predicate for A). Now Γ is not realized in M, because if $\bar{x}_1 \to \bar{b}_1$, $\bar{x}_2 \to \bar{b}_2$, $\bar{y} \to \bar{d}$ realized it then $\bar{d} \subseteq A$ and
$q_1 = q(\bar{x}) \cup \{\varphi(\bar{x},\bar{d})\}$ is consistent ($\bar{b}_1$ realized it)
$q_2 = q(\bar{x}) \cup \{\neg\varphi(x,\bar{d})\}$ is consistent ($\bar{b}_2$ realized it)

contradicting "$q \vdash tp(\bar{b},A)$."

So this holds if we replace q by some finite $q' \subseteq q$ hence by some formula $\psi_{\varphi,\bar{b}}(\bar{x},\bar{c}_\varphi) \in tp(\bar{b},A)$. So

$$\psi_{\varphi,\bar{b}}(\bar{x},\bar{c}_{\varphi,\bar{b}}) \vdash tp_\varphi(\bar{b},A), \quad \text{and} \quad \models \psi_{\varphi,\bar{b}}(\bar{b},\bar{c}_{\varphi,\bar{b}})$$

Similarly we can deduce the uniformity from the $|T|^+$-saturativity.

3.3A Notation: 1) Let l.a. stand for locally atomic, u.l.a. stand for uniformly locally atomic.

2) Let $A \subseteq_t C$ means that if $\varphi(\bar{z},\bar{x}) \in L$, $\bar{c} \in C$, $\bar{a} \in A$, $\mathfrak{C} \models \varphi[\bar{c},\bar{a}]$ then there is $\bar{c}' \in A$ such that $\mathfrak{C} \models \varphi[\bar{c}',\bar{a}]$.

3.4 Claim: 1)* If A is complete, unstable and $|T|^+$-saturated, then over A there is an m-type p of power $\leq |T|$ with no $|T|^+$-isolated extension.

2) If A is complete stable, λ-saturated and $\lambda > |T|$, then

(a) for every m-type p over A of cardinality $< \lambda$ there is an m-type q over A, $p \subseteq q$, $|q - p| \leq |T|$, q has a unique extension in $S^m(A)$ and it is in $S^m_*(A)$.

(b) over A there is a primary model N, so necessarily $N \cap P^{\mathbb{E}} = \cap P^{\mathbb{E}}$.

3)* If A is complete, $A \cap P^{\mathbb{E}}$ is λ-saturated and $P^M \subseteq A \subseteq M$ *then* M is λ-saturated.

Remark: We use "absolutely no two cardinal model" for 1) and 3)

3.5 Claim: Suppose A is complete, $A \subseteq_t B$ and $C \subseteq P^{\mathbb{E}}$. then $A \subseteq_t B \cup C$.

Proof: Let $\bar{a} \in A$, $\bar{b} \in B$, $\bar{c} \in C$, and suppose $\models \varphi[\bar{c}, \bar{b}, \bar{a}]$.

Let $\psi(\bar{y}, \bar{x}) = (\exists z_0, z_1, \cdots)[\varphi(z_0, z_1, \ldots, \bar{y}, \bar{x}) \wedge \bigwedge_{\ell} P(z_0)]$, so clearly $\models \psi[\bar{b}, \bar{a}]$, hence for some $\bar{b}' \in A$ $\models \psi[\bar{b}', \bar{a}]$. As A is complete, and $\bar{a}, \bar{b}' \in A$ clearly for some $c_0', c_1', \ldots, \in A$, $\models \varphi[c_0', c_1', \ldots, \bar{b}', \bar{a}]$.

This proves $A \subseteq_t B \cup C$.

3.6 Claim: If $tp(\bar{b}, A)$ is locally isolated, $A \subseteq_t B$ *then* $tp(\bar{b}, A) \vdash tp(\bar{b}, B)$. If A' is l.a. [u.l.a.] over A, $A \subseteq_t B$ then A' is l.a. [u.l.a.] over $A \cup B$.

Proof: Easy.

§4 On $\mathcal{P}^-(3)$- systems and $\mathcal{P}^-(3)$ non- structure when there are unstable $\mathcal{P}^-(3)$- systems.

4.1 Definition : We define what is a $\mathcal{P}^-(3)$-system. It is $\mathbf{S} = \langle A_s : s \in \mathcal{P}^-(3) \rangle$ such that :

1) $A_\emptyset, A_{\{1\}}, A_{\{2\}}, A_{\{1,2\}} \prec \mathbb{E} \restriction P$

2) The rest are $\prec \mathfrak{E}$.

3) $A_s \cap P^{\mathfrak{E}} = A_{s-\{0\}}$

4) $A_s \cap A_t = A_{s \cap t}$

5) $(A_{\{1,2\}}, A_{\{2\}}) \succ (A_{\{1\}}, A_\phi)$

6) $A_{\{1,0\}}$ is uniformly locally atomic over $A_{\{1\}} \cup A_{\{0\}}$ and $A_{\{2,0\}}$ is uniformly locally atomic over $A_{\{2\}} \cup A_{\{0\}}$

Now 6) follows by previous hypothesis, for T absolutely with no two cardinal model, (see 3.3). We say **S** is stable if $\bigcup_s A_s$ is stable. **S** has the existence property if $(\exists M \supseteq \bigcup_s A_s) P^M \subseteq \bigcup_s A_s$.

4.2 Fact: Being a $\mathcal{P}^-(3)$-system depends on the first theory only [of $(\bigcup_{s \in \mathcal{P}^-(3)} A_s, \cdots A_s \cdots)_{s \in \mathcal{P}^-(3)}]$ (because we have u.l.a. not just l.a.).

E.Question: Is there unstable $\mathcal{P}^-(3)$-system?

4.3 Theorem* : Suppose $\langle A_s^* : s \in \mathcal{P}^-(3) \rangle$ is unstable, $\lambda = \lambda^{<\lambda} > |T|$ and Q is the forcing of adding λ^{++}-Cohen subsets to λ(and $2^\lambda = \lambda^+$, $2^{\lambda^+} = \lambda^{++}$) and $\mu \geq \lambda^{++}$. Then in V^Q there are $2^{\lambda^{++}}$ non isomorphic models of T of power μ with the same P of power μ. [If e.g. $\mu^{<\lambda} = \mu$ then we can have λ-saturated models).

4.3A Remark: We do not try here to eliminate the set theory. We are more interested to show the dividing line is right.

4.4 Claim: Suppose for $\ell = 0,1$ $\langle A_s^\ell : s \in \mathcal{P}^-(3) \rangle$ is a $\mathcal{P}^-(3)$-system, $\langle A_s^\ell : s \in \mathcal{P}(\{1,2\}) \rangle$ is saturated of power $\lambda > |T|$, $\langle A_s^\ell : s \in \mathcal{P}(\{1,2\} \rangle$, $(\ell = 0,1)$ are elementarily equivalent and A_s^ℓ is saturated of power λ when $0 \in s$. Then the two systems are isomorphic.

Proof : Obviously there is an isomorphism g from $\langle A_s^0 : s \in \mathcal{P}(\{1,2\}) \rangle$ onto $\langle A^{s^1} : s \in \mathcal{P}(\{1,2\}) \rangle$. Now we know (see 1.3) that: as $A_{\{0\}}^\ell$ is saturated of power λ, it is unique over $A_{\{1\}} \cap P^{\mathfrak{E}} = A_\phi^\ell$. So we can extend $g \restriction A_\phi^0$ to a isomorphism g_0 from $A_{\{0\}}^0$ onto $A_{\{0\}}^1$. Now (by 1.6, 2.16) we know that

$tp_*(A^{\ell}_{\{0\}}, A^{\ell}_{\phi}) \vdash tp_*(A^{\ell}_{\{0\}}, P^{\mathfrak{C}})$ hence $tp_*(A^{\ell}_{\{0\}}, A^{\ell}_{\phi}) \vdash tp(A^{\ell}_{\{0\}}, A^{\ell}_{\{1,2\}})$ hence $g^0 \stackrel{def}{=} g_0 \cup g$ is an elementary mapping. We know (by condition 6 of Definition 4.1) that $A^{\ell}_{\{2,0\}}$ is u.l.a over $A^{\ell}_{\{2\}} \cup A^{\ell}_{\{0\}}$, hence it is λ-atomic over it, so as it is λ-saturated it is unique over $A^{\ell}_{\{2\}} \cup A^{\ell}_{\{0\}}$. Hence $g^0 \restriction (A^{\ell}_{\{2\}} \cup A^{\ell}_{\{0\}})$ can be extended to an isomorphism g_1 from $A^{\ell}_{\{2,0\}}$ onto $A^{1}_{\{2,0\}}$. As we know $tp_*(A_{\{0,2\}}, A_{\{2\}}) \vdash tp_*(A_{\{0,2\}}, P^{\mathfrak{C}})$ also $tp_*(A_{\{0,2\}}, A_{\{2\}}) \vdash tp_*(A_{\{0,2\}}, A_{\{1,2\}})$ hence $tp_*(A_{\{0,2\}}, A_{\{2\}} \cup A_{\{0\}}) \vdash tp_*(A_{\{0,2\}}, A_{\{1,2\}} \cup A_{\{0\}})$ [note $A_{\{1,2\}} \cup A_{\{0\}} \subseteq A_{\{0,2\}}$] so necessarily $g^1 \stackrel{def}{=} g_1 \cup g^0$ is an elementary mapping. Now $A^{\ell}_{\{1,0\}}$ is also λ-prime over $A_{\{1\}} \cup A_{\{0\}}$ so again there is an isomorphism g_2 extending $g^1 \restriction (A^{\ell}_{\{1\}} \cup A^{\ell}_{\{0\}})$ to an isomorphism from $A^{\ell}_{\{1,0\}}$ onto $A^{1}_{\{1,0\}}$. So it suffices to prove that $g_2 \cup g^1$ is an elementary mapping. As $A^{\ell}_{\{1,0\}}$ is u.l.a. over $A^{\ell}_{\{1\}} \cup A^{\ell}_{\{0\}}$ it suffices to prove $tp_*(A^{\ell}_{\{0,1\}}, A^{\ell}_{\{1\}} \cup A^{\ell}_{\{0\}}) \vdash tp_*(A^{\ell}_{\{0,1\}}, A^{\ell}_{\{1,2\}} \cup A^{\ell}_{\{0,2\}})$,

for this, by 3.5 (and see 3.3A) it suffices to prove:

(*) $A^{\ell}_{\{1\}} \cup A^{\ell}_{\{0\}} \subseteq_t A^{\ell}_{\{1,2\}} \cup A^{\ell}_{\{0,2\}}$

Let $\bar{c}_s \in A^{\ell}_s$, $\bar{b}_s \in A^{\ell}_{s \cup \{2\}}$ for $s = \phi, \{0\}, \{1\}$ be such that $\models \varphi(\ldots, \bar{c}_s, \ldots, \bar{b}_s, \ldots)_{s \in \mathcal{P}^-(2)}$. We shall show that there are $\bar{c}'_s \in A^{\ell}_s$, (for $s \in \mathcal{P}^-(2)$) such that $\models \varphi[\ldots, \bar{c}'_s, \ldots, \bar{b}_s, \ldots]_{s \in \mathcal{P}^-(2)}$. As we have already proved that $tp_*(A^{\ell}_{\{0,2\}}, A^{\ell}_{\{0\}} \cup A^{\ell}_{\{2\}}) \vdash tp_*(A^{\ell}_{\{0,2\}}, A^{\ell}_{\{1,2\}} \cup A^{\ell}_{\{0\}})$, w.l.o.g. for some ψ_1, ψ_2:

a) $\mathfrak{C} \models \psi_1[\bar{c}_{\phi}, \bar{c}_{\{1\}}, \bar{b}_{\phi}, \bar{b}_{\{1\}}, \bar{b}_{\{0\}}]$

b) $\mathfrak{C} \models \forall \bar{y}_{\phi}, y_{\{1\}}, \bar{x}_{\phi}, x_{\{1\}}, \bar{x}_{\{0\}}([\psi_1(\bar{y}_{\phi}, , \bar{x}_{\{1\}} \bar{x}_{\{1\}}, \bar{x}_{\phi}, \bar{x}_{\{0\}}) \rightarrow \psi_2(y_{\phi}, \bar{x}_{\phi}, \bar{x}_{\{0\}}])$

c) $\mathfrak{C} \models (\forall y_{\phi}, \bar{x}_{\phi}, \bar{x}_{\{0\}}) [\psi_2(\bar{y}_{\phi}, \bar{x}_{\phi}, \bar{x}_{\{0\}}) \rightarrow (\exists \bar{y}_{\{0\}}) \vartheta(\bar{y}_{\phi}, y_{\{0\}}, \bar{x}_{\phi}, \bar{x}_{\{0\}})]$

d) $\mathfrak{C} \models \forall \bar{y}_{\phi}, \bar{y}_{\{1\}}, \bar{y}_{\{0\}}, \bar{x}_{\phi}, \bar{x}_{\{1\}}, \bar{x}_{\{0\}})[\psi_1(y_{\phi}, \bar{y}_{\{1\}}, \bar{x}_{\phi}, x_{\{1\}}, \bar{x}_{\{0\}}) \wedge$

$$\vartheta(y_{\phi}, y_{\{0\}}, x_{\phi}, x_{\{0\}}) \rightarrow \varphi(\bar{y}_{\phi}, \bar{y}_{\{1\}}, \bar{y}_{\{0\}}, \bar{x}_{\phi}, \bar{x}_{\{1\}}, \bar{x}_{\{0\}})]$$

So in fact we have shown that w.l.o.g. $\bar{c}_{\{0\}}$ is empty [replace φ by ψ_1, (a) is the assumption; so suppose $\bar{c}'_{\phi} \in A^{\ell}_{\phi}$, $\bar{c}'_{\{1\}} \in A^{\ell}_{\{1\}}$ and $\models \psi_1[\bar{c}'_{\phi}, \bar{c}'_{\{1\}}, \bar{b}_{\phi}, \bar{b}_{\{1\}}, \bar{b}_{\{0\}}]$ hence by (b), $\models \psi_2[\bar{c}'_{\phi}, \bar{b}_{\phi}, b_{\{0\}}]$ and (c) $\models (\exists \bar{y}_{\{0\}}) \ \vartheta(\bar{c}'_{\phi}, \bar{y}_{\{0\}}, \bar{b}_{\phi}, \bar{b}_{\{0\}})$ and as $A^{\ell}_{\{0\}} \prec \mathfrak{C}$ for some $\bar{c}'_{\{0\}} \in A^{\ell}_{\{0\}}$, $\models \vartheta[\bar{c}'_{\phi}, \bar{c}'_{\{0\}}, \bar{b}_{\phi}, \bar{b}_{\{0\}})$, and by (d) we finish].

Then we can eliminate the use of $\bar{b}_{\{0\}}$ as $tp_{\Delta}(\bar{b}_{\{0\}}, P^{\mathfrak{C}})$ is isolated by some formula in $tp(\bar{b}_{\{0\}}, A_{\phi})$ (for Δ a finite set of formulas). At last we know that $(A_{\{1\}}, A_{\phi}) \prec (A_{\{1,2\}}, A_{\{2\}})$.

In fact we have prove

4.5 Claim : If $\langle A_s : s \in \mathcal{P}^{-}(3) \rangle$ is $\mathcal{P}^{-}(3)$-system, and $\langle A_s : s \in \mathcal{P}(\{1,2\}) \rangle$ is λ-saturated then S is λ-saturated.

Proof of 4.3: The Hypothesis 4.3: There is a $\mathcal{P}^{-}(3)$-system $\langle A^{*}_s : s \in \mathcal{P}^{-}(3) \rangle$ such that $\bigcup_s A^{*}_s$ unstable.

Assumptions: $\lambda = \lambda^{<\lambda} > |T|$, $2^{\lambda} = \lambda^{+}$, $2^{\lambda^{+}} = \lambda^{++}$.

We first define $A_i (i < \lambda^{++})$ increasing continuous, $A_i \prec \mathfrak{C} \restriction P$, $|A_i| = \lambda^{+}$, A_i w.l.o.g. a set of ordinals $< \lambda^{++}$ [$cf(i) \in \{0, 1, \lambda^{+}\} \Longrightarrow A_i$ is saturated]. For each $j < \lambda^{++}, cf(j) = \lambda^{+}$, $i = j+1$ we define $A^{i}_{\alpha}, A^{j}_{\alpha}$ for $\alpha < \lambda^{+}$ such that:

$A^{j}_{\alpha} \subseteq A^{i}_{\alpha}$, $|A^{i}_{\alpha}| = |A^{j}_{\alpha}| = \lambda$, $\bigcup_{\alpha < \lambda^{+}} A^{i}_{\alpha} = A_i$, $\bigcup_{\alpha < \lambda^{+}} A^{j}_{\alpha} = A_j$

$(A^{i}_{\alpha}, A^{j}_{\alpha})$ is an elementary chain (increasing-continuous) in α
$(A^{i}_{\alpha}, A^{j}_{\alpha}) \equiv (A^{*}_{\{1\}}, A^{*}_{\phi})$ and

[$cf(\alpha) = \lambda \Longrightarrow (A^{i}_{\alpha+1}, A^{j}_{\alpha+1}, A^{i}_{\alpha}, A^{j}_{\alpha}) \equiv (A^{*}_{\{1,2\}}, A^{*}_{\{2\}}, A^{*}_{\{1\}}, A^{*}_{\phi})$, and is saturated.]

We do it by induction on i,
For i = 0 , or i limit: no problem.
i = j+1, cf j :≠ λ^{+}: no problem
i = j+1, cf j = λ^{+}: no real problem. First we define by induction on α, $(A^{i}_{\alpha}, A^{j}_{\alpha}) \equiv (A^{*}_{\{1\}}, A^{*}_{\phi})$ a continuous increasing (in α) chain; [$cf\ \alpha \in \{0, 1, \lambda\} \Longrightarrow (A^{i}_{\alpha}, A^{j}_{\alpha})$ is saturated], so that $\bigcup_{\alpha < \lambda^{+}} (A^{i}_{\alpha}, A^{j}_{\alpha})$ will be saturated:

for $\alpha = 0$, or α *limit* or $\alpha = \beta + 1$, $cf\ (\beta) \neq \lambda$: no problem arise and take care of the saturation of the union

$\alpha = \beta+1, cf\ \beta = \lambda$: Let $(A_{\{1,2\}}, A_{\{2\}}, A_{\{1\}}, A_{\phi})$ be a saturated model of power λ of the theory of $(A^{*}_{\{1,2\}}, A^{*}_{\{2\}}, A^{*}_{\{1\}}, A^{*}_{\phi})$. So $(A_{\{1\}}, A_{\phi})$ and $(A^{i}_{\beta}, A^{j}_{\beta})$ are saturated models of the same power and theory; hence isomorphic, so w.l.o.g. equal and let

$$A^i_\alpha = A_{\{1,2\}} \qquad A^j_\alpha = A_{\{2\}}$$

Now $\bigcup_{\alpha<\lambda^+} A^j_\alpha$ is a saturated model of the theory of A^*_ϕ(= the theory of A^j_α) and has power λ^+, so it is isomorphic to A_i and w.l.o.g. they are equal. So we have defined A_i.

* * *

Now we define by induction on $i < \lambda^{++}$

$M_i \prec \mathfrak{C}$, such that:

a) $M_i \cap P^{\mathfrak{C}} = A_i$, M_i increasing continuous.

b) M_i is λ-constructible over A_i,

c) when $cf\ i \in \{0,1,\lambda^+\}$ M_i is λ-saturated and

d) if $j < i$ then M_i is λ-atomic over $M_j \cup A_i$;

We will define the M_i's in some forcing extension V^Q of V: but Q is λ-complete: so (when $cf\ i \in \{0,1,\lambda^+\}$) M_i is isomorphic over A_i to some $M_i' \in V$ [as over A_i there is in V a λ-prime model M_i' in fact a λ-primary one and this property is still true in V^Q. This property is also satisfied by M_i over A_i; so they are isomorphic: use the uniqueness of the λ-primary model (see [Sh 1], Ch. II, §5).

Specifically, Q will be "adding λ^{++}-Cohen subsets of λ, $\langle r^\alpha : \alpha < \lambda^+ \rangle$". For every sequence $\bar{r}, \bar{r} = \langle r_{i,\alpha} : i < \lambda^{++}, \alpha < \lambda^+ \rangle$ (where for some $h \in V$, $r_{i,\alpha} = r^{h(i,\alpha)}$, h one to one) we shall define a model $\overline{M}^{\bar{r}}$. For a while we suppress the superscript $\bar{r}$.

Case I: $i = 0$: by the proof of the existence of a λ-primary model over any λ-saturated $A \prec \mathfrak{C} \restriction P$, $|A| = \lambda^+$ (see 2.14).

Case II: *i limit:* The only problematic point is "M_i is λ-constructible over A_i, and M_i is λ-atomic over $M_j \cup A_i$ for $j < i$". Let $j < i$, every $\bar{c} \in M_i$

belongs

to M_ξ for some ξ, $j < \xi < i$, so by the induction hypothesis $tp(\bar{c}, A_\xi \cup M_j)$ is λ-isolated, but $M_j \cup A_\xi \cup \bar{c}$ is compete hence $tp(\bar{c}, A_\xi \cup M_j) \vdash tp(\bar{c}, A_i \cup M_j)$ so the latter is λ-isolated too. So M_i is λ-atomic over $M_j \cup A_i$.

Now each M_j $(j < i)$ is λ-constructible over A_j, hence over A_i. So (see [Sh 1 IV §3] $M_j = \bigcup_{\alpha<\lambda^+} M_{j,\alpha}$, $\|M_{j,\alpha}\| = \lambda$, $M_{j,\alpha}$ increasing continuous in α and M_j is λ-constructible hence λ-atomic over $A_j \cup M_{j,\alpha}$ and even over $A_i \cup M_{j,\alpha}$. Let $i = \bigcup_{\alpha<\lambda^+} W_\alpha$, $|W_\alpha| \le \lambda$, W_α increasing continuous, W_α with no last element. Let $N_\alpha = \bigcup_{j\in W_\alpha} M_{j,\alpha}$, so clearly $\|N_\alpha\| \le \lambda$. Let

$$C_0 = \{\alpha < \lambda^+ : \forall j < \xi \in W_\alpha,\ M_{\xi,\alpha} \cap M_j = M_{j,\alpha}\}$$

Clearly C_0 is a closed unbounded subset of λ^+.

Now for every $j < \xi < i$, M_ξ is λ-atomic over $M_j \cup A_\xi$ hence (as usual) over $M_j \cup A_i$, and for every $\bar{c} \in M_\xi$ there is $\alpha(\bar{c}, j) < \lambda^+$ such that $tp(\bar{c}, M_{j,\alpha(\bar{c},j)} \cup (M_{\xi,\alpha(\bar{c},j)} \cap A_\xi)) \vdash tp(\bar{c}, M_j \cup A_i)$ (are λ-isolated). Clearly $C_1 = \{\alpha \in C_0 : \forall \bar{c}(\forall j, \xi \in W_\alpha)[j < \xi \wedge \bar{c} \in M_{\xi,\alpha} \rightarrow \alpha(\bar{c}, j) < \alpha]\}$ is closed unbounded. It suffices to prove that for every $\alpha \in C_1$, $N_{\alpha+1}$ is λ-atomic over $N_\alpha \cup A_i$ (hence λ-constructible). (as we know N_0 is λ-atomic over N_i). First we prove that for every $j \in W_\alpha$, M_j is λ-atomic over $N_\alpha \cup A_i$; let $\bar{d} \in M_j$ then as $\alpha \in C_1$, $tp(\bar{d}, M_{j,\alpha} \cup A_j)$ is λ-atomic hence $tp(\bar{d}, M_{j,\alpha} \cup A_i)$ is λ-atomic, so it suffices to prove $tp(\bar{d}, M_{j,\alpha} \cup A_i) \vdash tp(\bar{d}, \bar{c} \cup M_{j,\alpha} \cup A_i)$ for every $\bar{c} \in N_\alpha$. For any such $\bar{c}$, as W_α has no last element, for some ξ $\bar{c} \in M_{\xi,\alpha}$ $j < \xi \in W_\alpha$. Now $\alpha(\bar{c}, j) < \alpha$, hence $tp(\bar{c}, M_{j,\alpha(\bar{c},j)} \cup (M_{\xi,\alpha(\bar{c},j)} \cap A_\xi)) \vdash tp(\bar{c}, M_j \cup A_i)$ as $\bar{d} \in M_j$, this implies $tp(\bar{c}, M_{j,\alpha} \cup A_i) \vdash tp(\bar{c}, M_{j,\alpha} \cup A_i \cup \bar{d})$ and by symmetry we get the conclusion. So we have proved that M_j is λ-atomic over $N_\alpha \cup A_i$, hence $\bigcup_{j\in W_\alpha} M_j$ is λ-atomic over $N_\alpha \cup A_i$, but $\bigcup_{j \in W_\alpha} M_j$ is $M_{\sup(W_\alpha)}$ and so we have proved it if $\sup(W_\alpha) = i$. Now if $\zeta \overset{def}{=} \sup W_\alpha < i$, then remember that we had proved that M_i is λ-atomic over $M_\zeta \cup A_i$; as we have just proved that M_ζ is λ-atomic over

$N_\alpha \cup A_i$, together we get that M_i is λ-atomic over $N_\alpha \cup A_i$.

Case III: $i = j + 1, cf\ j < \lambda^+$.

As M_j is λ-constructible over A_j, we can find $M_{j,\alpha}$, and $A_{i,\alpha}$ for $\alpha < \lambda^+$ such that, $M_j = \bigcup_{\alpha<\lambda^+} M_{j,\alpha}$ where $M_{j,\alpha}$ is increasing continuous (in α) $\|M_{j,\alpha}\| \le \lambda$, M_j λ-atomic over $M_{j,\alpha} \cup A_j$ (hence over $M_{j,\alpha} \cup A_i$), and $|A_{i,\alpha}| \le \lambda$, $A_i = \bigcup_{\alpha<\lambda^+} A_{i,\alpha}$. $A_{i,\alpha}$ increasing continuous in α, and $(A_{i,\alpha}, M_{j,\alpha}, A_{j,\alpha}) \prec (A_i, M_j, A_j)$ where $A_{j,\alpha} = A_{i,\alpha} \cap A_j = M_{j,\alpha} \cap A_j$, and when $cf\ \alpha \in \{0,1,\lambda\}$. $(A_{i,\alpha}, A_{j,\alpha})$ is λ-saturated, also when $cf\ \alpha \in \{0,1,\lambda\}$, $(A_{i,\alpha}, M_{j,\alpha}, A_{j,\alpha}) \prec_{L_{\lambda\lambda}} (A_i, M_j, A_j)$.

We define by induction on α, $M_{i,\alpha}$ such that $A_{i,\alpha} \cup M_{j,\alpha} \subseteq M_{i,\alpha}$, $P^{M_{i,\alpha}} = A_{i,\alpha}$, $[cf\ \alpha \in \{0,1\lambda\} \rightarrow M_{i,\alpha}$ is λ-saturated$]$, $M_{i,\alpha}$ increasing continuous in α, and $M_{i,\alpha}$ is λ-atomic over $M_{i,\alpha} \cup M_{j,\alpha}$ and also over $A_{i,\alpha} \cup M_j$. For the last demand note that

(*) when $cf\ \alpha \in \{0,1,\lambda\}$, as $(A_{i,\alpha}, M_{j,\alpha}, A_{j,\alpha}) \prec_{L_{\lambda\lambda}} (A_i, M_j, A_j)$ it suffices to prove that $M_{i,\alpha}$ is λ-atomic over $A_{i,\alpha} \cup M_{j,\alpha}$.

So for $\alpha = 0$ it is easy, by the last sentence, for α-limit there is no problem. For $\alpha = \beta + 1$, over $A_{i,\alpha} \cup M_{j,\alpha}$ there is a λ-atomic λ-saturated model $M_{i,\alpha}$, but why $M_{i,\beta} \subseteq M_{j,\alpha}$? As the previous is λ-atomic over $A_{i,\alpha} \cup M_{j,\alpha}$ ([prove it as you have proved (*) and for β limit we use $M_\beta = \bigcup_{\gamma<\beta} M_{\gamma+1}$) and as $\|M_{i,\beta}\| \le \lambda$, clearly $M_{i,\beta}$ is λ- constructible over $A_{i,\alpha} \cup M_{j,\alpha}$, and we can embed it into $M_{i,\alpha}$, over $A_{i,\alpha} \cup M_{j,\alpha}$ and so by renaming we can finish.

So $M_i \overset{def}{=} \bigcup_{\alpha<\lambda^+} M_{i,\alpha}$ is λ-atomic over $M_j \cup A_i$ (hence over $M_\xi \cup A_i$ for $\xi < i$ (see [Sh 1] ch. IV §3]) and is λ-saturated. We still have to show that it is λ-constructible over A_i. For this it suffices to prove $M_{i,\alpha+1}$ is λ-atomic over $M_{i,\alpha} \cup A_{j,\alpha+1}$ which we could have guaranteed this easily in the construction. More exactly, $M_{i,\alpha+1}$ is a λ-saturated model of cardinality λ extending $M_{i,\alpha} \cup A_{j,\alpha+1}$; Now if Γ is a set of $\le \lambda$ types over $M_{i,\alpha} \cup A_{j,\alpha+1}$ each with no support of power $<\lambda$ (i.e. no type q over $M_{i,\alpha} \cup A_{j,\alpha+1}$, (consistent), $|q| < \lambda$, $q \vdash p$ where p is the type from Γ), then there is a λ-saturated $M \supseteq A_{i,\alpha} \cup M_{j,\alpha+1}$, M omitting every $p \in \Gamma$. Now the other demands on $M_{i,\alpha+1}$ are of the form: omits some type; and to prove those types have no support $< \lambda$, it suffices to find a (λ-saturated M, $M \supseteq A_{i,\alpha} \cup M_{j,\alpha+1}$) omitting such a type for each $p \in \Gamma$ separately.

Case IV: $i = j+1, cf\ j = \lambda^+$.

We act exactly as in Case III, with one additional feature. When $\alpha = \beta + 1, cf\ \beta = \lambda$, we demand

$$(**) \qquad \langle M_{j,\alpha}, M_{i,\alpha}, M_{j,\beta}, A_{i,\alpha}, A_{j,\alpha}, A_{i,\beta}, A_{j,\beta} \rangle$$

$$= \langle A^*_{\{0,2\}}, A^*_{\{0,1\}}, A^*_{\{0\}}, A^*_{\{1,2\}}, A^*_{\{2\}}, A^*_{\{1\}}, A^*_{\phi} \rangle$$

[Remember $A_{j,\gamma}, A_{i,\gamma} (\gamma < \lambda^+)$ were defined in the first part of the proof, so that the relevant part of (**) holds. We then can define $M_{j,\gamma} (\gamma \geq 0)$, λ-saturated of power λ, $M_{j,\gamma} \cap P^{\mathbb{E}} = A_{j,\gamma}$, and $M_{j,\gamma+1}$ is λ-atomic over $A_{j,\gamma+1} \cup M_{j,\gamma}$, by 2.14 w.l.o.g. $M_j = \bigcup_{\gamma} M_{j,\gamma}$. Now we defined by induction on γ, $M_{i,\gamma}$, λ-atomic over $M_{j,\gamma} \cup A_{i,\gamma}$. Clearly there is a λ-saturated model of cardinality λ elementarily equivalent to $\langle A^*_{\{0,2\}}, A^*_{\{0,1\}}, A^*_{\{0\}}, A^*_{\{1,2\}}, A^*_{\{2\}}, A^*_{\{1\}}, A^*_{\phi} \rangle$, and by 4.4 it is isomorphic to $\langle M_{j,\alpha}, M_{i,\alpha}, M_{j,\beta}, A_{i,\alpha}, A_{j,\alpha}, A_{i,\beta}, A_{j,\beta} \rangle$ so (**) holds].

So the left system is unstable so by 3.5 there is an m-type p over it of power $< \lambda$ with no λ-isolated extension over $M^j_\alpha \cup M^i_\beta \cup A^i_\alpha$, so in the construction we have a perfect (i.e. homeomorphic to M) set of possibilities an we use $r_{j,\beta}$ to decide (except here we do not use the Cohen sets, though once used we may continue to use it).

The non isomorphism is as in previous proofs.

Remark: We could simplify the proof of 4.3 by a more extensive use of 0.1.

§5 General system and relevant symmetry.

We change slightly the thing we analyze - we shall analyze "the possible existence of a λ-prime model over any $A \prec P^{\mathfrak{C}}$". Remember

Hypothesis: Every formula is equivalent to a relation.

In this section we shall deal with systems of the following kind:

5.1 Definition : A I-system is $\mathbf{S} = \langle A_s : s \in I \rangle$ ($I = I(\mathbf{S})$) where

1) for some $n = n(I) = n(\mathbf{S})$, $\mathcal{P}(\{1, \ldots, n-1\}) \subseteq I \subseteq \mathcal{P}(n)$, I close under subsets

2) $A_s \cap A_t = A_{s \cap t}$

3) a) if $0 \notin s$, then $A_s \prec \mathfrak{C} \restriction P$, b) if $0 \in s$, $A_s \prec \mathfrak{C}$

4) $\langle A_s : s \in \mathcal{P}(\{1, \ldots, n-2\}) \rangle \prec \langle A_{s \cup \{n-1\}} : s \in \mathcal{P}(\{1, \ldots, n-2\}) \rangle$ are both systems (so the definition is by induction on n).

5) if $0 \in s$, A_s is u.l.a. over $\bigcup_{t \subset s} A_t$.

Remark: This is useful when no two cardinal models exist.

5.2 Definition : 1) A system $\mathbf{S}$ is stable if $\bigcup_{s \in I(\mathbf{S})} A_s^{\mathbf{S}}$ is

2) A system has the existence property if there is M, $P^M \subseteq \bigcup_s A_s^{\mathbf{S}} \subset M$.

3) The I-goodness holds if every I-system is stable.

4) $n^*(T)$ is sup $\{n+1 : \mathcal{P}^-(n)\text{-goodness holds}\}$ (so $n^*(T) \leq \omega$).

5) $n^{**}(T)$ is sup $\{n+1$: every $\mathcal{P}^-(n)$-system has the atomicity property $\}$.
where

6) $\langle A_s : s \in I \rangle$ has the atomicity property if for every $|T|^+$-saturated $\langle A_s^+ : s \in I \rangle \equiv \langle A_s : s \in I \rangle$, and m-type p over $\bigcup_{s \in I} A_s^+$ of cardinality

$\leq |T|$, has a $|T|^+$-isolated extension over $\bigcup_{s\in I} A_s$.

5.3 Lemma : 1) Being an I-system depends only on its first order theory.

2) Having the atomicity property (for an I-system) depends only on its first order theory.

3) If $\langle A_s : s \in I\rangle$ is a system, $n(T) > 0$, then so are $\langle A_s : s \in I \cap \mathcal{P}(n(I)-1)\rangle$ and $\langle A_{s\cup\{n(I)-1\}} : s \cup\{n(I)-1\} \in I,\ (n(I)-1) \not\in s\rangle$.

4) If $J \subseteq I$ satisfies (1) of 5.1 then $\langle A_s : s \in J\rangle$ is a J-system.

5)* If every model is stable (i.e., $|S^m(M)| \leq \|M\|^{|T|}$) then $n^*(T) = n^{**}(T)$, in fact stability and atomicity of $\mathcal{P}^-(n)$-systems are equivalent. (see 3.4(2)(a)). (Without 0.1 we get: stability implies atomicity.)

5.4 Lemma : For any system $\langle A_s : s \in I\rangle$, ($n = n(I)$):

a) if $0 \in s \in I$ then $tp_*(A_s, \bigcup_{t\subset s} A_t) \vdash tp_*(A_s, \bigcup\{A_t : t \in I, s \not\subseteq t\})$; moreover for every $\varphi(\bar{x},\bar{y})$ and $\bar{c} \in A_s$ for some $\psi_\varphi(\bar{x},\bar{b}_{\varphi,\bar{c}}) \in tp(\bar{c}, \bigcup_{t\subset s} A_t)$, $\psi_\varphi(\bar{x},\bar{b}_{\varphi,\bar{c}}) \vdash tp_\varphi(\bar{c}, \bigcup\{A_t : t \in I, s \not\subseteq t\})$.

b) $\bigcup_{\substack{(n-1)\not\in s\\ s\in I}} A_s \subseteq_t \bigcup_{s\in I} A_s$, in fact: for $\bar{b}_t \in A_t (t \in I)$ such that $\models \varphi(\ldots\bar{b}_t,\ldots)$ we can find $\bar{b}_t^+ \in A_{s-\{n-1\}}$, such that $[(n-1) \not\in t \Rightarrow \bar{b}_t^+ = \bar{b}_t]$, and $\models\varphi(\ldots,b_t^+\ldots)$.

Proof : The proof is by simultaneous induction on $|I|$ (for all systems and both a) and b)). The proof is splitted to cases.

Proof of a):

Case 1: There is $t \in I$, $s \subset t$.

Then we can reduce the problem to one on $I^+ \subset I$ and use the induction hypothesis. So if not Case 1 $\{t : t \in I, s \not\subseteq t\} = I - \{s\}$.

Case 2 : not Case 1 and $(n-1) \not\in s$.

Let $\bar{c} \in A_s$, $\varphi(\bar{x},\bar{y})$ be a formula.

Let $J = \{t \in I : (n-1) \notin t\}$, then by the induction hypothesis [as $|J| < |J \cup \{n-1\}| \le |I|$, because $\{n-1\} \notin J$, (and $\{n-1\} \in I$, as I is downward closed and $n = n(I)$). Note that $n-1 > 0$ as $n-1 \notin s, 0 \in s$] for some $\psi_\varphi(\bar{x},\bar{b}_{\varphi,\bar{c}}) \in tp_*(A_s, \bigcup_{t \subset s} A_t)$, $\psi_\varphi(\bar{x},\bar{b}_{\varphi,\bar{c}}) \vdash tp_\varphi(\bar{c}, \bigcup_{\substack{t \neq s \\ t \in J}} A_t)$.

So for no $\bar{d} \in \bigcup_{\substack{t \neq s \\ t \in J}} A_t$. $\models (\exists \bar{x})[\psi_\varphi(\bar{x},\bar{b}_{\varphi,\bar{c}}) \wedge \varphi(\bar{x},\bar{d})] \wedge (\exists \bar{x})[\psi_\varphi(\bar{x},\bar{b}_{\varphi,\bar{c}}) \wedge \neg \varphi(\bar{x},\bar{d})]$. Applying the induction hypothesis to $I-\{s\}$ (for (b)) we see that $\bigcup\{A_t : t \in J-\{s\}\} \subseteq_t \bigcup\{A_t : t \in I-\{s\}\}$. So also in $\bigcup\{A_t : t \in I-\{s\}\}$ we cannot find $\bar{d}$ as above. So $\psi_\varphi(\bar{x},\bar{b}_{\varphi,\bar{c}}) \vdash tp_\varphi(\bar{c}, \bigcup\{A_t : t \in I-\{s\}\})$, as required.

Case 3: Not 1 nor 2 and there is v, a maximal member of I, $0 \in v, v \neq s, (n-1) \notin v$. So v,s are $\subseteq$- incomparable.

By using the induction hypothesis for $I-\{v\}, s$ and case 2 for I,v we see that

$$tp_*(A_s, \bigcup_{t \subset s} A_t) \vdash tp_*(A_s, \bigcup\{A_t : t \in I, t \neq v, s\})$$
$$tp_*(A_v, \bigcup_{t \subset v} A_t) \vdash tp_*(A_s, \bigcup\{A_t : t \in I, t \neq v\})$$

Together we get the first close of (a). As for the second: we can treat our system as an $|I|$-sorted model, find a $|T|^+$-saturated elementary extension, so also there we get the first close of (a). By saturativity we get the ψ_φ and note that its property is preserve by elementary equivalence.

Case 4: For some $t \in I$, $0 \notin t$ and $t \cup \{0\} \notin I$.

Let $J_0 = \{v \in I : 0 \notin v\}$, $J_1 = \{v \in I : 0 \notin v, v \cup \{0\} \in I\}$, $J_2 = \{v \in I : 0 \in v\}$. We shall prove that $tp(\bigcup_{u \in J_2} A_u, \bigcup_{u \in J_1} A_u) \vdash tp(\bigcup_{u \in J_2} A_u, \bigcup_{u \in J_0} A_u)$ (by [Sh 1, ch. IV §2 §3], this suffices for the first phrase of (a),) then proceed as in Case 3.

For each $v \in J_2$, as $tp_*(A_v, \bigcup_{u \subset v} A_u)$ is u.l.a. by the induction hypothesis $\bigcup_{u \subset v} A_u \subset_t \bigcup \{A_u : v \not\subset u,\ u \in J_2 \cup J_1\}$, hence by 3.5 $\bigcup_{u \subset v} A_u \subset_t \bigcup \{A_u : v \not\subset u, u \in I\}$, together we get $tp_*(A_v, \bigcup_{u \subset v} A_u) \vdash tp(A_v, \bigcup\{A_u : v \not\subset u, u \in I\})$. By [Sh 1, IV 3.3] this gives $tp(\bigcup_{v \in J_2} A_v, \bigcup_{u \in J_1} A_u) \vdash tp(\bigcup_{v \in J_2} A_v, \bigcup_{u \in J_0} A_u)$.

Case 5: not cases 1,2,3,4.

So $(n-1) \in s$ [as not Case 2] and $(\forall t \in I)(t \cup \{n-1\} \in I)$, [if t is a counterexample, as I is downward closed w.l.o.g. t is maximal in I; as $t \cup \{n-1\} \not\in I$ clearly $t \not\subset s$, $n-1 \not\in t$, by "not case 4" $t \cup \{0\} \in I$ hence by t's maximality $0 \in t$, and we get Case 3, contradiction]. So $I = J \cup \{t \cup \{n-1\} : t \in J\}$ where $J = \{t \in I : (n-1) \not\in I\}$. We apply the induction hypothesis to $\langle A_{s \cup \{n-1\}} : s \in J \rangle$, $s - \{n-1\}$ (remember 5.3(3)) so

$$tp_*(A_s, \bigcup\{A_t : t \subset s, (n-1) \in t\}) \vdash tp_*(A_s, \bigcup\{A_t : t \neq s, s - \{n-1\}\})$$

hence the first close of (a) follows (by Ax VII of [Sh 1, ch IV §1]) and we prove (a) as in the Case 3.

Proof of (b) of 5.4: Now we prove (b) of 5.4. Let $J = \{t \in I : (n-1) \not\in t\}$.

First replace our system by a $|T|^+$-saturated one. Then by increasing the $\bar{b}_s$ to sequences of length $< |T|^+$ we can assume for each $s \in I$: if $0 \in s$ then $tp(\bar{b}_s, \bigcup_{t \subset s} \bar{b}_t) \vdash tp(\bar{b}_s, \bigcup\{A_t : t \in I, s \not\subset t\})$. Now we define the $\bar{b}_s^+$. If $(n-1) \not\in s$ let $\bar{b}_s^+ = \bar{b}_s$. Next choose $\langle \bar{b}_s^+ : s \in I, 0 \not\in s, (n-1) \in s \rangle$, so that $\bar{b}_s^+ \in A_{s-\{n-1\}}$ and in the model $\langle A_{s \cup \{n-1\}} : s \in \mathcal{P}(\{1, \ldots, n-2\}) \rangle$ it realizes over $\bigcup\{\bar{b}_s : s \in \mathcal{P}(\{1, \ldots, n-2\})\}$ the same type as $\langle \bar{b}_s : s \in I, 0 \not\in I, (n-1) \in s \rangle$ (possible by (4) of Definition 5.1). For the others, define by induction on $|s|$, $\bar{b}_s^+$ such that $tp(\cdots {}^\frown \bar{b}_t {}^\frown \cdots)_{t \subset s} = tp(\cdots {}^\frown \bar{b}_t^+ {}^\frown \cdots)_{t \subset s}$, and simultaneously prove that the mapping $\bar{b}_s \to \bar{b}_s^+$ defined so far is elementary (for $\mathfrak{C}$).

5.5 Conclusion: 1)* Suppose $\lambda = \lambda^{<\lambda} > |T|$, and $\langle A_s^\ell : s \in I \rangle$ is a system, $\langle A_s^\ell : 0 \not\in s \in I \rangle$ is λ-saturated, each A_s^ℓ is λ-saturate and of power λ and

$\langle A_s^0 : 0 \notin s \in I \rangle \equiv \langle A_s^1 : 0 \notin s \in I \rangle$. Then the two systems are isomorphic.

2) If in (1) we do not assume 1.0, we need $(A_s^\ell, c)_{c \in \bigcup_{t \subset s} A_t^\ell}$ is λ-saturated when $0 \in s \in I$.

5.6 Conclusion: 1) If $\langle A_s : s \in I \rangle$ is an I-system, $\mathcal{P}(\{1, \ldots, n-1\}) \subseteq J \subseteq I$ then $\bigcup_{s \in I} A_s$ is u.l.a. over $\bigcup_{s \in J} A_s$.

2) If $\langle A_s : s \in I \rangle$ is an I-system, then for $s \in I$, $A_s \cap P^{\mathfrak{C}} = A_{s-\{0\}}$.

§6 A proof of the existence property.

6.0 Hypothesis: $n^{**}(T) = \omega$.

6.1 Theorem * : Suppose T is countable and $\langle A_s : s \in \mathcal{P}^-(n) \rangle$ is a system satisfying:

(*) $0 \in s \in \mathcal{P}^-(n) \Longrightarrow A_s$ is $\mathbf{F}^\ell_{\aleph_0}$ - constructible over $\bigcup_{t \subset s} A_t$.

Then there is a model M $\mathbf{F}^\ell_{\aleph_0}$-constructible over $\bigcup_s A_s$, u.l.a. over it, and $P^M \subseteq \bigcup_s A_s$. So the existence property holds for such systems.

Proof : The proof is broken to some claims.

6.2 Claim: If $A \subseteq_t C$, B is $\mathbf{F}^\ell_{\aleph_0}$-constructible over A, then B is $\mathbf{F}^\ell_{\aleph_0}$-constructible over C (by the same sequence), $tp_*(B,A) \vdash tp_*(B,C)$, and $A \cup B \subseteq_t C$.

Proof : See [Sh 1 , Ch. XII].

6.3 Claim *: If M is $\mathbf{F}^\ell_{\aleph_0}$-constructible over $\bigcup A_s \subseteq M$, $\langle A_s : s \in \mathcal{P}^-(n) \rangle$ is a system *then* M is u.l.a. over $\bigcup_s A_s$.

Proof : W.l.o.g. (by easy set theory) for some $\lambda > |\bigcup_s A_s| + |T|$,

$\lambda = \lambda^{<\lambda}$ so let $\langle A_s' : s \in \mathcal{P}^-(n)\rangle$ be a saturated elementary extension of $\langle A_s : s \in \mathcal{P}^-(n)\rangle$. By 6.2 $tp_*(M, \bigcup_s A_s')$ is λ-isolated so there is a λ-primary N, $M \cup \bigcup_s A_s' \subseteq N$. Hence N is u.l.a. over $\bigcup A_s$ see 3.3 and we finish by the next Fact (6.4).

6.4 Fact: If $A \prec C$, and B is u.l.a. over C, $tp_*(B,A) \vdash tp_*(B,C)$ then B is u.l.a. over A (witnessed in the same way).

Remark: Note that we assume $A \prec C$, i.e. $\mathfrak{C} \restriction A \prec C \restriction \mathfrak{C}$, not just $A \subseteq_t C$.

Proof : Let $\bar b \in B$, $\varphi \in L$, then for some $\psi_\varphi(\bar x, \bar c_{\varphi,\bar b}) \in tp(\bar b, C)$ $\psi_\varphi(\bar x, \bar c_{\varphi,\bar b}) \vdash tp_\varphi(\bar b, C)$. As $tp_*(B,A) \vdash tp_*(B,C)$ there is $\vartheta(\bar x, \bar a) \in tp_*(\bar b, A)$, $\vartheta(\bar x, \bar a) \vdash \psi_\varphi(\bar x, \bar c_{\varphi,\bar b})$. So

$$\models (\forall \bar x)[\vartheta(\bar x, \bar c_{\varphi,\bar b}) \rightarrow \psi_\varphi(\bar x, \bar c_{\varphi,\bar b})] \wedge$$
$$(\forall \bar y \in C)[(\forall \bar x)(\psi_\varphi(\bar x, c_{\varphi,b}) \rightarrow \varphi(\bar x, \bar y)) \vee (\forall \bar x)(\psi_\varphi(\bar x c_{\varphi,\bar b}) \rightarrow \neg \varphi(\bar x, \bar y))]$$

so there is $\bar c'_{\varphi,b} \in A$ with those properties.

6.5 Claim: If $\langle A_s : s \in \mathcal{P}^-(n)\rangle$ is a system, satisfying (*) (from 6.1) $\lambda = \sum_s |A_s| > |T|$ then we can define $\langle A_s^\alpha : s \in \mathcal{P}^-(n)\rangle$ $(\alpha < \lambda)$ such that

(1) $\sum_s |A_s^\alpha| = |\alpha| + |T|$

(2) $\langle A_s^\alpha : s \in \mathcal{P}^-(n)\rangle \prec \langle A_s : s \in \mathcal{P}^-(n)\rangle$

(3) $\langle A_s^\alpha : s \in \mathcal{P}^-(n)\rangle$ is increasing continuous in α.

(4) $\langle A_s^\alpha : s \in \mathcal{P}^-(n)\rangle$ is a system, as well as $\langle A_s^\alpha : s \in \mathcal{P}^-(n+1), s \neq \{0, \ldots, n-1\}\rangle$ (where $A^\alpha_{s \cup \{n\}} = A_s^{\alpha+1}$ for $s \in \mathcal{P}^-(n)$), satisfying (*) in both cases.

Proof : Easy. [Sh 1, ch. IV, §3]

Proof of 6.1: We prove it by induction on $\lambda = \sum_s |A_s|$ (for all n simultaneously).

Easily of the three properties demanded of M in 6.1 the first implies the second (by 6.3) and the third (apply u.l.a. for the formula $x=y$). Remember T is countable.

Case 1: $\lambda = \aleph_0$.

So $A \stackrel{def}{=} \bigcup_s A_s$ is countable. By Hypothesis 6.0, easily for every $\varphi(\bar{x},\bar{a})$ $\bar{a} \in A$, $\models \exists \bar{x} \varphi(\bar{x},\bar{a})$, and $\varphi_1(\bar{x},\bar{y})$ there is $\vartheta(\bar{x},\bar{a}_1)$, $\bar{a}_1 \in A$, and $\models (\exists \bar{x})[\varphi(\bar{x},\bar{a}) \wedge \vartheta(\bar{x},\bar{a}_1)]$ and $\vartheta(\bar{x},\bar{a}_1) \vdash tp_{\varphi_1}(\bar{c},A)$ for some $\bar{c}$. [otherwise replace $\langle A_s : s \in \mathcal{P}^-(n)\rangle$ by an elementarily equivalent $|T|^+$-saturated system and get contradiction to 5.2(6)]. So we can define by induction on n, $\varphi_n(x_0, \ldots, x_n, \bar{a}_n)$, $\bar{a}_n \in A$ such that $\models (\exists x_0, \ldots, x_n)\varphi_n(x_0, \ldots, x_n, \bar{a}_n)$, $\models (\forall x_0, \ldots, x_{n+1})$ $(\varphi_{n+1}(x_0, \ldots, x_{n+1}, \bar{a}_{n+1}) \rightarrow \varphi_n(x_0, \ldots, x_n, \bar{a}_n))$ and for every $\psi = \psi(x_0, \ldots, x_n; \bar{y})$ for some $k > n$ and $c_0, \ldots, c_n$ $(\exists x_{n+1} \cdots x_k)$ $\varphi_n(x_0, \ldots, x_k, \bar{a}_k) \vdash tp(\langle c_0, \ldots, c_n\rangle A)$ $\{\varphi_n(x_0, \ldots, x_n, \bar{a}_n) : n < \omega\}$ is complete over A (in $\{x_n : n < \omega\}$) and is the complete diagram over A of a model as required (remember Ax VII (of [Sh 1 , Ch. IV. §1]. holds for $\mathbf{F}^{\ell}_{\aleph_0}$.)

Case 2: $\lambda > |T|$.

Define $A_s^\alpha(\alpha < \lambda)$ by 6.5. We now define by induction on α, a model M_α, so that M_α is $\mathbf{F}^{\ell}_{\aleph_0}$-constructible over $\bigcup_s A_s^\alpha$, $\bigcup_s A_s^\alpha \subseteq M_\alpha$, also if α is limit $M_\alpha = \bigcup_{\beta<\alpha} M_\beta$, and $\alpha = \beta + 1$ M_α is $\mathbf{F}^{\ell}_{\aleph_0}$-constructible over $\bigcup_s A_s^\alpha \cup M_\beta$. We should prove for each α, that $\langle A_s^\alpha : \alpha \in \mathcal{P}^-(n+1)\rangle$ is a system where $A_n^\alpha = M_\alpha$, this follows by 6.5(4) and noting M_n is u.l.a. over $\bigcup\{A_s^\alpha : s \in \mathcal{P}^-(n)\}$ by 6.3.

6.6 Theorem * : Suppose T is countable. If M,N are $\aleph_1$-saturated, with $M \restriction P = N \restriction P$ then M,N are isomorphic over P^M. (by 3.4(3) the $\aleph_1$-saturation of $M \restriction P$ implies that of M).

Proof : Over P^M there is a $\mathbf{F}^{\ell}_{\aleph_0}$ primary model M^+, so M^+ is $\mathbf{F}^{t}_{\aleph_1}$-primary and $\mathbf{F}^{t}_{\aleph_1}$-prime. So it can be elementarily embedded into M over P^M hence its image is equal to M. Similarly for N.

This theorem is made more interesting by the following (not using 6.0 anymore):

6.7 Fact: Assume for every M, P^M is stable.

If there are $M_0 \prec M, P^M \subset M_0 \neq M_1$ then for every $\lambda \geq |T| + |\delta|$ we can find M and $a_i \in M (i < \delta)$ such that :

(a) $i \neq j \Longrightarrow a_i \neq a_j$, $\|M\| = \lambda = |P^M|$,

(b) for $i < j$ $\ tp(a_j, P^M \cup \{a_\alpha : \alpha < i\}) = tp(a_i, P^M \cup \{a_\alpha : \alpha < i\})$.

(c) moreover for every $\bar{b} \in M$ there is $i(\bar{b}) < \delta$ such that: if $i(b) \leq i < j$, $tp(a_j, P^M \cup \bar{b} \cup \{a_\alpha : \alpha < i\}) = tp(a_i, P^M \cup \bar{b} \cup \{a_\alpha : \alpha < i\})$

Proof : W.l.o.g. M_0 is λ^+-saturated. Let $a \in M_1 - M_0$ and define by induction on $i < \delta$, $N_i \prec M_0$, $\|N_i\| = \lambda$ and a_i such that $\bigcup_{j<i} N_j \cup \{a_j : j < i\} \subset N_i$ and $a_i \in M_0$ realizes $tp(a, N_i)$. By claim 2.16 (or 1.4) $\bigcup_{i<\delta} N_i$, $\{a_i : i < \delta\}$ are as required.

6.8 Lemma : 1) Under the assumption of 6.7, if the conclusion of 6.1 holds then when $|T| < \lambda < \mu$ there is a model M^*, $\|M^*\| = |P^{M^*}| = \mu$, so that there are $a_i (i < \delta)$ as there (for M^*) when $\delta = \lambda$ but not when $\lambda < cf\ \delta \leq \mu$.

2) If 1.0 fails, λ regular $2^\lambda = \lambda^+$, then we can find M, $a_i (i < \lambda)$ as in 6.7, P^M is saturated, $\|M\| = \|P^M\| = \lambda^+$.

Proof : 1) Let M ,$a_i (i < \lambda)$ be as there, choose $A, P^M \in A \prec \mathfrak{C} \restriction P$, $|A| = \lambda$ and let M^* be $\mathbf{F}^{\ell}_{\aleph_0}$-constructible over $M \cup A$. By the P^M's stability, $a_i (i < \lambda)$ has the property in M^* too. Suppose $\langle c_i : i < \delta \rangle$ has the property (in M, i.e. a),b),c) of 6.7) too and $\lambda < cf\ \delta$. By [Sh 1 Ch IV §3] we can find $N \prec M^*$, $M \subset N$, $\|N\| = \lambda$, N closed enough (under history of the construction and the function $\bar{b} \to c_{i(\bar{b})}, c_i \to c_{i+1}$), so that M^* is $\mathbf{F}^{\ell}_{\aleph_0}$-constructible over $N \cup A$, and $\langle a_i : a_i \in N \rangle$ has the property in N and $cf(\sup\{i : a_i \in N\}) > |T|$. Then

$tp(a_{\sup\{i:a_i \in N\}}, N \cup A)$ is not $\aleph_1$- isolated, contradiction.

2) Left to the reader.

§7 Manipulations with systems for an arbitrary theory.

7.0 Discussion: We are dealing with several kinds of I-systems, so we shall use the name "$I-x-$system", x a latin letter to differentiate. For Definition 5.1 we use $x=a$ and say it is for $\mathbb{E}$ or for T.

7.1 Definition: We call $\mathbf{S} = \langle A_s : s \in I \rangle$ an $I-b-$system for T if:

1) for some $n = n(I) = n(S)$, $I \subseteq \mathcal{P}(n)$, $I \not\subseteq \mathcal{P}(n-1)$, I closed under subsets.

2) $\langle A_s : s \cup \{n-1\} \in I, n-1 \notin s \rangle \prec \langle A_{s \cup \{n-1\}} : s \cup \{n-1\} \in I, n-1 \notin s \rangle$

3) each A_s is a model of T.

4) $\langle A_s : s \in J \rangle$ is a I-system when $J = \{s \in I : (n-1) \notin s\}$.

5) If $n-1 \in t \in I$ then $A_t \cap (\bigcup\{A_s : n-1 \notin s \in I\}) \subseteq \bigcup\{A_s : n-1 \notin s \in I, s \cup \{n-1\} \in I\}$.

7.2 Fact: 1) For $\mathbf{S}$ to be an $I-b-$system for T depends on its first order theory only.

2) If $\langle A_s : s \in I \rangle$ is an $I-b-$system then $A_s \cap A_t = A_{s \cap t}$ for any $s, t \in I$.

Proof: 1) Check.

2) Prove it by induction on n. If $n-1 \notin s \cup t$-trivial using condition 4). If $n-1 \in s \cap t$, by condition (4) and the induction hypothesis $A_{s-\{n-1\}} \cap A_{t-\{n-1\}} = A_{s \cap t-\{n-1\}}$ and use condition (2). If $n-1 \in s, n-1 \notin t$, then $s \cap t = (s - \{n-1\}) \cap t$, and again $A_{s-\{n-1\}} \cap A_t = A_{s \cap t}$ by condition (4), and by (2) $A_{s-\{n-1\}} = A_s \cap (\bigcup \{A_v : v \cup \{n-1\} \in I, n-1 \neq I\}$, and we finish by (5).

7.3 Fact: $\langle A_s : s \in \mathcal{P}(n)\rangle$ is a $\mathcal{P}(n)-b$-system for $\mathbb{E}\restriction P$ *iff* $\langle A_{s-1} : s \in \mathcal{P}(n+1), 0 \notin s\rangle$ is a $\mathcal{P}(\{1,\ldots,n\})-a$-system (for an integer k let $s-k = \{i-k : i \in k\}$ $s+k$ is defined similarly.)

7.4 Fact: If $n > 0$, $\langle A_s : s \in \mathcal{P}(n)\rangle$ is a $\mathcal{P}(n)-b$-system for T and for $s \in \mathcal{P}(n-1)$ $B_s = (A_{(s+1)\cup\{0\}}, A_{s+1})$ then $\langle B_s : s \in \mathcal{P}(n-1)\rangle$ is a $\mathcal{P}(n-1)-b$-system for $T_1 = Th(A_{\{0\}}, A_\phi)$

Proof: We prove it by induction on n

n=1: so $\mathcal{P}(n-1) = \mathcal{P}(0) = \{\phi\}$, so $\langle B_s : s \in \mathcal{P}(n-1)\rangle$ consistent of one model, of T_1 of course:

n+1:

Condition: 1) is trivial.

Condition: 2) We should prove

$$\langle B_s : s \in \mathcal{P}(n-1)\rangle \prec \langle B_{s\cup\{n-1\}} : s \in \mathcal{P}(n-1)\rangle$$

(looking what I is).

This is equivalent to

$$\langle\langle A_{(s+1)\cup\{0\}}, A_{s+1}\rangle : s \in \mathcal{P}(n-1)\rangle \prec \langle\langle A_{(s+1)\cup\{0,n\}}, A_{(s+1)\cup\{n\}}\rangle : s \in \mathcal{P}(n-1)\rangle$$

which is equivalent to

$$\langle A_s : s \in \mathcal{P}(n)\rangle \prec \langle A_{s\cup\{n\}} : s \in \mathcal{P}(n)\rangle$$

which holds as $\langle A_s : s \in \mathcal{P}(n+1)\rangle$ is a $\mathcal{P}(n+1)-b$-system.

Condition: 3) we know $\langle A_s : s \in \mathcal{P}(n)\rangle$ is a $\mathcal{P}(n)-b$-system hence by the induction hypothesis for $s \in \mathcal{P}(n-1)$, $(A_{(s+1)\cup\{0\}}, A_{(s+1)}) \equiv (A_{\{0\}}, A_\phi)$. As we have proved condition 2), for $s \in \mathcal{P}(n-1)$ $B_s \prec B_{s\cup\{n-1\}}$ i.e. $(A_{(s+1)\cup\{0\}}, A_{s+1}) \equiv (A_{(s+1)\cup\{0,n-1\}}, A_{(s+1)\cup\{n-1\}})$, so the condition holds for $s \cup \{n-1\}$ when $s \in \mathcal{P}(n-1)$.

So it holds for every $s \in \mathcal{P}(n)$, as required.

Condition: 4) Easy.

Condition 5): Obvious, (by 5) for $\langle A_s : s \in \mathcal{P}(n+1)\rangle$).

7.5 Lemma: 1) Suppose $\langle M_s : s \in \mathcal{P}(n)\rangle$ is a $\mathcal{P}(n)-b$-system for T, $\lambda = \sum_s \|M_s\|$, $\lambda > |T|$. Then we can find $\langle M_{s,\alpha} : s \in \mathcal{P}(n)\rangle$ $(\alpha < \lambda)$ such that:

(i) $\langle M_{s,\alpha} : s \in \mathcal{P}(n)\rangle \prec \langle M_s : s \in \mathcal{P}(n)\rangle$

(ii) $\|M_{s,\alpha}\| = |T| + |\alpha|$.

(iii) Let for $\alpha < \lambda^+, s \in \mathcal{P}(n)$,

$$M_s^\alpha = M_s^\alpha, M^\alpha_{s\cup\{n\}} = M_s^{\alpha+1}$$

Then $\langle M_s^\alpha : s \in \mathcal{P}(n+1)\rangle$ is a $\mathcal{P}(n+1)-b$-system for T.

2) If $\langle M_s : s \in \mathcal{P}(n)\rangle$ is κ-saturated $(\forall \alpha < \lambda)[|\alpha|^{<\kappa} < \lambda]$, $2^{|T|} < \lambda$ then we can demand $\langle M_{s,\alpha} : s \in \mathcal{P}(n)\rangle$ is κ-saturated when $cf\ \alpha \in \{0,1\}$ or $cf\ \alpha \geq \kappa$, but then $\|M_{s,\alpha}\| \leq (|T| + |\alpha|)^{<\kappa}$ (if we ask just for κ-compact then $\|M_{s,\alpha}\| \leq (|T| + |\alpha|)^{<\kappa}$.

(2a) We can even demand this for each $I \subseteq \mathcal{P}(n)$ separately.

3) If $\lambda = \kappa^+$, $\kappa = \kappa^{<\kappa} > |T|$, and $\langle M_s : s \in \mathcal{P}(n)\rangle$ is saturated, then we can also demand $\langle M_s^\alpha : s \in \mathcal{P}(n+1)\rangle$ is saturated when $cf\ \alpha \notin [\aleph_0, \kappa)$. (but $\|M_{s,\alpha}\| = \kappa$) We can, except for some unbounded non stationary subset determine its theory as that of $\langle N_s : s \in \mathcal{P}(n+1)\rangle$ a $\mathcal{P}(n+1)-b$-system, provided that $\langle N_s : s \in \mathcal{P}(n)\rangle \equiv \langle M_s : s \in \mathcal{P}(n)\rangle$.

Proof: 1) Easy, 2) Easy, 3) See proofs in §4.

7.6 Lemma: Suppose $\lambda = \lambda^{<\lambda}$, and $2^{\lambda^{+\ell}} = \lambda^{+\ell+1}$ for $\ell < n$. Suppose $\langle A_s^* : s \in \mathcal{P}(n)\rangle$ is a $\mathcal{P}(n)-b$-system for T. Let $J = J_{\lambda,n} \stackrel{\text{def}}{=} \{\eta : \eta$ a sequence of ordinals of length $\leq n, \eta(\ell) < \lambda^{+(n-\ell)}\}$.

Then we can define models $M_{\eta,t}$ $(\eta \in J,\ t \in \mathcal{P}(\ell(\eta)))$ of T such that:

(i) $M_{\eta,t}$ is a model of T of power $\lambda^{+(n-\ell(\eta))}$, it is saturated provided that $(\forall \ell < \ell(\eta))[cf\ \eta(\ell) \in \{0, 1, \lambda^{+(n-\ell)}\} \vee \ell \in t]$.

(ii) if $\eta \in J, t \in \mathcal{P}(\ell(\eta))$ and $\ell(\eta) < n$, then

a) $M_{\eta,t} = \bigcup \{M_{\eta\frown<i>,t} : i < \lambda^{(n-\ell(\eta))}\}$

b) if $\eta\frown<\delta> \in J$ δ limit then $M_{\eta\frown<\delta>,t} = \bigcup_{i<\delta} M_{\eta\frown<i>,t}$

c) $M_{\eta\frown<i>,t\cup\ell(n)} = M_{\eta\frown<i+1>}$.

(iii) for each $\eta \in J, S^{\eta} \stackrel{def}{=} \langle M_{\eta,t} : t \in \mathcal{P}(\ell(\eta))\rangle$ is a $\mathcal{P}(\ell(\eta))-b$ –system.

(iv) if $(\forall \ell < \ell(\eta))[cf\ \eta(\ell) \in \{0,1,\lambda^{+(n-\ell)}\}]$ then S_{η} is saturated and its theory is that of $\langle A_s^{*} : s \in \mathcal{P}(\ell(\eta))\rangle$.

Proof: We prove it by induction on n for all possible T, $\langle A_s^{*} : s \in \mathcal{P}(n)\rangle$. For each $n > 0$ we define by induction on $\alpha < \lambda^{+n}$ the models $M_{<\alpha>,\phi}$ and $M_{\eta,t}(\eta(0) < \alpha,)$ $\eta \in J, t \in \mathcal{P}(\ell(\eta))$, such that when $cf\ \alpha \in \{0,1,\lambda^{+(n-1)}\}$, $M_{<\alpha>,\phi}$ is a saturated, of cardinality $\lambda^{+(n-1)}$ $M_{<\alpha>} \models T$, $M_{<\alpha>,\{1\}}$ is saturated, $\langle M_{<\alpha>} : \alpha < \lambda^{+n}\rangle$ an elementary chain, for α limit, $M_{<\alpha>} = \bigcup_{\beta<\alpha} M_{<\beta>}$, for $\alpha = \beta + 1$ $M_{<\alpha>,\phi} = M_{<\beta>,\{1\}}$.

For α limit or zero - no problem. For $\alpha = \beta + 1$, $cf\ \beta \notin \{0,1,\lambda^{+(n-1)}\}$, we let $M_{<\alpha>,\{1\}}$ be a saturated elementary extension of $M_{<\beta>}$ of power $\lambda^{+(n-1)}$ and then use 7.5 (2a). For $\alpha = \beta + 1, \beta=0$ for $M_{<\alpha>}$ there is no problem and then use 7.5. For $\alpha = \beta+1$, $cf\ \beta \in \{1,\lambda^{+(n-1)}\}$, $M_{<\beta>,\phi}$ is saturated. We use the induction hypothesis for $n-1$, and T_1 from 7.4 (starting there with $\langle A_s^{*} : s \in \mathcal{P}(n)\rangle$). Getting $(M^1_{\eta,t}, M^0_{\eta,t})$ $\eta \in J_{\lambda,n-1}$. So $M^0_{<\alpha>,\phi}$ is a model of T of power $\lambda^{+(n-1)}$, saturated hence $\cong M_{<\beta>,\phi}$ so w.l.o.g. it is $M_{<\beta>}$, let $M_{<\alpha>,\phi} = M^1_{<>} = M_{<\beta>,\{0\}}$, $M_{<\alpha>\frown\eta(t-\{0\})-1}$ is $M^0_{\eta,t}$ if $0 \notin t$, and is $M^1_{\eta,(t-\{0\})-1}$ if $0 \in t$.

7.7. Lemma. Suppose $\lambda^{<\lambda} > |T|$, $2^{\lambda^{+\ell}} = \lambda^{+\ell+1}$ for $\ell < n$ and $\langle A_s^{*} : s \in \mathcal{P}(n)\rangle$ a $\mathcal{P}(n)-b$ –system for T. Let

$W(\lambda) = \{\delta < \lambda^{+} : cf\ \delta = \lambda\}$ $\quad W^{+}(\lambda) = \{\delta+1 : \delta \in W(\lambda)\}$, and $W^{*}(\lambda) = W(\lambda) \cup W^{+}(\lambda)$.

$J'_{\lambda,n} = \{\eta : \eta$ a sequence of ordinals of length $\leq n$, $\eta(\ell) < \lambda^{+(n-\ell)}$,

$$[\ell+1 < \ell(\eta) \Longrightarrow \eta(\ell) \in W^*(\lambda^{+(n-\ell-1)})]\}$$

$J^a_{\lambda,n} = \{\eta \in J'_{\lambda,n}$: for every $\ell+1 < \ell(\eta)$, $\eta(\ell) \in W(\lambda^{+(n-\ell)})\}$.

$J''_{\lambda,n} = \{\eta \in J'_{\lambda,n}: cf\ (\eta(\ell(\eta)-1)) \in W^*(\lambda^{+(\ell(\eta)-1)})\}$.

Then we can define models $M_\eta (\eta \in J'_{\lambda,n})$ of T such that:

(i) M_η is a model of T of power $\lambda^{+(n-\ell(\eta))}$

(ii) if $\eta \in J''_{\lambda,n}$ then M_η is saturated.

(iii) if $\eta \in J'_{\lambda,n}$ is not maximal then
$i < j \Longrightarrow M_{\eta\frown\langle i\rangle} \prec M_{\eta\frown\langle j\rangle}$; $M_\eta = \bigcup_i M_{\eta\frown\langle i\rangle}$; for δ limit $M_{\eta\frown\langle\delta\rangle} = \bigcup_{i<\delta} M_{\eta\frown\langle i\rangle}$.

(iv) For each $\eta \in J^a_{\lambda,n}$ we define a $\mathcal{P}(\ell(\eta))-b$-system S^η: $S^\eta = \langle M^{S^\eta}_t : t \in \mathcal{P}(\ell(\eta))\rangle$, $M^{S^\eta}_t = M_{\nu(\eta,t)}$ where $\ell(\nu(\eta,t)) = \ell(\eta)$ and

$$\nu(\eta,t)(\ell) = \begin{cases} \eta(\ell) & \text{if } \ell < \ell(\eta) \quad \ell \notin t \\ \eta(\ell)+1 & \text{if } \ell < \ell(\eta) \quad \ell \in t \end{cases}$$

We shall want:

(iv) If $\eta \in J''_{\lambda,n}$, S^η is saturated and $\equiv\langle A^*_s : s \in \mathcal{P}(\ell(\eta))\rangle$.

Proof: Like 7.6, only simpler.

§8 The structure theory we can still get when $k < n^{**}(T)$

8.1 Claim: If $A \subseteq^t_\lambda C$, and B is $\mathbf{F}^t_\lambda$-constructible over A, then B is $\mathbf{F}^t_\lambda$-constructible over C (by the same construction) and $tp_*(B,A) \vdash tp(B,C)$.

Proof: See [Sh 1, Ch. XI].

Remark: 1) $A \subseteq^t_\lambda C$ if every m-type of power $< \lambda$ over A realized in C is realized in A.

2) The same holds for $\subseteq^{\ell}_\lambda$, but we ignore this distinction (important for $\lambda = |T|$).

3) Remember M is λ-compact if every m-type over M of power $< \lambda$, finitely satisfiable in M is realized in M.

8.2 Claim*: If M is $\mathbf{F}^t_\lambda$-constructible over $\bigcup_s A_s \subseteq M$, $\langle A_s : s \in \mathcal{P}^-(n) \rangle$ a λ-compact $\mathcal{P}^-(n)$-a-system and $n < n^{**}(T)$ *then* M is u.l.a. over $\bigcup A_s$.

Proof: W.l.o.g. for some $\mu > \Sigma |A_s| + |T|$, $\mu = \mu^{<\mu}$ and let $\langle A'_s : s \in \mathcal{P}^-(n) \rangle$ be a μ-compact elementary extension of $\langle A_s : s \in \mathcal{P}^-(n) \rangle$ which has power μ. As $\langle A_s : s \in \mathcal{P}^-(n) \rangle$ is λ-compact clearly $\bigcup A_s \subseteq^t_\lambda \bigcup_s A'_s$ (in case of saturation instead compactness - even $\subseteq^s_\lambda$) so by 8.1 M is $\mathbf{F}^t_\lambda$-constructible over $\bigcup_s A'_s$, so $tp(M, \bigcup_s A_s) \vdash tp(M, \bigcup_s A'_s)$ hence there is a μ-primary model N over $\bigcup_s A'_s$, $M \subseteq N$. We know (see 3.3) N is u.l.a. over $\bigcup_s A'_s$. So for every $\bar{c} \in M$ and φ there is a $\psi = \psi(x, \bar{b}) \in tp(\bar{c}, \bigcup A'_s)$. $\psi \vdash tp_\varphi(\bar{c}, \bigcup_s A'_s)$. But we know $tp(\bar{c}, \bigcup_s A_s) \vdash tp(\bar{c}, \bigcup_s A'_s)$ hence for some $\vartheta \in tp(\bar{c}, \bigcup_s A_s)$ $\vartheta \vdash \psi$. So $\vartheta \in tp(\bar{c}, \bigcup A_s), \vartheta \vdash tp_\varphi(\bar{c}, \bigcup_s A_s^{(0)})$. We get M is l.a. over $\bigcup A_s$. But we want u.l.a.

This follows from 6.4.

8.3 Claim*: Let $\mathbf{S} = \langle A_s : s \in I \rangle$ be an I-a-system and $\lambda > |T|$. $\mathbf{S}$ is λ-saturated *iff* $\langle A_s : s \in I \cap \mathcal{P}(\{1, \ldots, n(I)-1\}) \rangle$ is λ-saturated and each $M_s (s \in I,\ 0 \in s)$ is λ-saturated.

Proof: $\Longrightarrow$ trivial.

$\Longleftarrow$: We prove it by induction on $|I|$. Let $p = p\ (x_0, \ldots, x_{m-1})$ be an m-type over $\mathbf{S}$. $|\mathrm{Dom}\ p| < \lambda$ and p is finitely satisfiable in $\mathbf{S}$. If $I = \mathcal{P}(\{1, \ldots, n(I)-1\})$ this is trivial. Otherwise choose $t \in I$, $0 \in t$, t maximal, and let $J = I - \{t\}$. W.l.o.g.

$$\{x_0 \in A_t - \bigcup_{s \subset t} A_s, \ldots, x_{k-1} \in A_t - \bigcup_{s \subset t} A_s,$$
$$x_k \notin A_t - \bigcup_{s \subset t} A_s, \ldots, x_{m-1} \in A_t - \bigcup_{s \subset t} A_s\} \subseteq p.$$

As A_t is u.l.a. over $\bigcup_{s \subset t} A_s$ (and 5.1) *there is* $\psi(\bar{x}, \bar{y}) \in L(T)$ such that for every

$a \in A_t - \bigcup_{s \subset t} A_s$, for some $\bar{b} \in \bigcup_{s \subset t} A_s$, $\models \psi[a, \bar{b}]$ and $\psi(x, \bar{b}) \vdash \{x \neq e : e \in \bigcup\{A_s : s \in J\}$ (this in $\mathbb{E}$, so w.l.o.g. ψ is atomic, we shall not mention such things). So $p \cup \{\psi(x_0, \bar{y}_0), \ \bar{y}_0 \subseteq \bigcup_{s \in J} A_s, \ (\forall z \in \bigcup_{s \in J} A_s)(\neg\psi(z, \bar{y}_0)\}$ is finitely satisfiable in S. So w.l.o.g. for $\ell < k$ $(\exists \bar{y})[\psi(x_\ell, \bar{y}) \wedge \bar{y} \subseteq \bigcup_{s \in J} A_s \wedge (\forall z \in \bigcup_{s \in J} A_s) \neg\psi(z, \bar{y})] \in p$. Now let $\bar{x}^0 = \langle x_0, \ldots, x_{k-1} \rangle$ $\bar{x}^1 = \langle x_k, \ldots, x_{m-1} \rangle$, and

$$p^1 = p \cup \{(\forall \bar{z} \subseteq \bigcup_{s \in I} A_s)[\varphi(\bar{x}^0, \bar{z}) \equiv \psi_\varphi(\bar{z}, \bar{y}_\varphi)] \wedge \bar{y}_\varphi \subseteq \bigcup_{s \subset t} A_s : \varphi \in L\}.$$

Let p^2 be the closure of p^1 under conjunctions. Let $p^3 = \{(\exists \bar{x}^0)\vartheta : \vartheta \in p^2\}$. By the induction hypothesis p^3 is realized say by $\bar{x}^1 \to \bar{a}^1, \bar{y}_\varphi \to \bar{b}_\varphi$ for $\varphi \in L$ (you may argue that p^3 has $|T|$ variable not some $m' < \omega$, but λ-compactness implies this). Now we can find $\bar{a}^0$ realizing $\{\psi_\varphi(\bar{x}^0, \bar{b}_\varphi) : \varphi \in L\}$. Still we do not know that $\bar{a}^\frown\bar{a}^1$ realizes p - it may contain formulas which are not atomic. But our conclusion follows from:

8.4 Claim: Let $\langle A_s : s \in I \rangle$ be an I–a–system, $0 \in t \in I$, t maximal. Let $\langle \psi_\varphi : \varphi \in L \rangle$ wittness the u.l.a. of A_t over $\bigcup_{s \subset t} A_s$ $\bar{d}^1, \bar{d}^2 \in \bigcup_{s \subset t} A_s$, $\bar{c}^1 \bar{c}^2 \in A_s - \bigcup_{s \subset t} A_s$, $\models \psi_\varphi(\bar{c}^\ell, b^\ell_\varphi)$, $\bar{b}^\ell_\varphi \in \bigcup_{s \subset A} A_s$, $(\cdots, \bar{b}^2_\varphi \cdots, \bar{d}^2) = tp(\cdots, b^1_\varphi \cdots, \bar{d}^1)$ *then* in $\langle A_s : s \in I \rangle$ the sequences $\bar{c}^1 \frown \bar{d}^1$, $\bar{c}^2 \frown \bar{d}^2$ realizes the same type.

Proof: Again as in the previous claim; then some automorphism of $\langle A_s' : \in J \rangle$ take $\bar{d}^1$ to $\bar{d}^2$ and b^1_φ to b^2_φ. Then there is an automorphism of N embedding it taking $\bar{c}^1$ to $\bar{c}^2$.

8.5 Claim: Suppose $\langle A_s : s \in \mathcal{P}^-(n) \rangle$ is an I–a–system, λ-compact, and $\mu = \Sigma|A_s| > |T|$.

Then we can find $\langle A^\alpha_s : s \in \mathcal{P}^-(n) \rangle$ $(\alpha < \mu)$ such that

(1) $|A^\alpha_s| < \mu$, if $\mu = \chi^+$, $|A^\alpha_s| = \chi$,

(2) $\langle A^\alpha_s : s \in \mathcal{P}^-(n) \rangle \prec \langle A_s : s \in \mathcal{P}^-(n) \rangle$.

(3) If $J \subseteq \mathcal{P}^-(n)$, $\langle A_s : s \in J \rangle$ is λ-compact $(\forall \alpha < \mu)$ $[|\alpha|^{<\lambda} < \mu]$ then $\langle A_s^{\alpha+1} : s \in J \rangle$ is λ-compact (also for $\alpha = -1$).

(4) If A_s is λ-constructible over $\bigcup_{t \subset s} A_t$ then there is an $\mathbf{F}_\lambda^t$-construction $\langle a_i^s, B_i^s : i < \mu \rangle$ of A_s over $\bigcup_{t \subset s} A_t$ such that for each α for some $j(\alpha)$,

$$A_s^\alpha - \bigcup_{t \subset s} A_t^\alpha = \{a_i^s : i < j(\alpha)\}, \ (\forall i < j(\alpha) B_i^s \subseteq \bigcup_{t \subset s} A_t^\alpha$$

Proof: Easy (for (4) see [Sh 1 Ch IV §3]

8.6 Claim *: A complete set A is stable iff it has the atomicity property provided.

Proof: W.l.o.g. A is saturated of power μ, $\mu = \mu^{<\mu} > |T|$. Now easily stability implies atomicity. So assume atomicity for A, so there is M, λ-primary over A . Let $(M', A') \equiv (M, A)$ be saturated of power μ, so w.l.o.g. $A = A'$ and $M \prec M'$. By the hypothesis 1.0 $M = M'$. Hence M' is atomic over A, so by the saturation M' is u.l.a. over A. Also for every $p \in S_*^m(A)$ there is a λ-saturated $M'' \supseteq A \supseteq P^{M''}$ realizing p, but as again w.l.o.g. $M = M''$, p is λ-isolated, hence $\mathbf{F}_{\aleph_0}^a$-isolated. From here atomicity is easy.

8.7 Lemma: Suppose $|T| \mu < \lambda = \lambda^{<\lambda}$, $2^{\lambda^{+\ell}} = \lambda^{+\ell+1}$ for $\ell < k+n$.

1)* In the definition of an I–a–system we can omit "A_s is u.l.a. over $\bigcup_{t \subset s} A_t$ for $s \in I, 0 \in s$" when $|s| < n^{**}(T)$.

2) $n^*(T) = n^{**}(T)$.

3) If for $\ell = 1,2$ $\langle A_s^\ell : s \in I \rangle$ is a $\mathcal{P}^-(n)$–a–system, $\langle A_{s+1}^\ell : s+1 \in I \rangle$ is saturated of power μ with first order theory not depending on ℓ, $n(I) < n^{**}(T)$ *then* $\langle A_s^\ell : s \in I \rangle \cong \langle A_s^2 : s \in I \rangle$.

4)* If $k+n < n^{**}(T)$, $\langle A_s : s \in \mathcal{P}^-(n) \rangle$ and $\mathcal{P}^-(n)$–a–system, $|A_s| \leq \lambda^{+k}$, $A_{\{1,\ldots,n-1\}}$ is λ- saturated *then* over $\bigcup_s A_s$ there is a λ-primary model $M, P^M = A_{\{1,\ldots,n-1\}}$.

Proof: 1) By 3.3,

2) See 8.6.

3) We prove by induction on n (similar proof occurs previoulsy) we start with an isomorphism from $\langle A^1_{s+1}: s+1 \in I \rangle$ onto $\langle A^2_{s+1}: s+1 \in I \rangle$ and extend it step by step. For this we have to prove $A_t (0 \in t \in I)$ is μ-primary over $\bigcup_{s \subset t} A_s$, for this it suffices to prove it is u.l.a. over $\bigcup_{s \subset t} A_s$, which follows by 3.3 if we have proved 2).

4) We prove it by induction on k. For $k = 0$, $\bigcup_s A_s$ is stable, so there is a λ-primary model over it but

8.8 Claim: If A is complete, $A \cap P^{\mathfrak{C}}$ is λ- compact, $p \in S^m(A)$ is λ-isolated *then* $p \in S_*^m(A)$.

For $k+1$: Use 8.5 to get $A^\alpha_s (\alpha < \lambda^{+(k+1)}$. Now we define by induction on $\alpha, A^\alpha_{P(n)}$ so that

(i) $A^0_{P(n)}$ is λ-primary over $\bigcup \{A^0_s : s \in P^-(n)\}$.

(ii) $A^{\alpha+1}_{P(n)}$ is λ-primary over $\bigcup \{A^{\alpha+1}_s : s \in P^-(n)\} \cup A^\alpha_{P(n)}$.

(iii) $A^\delta_{P(n)} = \bigcup_{\alpha<\delta} A^\alpha_{P(n)}$.

(iv) $A^\alpha_{P(n)}$ is u.l.a. over $\bigcup \{A^\alpha_s : s \in P^- n)\}$ and is a model,

(v) $A^\alpha_{P(n)} \cap P^{\mathfrak{C}} = A^\alpha_{P(n-1)+1}$ (and $tp_*(A^\alpha_{P(n)}, \bigcup \{A_s : s \in P^-(n)\}$ $\vdash tp_*(A^\alpha_{P(n)}, \bigcup \{A_s : s \in P^-(n)\})$. The induction step (for α) is by the induction hypothesis for k (as $|A^{\alpha+1}_s| \leq \lambda^{+k}$) and 7.7 for α successor, and remember 7.5(3).

§9 Non structure when $n^{**}(T) < \omega$ and there is no two cardinal model

9.0 Hypothesis : $P^N \subseteq M \prec N \Longrightarrow M = N$; every formula is equivalent to a relation (for T).

9.1 Main Theorem : Suppose $\lambda = \lambda^{<\lambda}$, $2^{\lambda^{+\ell}} = \lambda^{+\ell+1}$ for $\ell < n \overset{def}{=} n^{**}(T)$, Q

is the forcing of adding λ^{+n} Cohen subset to V say $\langle r_\eta : \eta \in J'_{\lambda,n} \rangle$. (see 7.7).

Then in V^Q there are $2^{(\lambda^{+n})}$ model M_i, $\|M_i\| = |P^{M_i}| = \lambda^{+n}$ which pairwise are not isomorphic over P^M; really we can make $\|M_i\| = |P^{M_i}| = \mu$, for any $\mu \geq \lambda^{+n}$.

Proof : Let $\langle A_s : s \in \mathcal{P}^-(n) \rangle$ be an I—a—system which is unstable. Working in V let $A_\eta(\eta \in J'_{\lambda,n})$ be as in 7.7 [A_η standing for M_η $\langle A^\bullet_{s+1} : s \in \mathcal{P}^-(n-1) \rangle$ for $\langle A^\bullet_s : s \in \mathcal{P}^-(n-1) \rangle$) and Th $(\mathbb{E} \restriction P)$ for T]. Define a well ordering $<^*$ on $J'_{\lambda,n}$: $\eta \leq^* \nu$ iff $\eta = \nu \restriction \ell(\eta)$ or $(\exists \ell)[\eta \restriction \ell = \nu \restriction \ell \wedge \eta(\ell) < \nu(\ell)]$. For $A \subseteq J'_{\lambda,n}, A \in V$, we now define for each n by induction on $<^*$ a model N^A_η such that

(i) $N^A_\eta \cap P^{\mathbb{E}} = A_\eta$, $N^A_\eta \prec \mathbb{E}$.

(ii) if $\eta \in J'_{\lambda,n}$ is not maximal then $[i < j \Longrightarrow N^A_{\eta\frown\langle i\rangle} \prec N^A_{\eta\frown\langle j\rangle}]$ for δ limit $N^A_{\eta\frown\langle\delta\rangle} = \bigcup_{i<\delta} N^A_{\eta\frown\langle i\rangle}$ and $N^A_\eta = \bigcup_i N^A_{\eta\frown\langle i\rangle}$:

(iii) if $s \subseteq t \subseteq \mathcal{P}(\ell(\eta))$ then $N_{\nu(\eta,s)} \subseteq N^A_{\nu(\eta,t)}$.

(iv) The construction of $\langle N_\eta : \eta <^* \nu \rangle$ is done in $V[\langle r_\eta : \eta <^* \nu, \eta \in A \rangle]$.

(where by renaming assume Q odd the sets $\langle r_\eta : \eta \in J'_{\lambda,n} \rangle$, r_η a function from λ to $\{0,1\}$.

There are no particular problems (especially if you have read §4).

References.

[Sh1]

S. Shelah, Classification Theory and the number of non isomorphic models, North Holland Publ. Co. 1978.

[Sh2]

———, Stability over a predicate, *Notre Dame J. of Formal Logic*, to appear.

[Sh3]

------, Classifcation theory for non-elementary Classes I, the number of uncountable models $\psi \in L_{\omega_1,\omega}$, Part A: *Israel J. Math.* 46 (183), 212-240., Part B: *Israel J. Math.* 46 (1983), 241-273.

[Sh4]

-------, Classification theory completed for countable theories, *North Holland Pub. Co.* (1985?).

[Sh5]

-------, Models with second order properties III. Omitting types in λ^+ for $L(Q)$. Proc. of a workshop in Berlin, July, 1979, *Archiv fur Math Logik*, 21 (1981), 1-11.

[Sh6]

------, Proper Forcing, *Springer Lecture Notes*, 940 (1982).

[Sh7]

------, Classification over a predicate II, in preparation.

Existence of Endo-Rigid Boolean Algebras

In [Sh 2] we answering a question of Monk have explicated the notion of "a Boolean algebra with no endomorphisms except the ones induced by ultrafilters on it" (see §2 here) and prove the existence of one with character density $\aleph_0$, assuming first $\diamondsuit_{\aleph_1}$ and then only CH. The idea was that if h is an endomorphism of B, not among the "trivial" ones, then there are pairwise disjoint $d_n \in B$ with $h(d_n) \not\subset d_n$. Then we can, for some $S \subseteq \omega$, add an element x such that $d_n \leq x$ for $n \in S$, $x \cap d_n = 0$ for $n \notin S$ while forbidding a solution for $\{y \cap h(d_n) = h(d_n) : n \in S\} \cup \{y \cap h(d_n) = 0 : n \notin S\}$. Further analysis showed that the point is that we are omitting positive quantifier free types. Continuing this Monk succeeds to prove in *ZFC* the existence of such Boolean algebras of cardinality $2^{\aleph_0}$ and density character $2^{\aleph_0}$. In his proof he

(a) replaces some uses of the countable density character by the $\aleph_1$-chain condition

(b) generally it is hard to omit $< 2^{\aleph_0}$ many types but because of the special character of the types and models involve, using $2^{\aleph_0}$ almost disjoint subsets of ω, he succeeds in doing this

(c) for another step in the proof (ensuring indecomposability - see Definition 2.1) he (and independently by Nyikos) find it is in fact easier to do this when for every countabe $I \subseteq B$ there is $x \in B$ free over it.

The question of the existence of such Boolean algebras in other cardinalities remains open (See [DMR] and a preliminary list of problems for the handbook of Boolean Algebras by Monk).

We shall prove (in ZFC) the existence of such B of density character λ and cardinality $\lambda^{\aleph_0}$ whenever $\lambda > \aleph_0$. We then conclude answers to some other questions from Monk's list, (combine 3.1 with 2.5). We use a combinatorial method from [Sh 3], [Sh 4] , it is represented in section 1.

In [Sh 1], [Sh 6] (and [Sh 7]) the author offers the opinion that the combinatorial proofs of [Sh 1], Ch. VIII (applied there for general first order theories) should be useful for proving the existence of many non-isomorphic, and/or pairwise non-embeddable structure which has few (or no) automorphism or endomorphism or direct decomposition etc. As an illumination in [Sh 6] a rigid Boolean algebra in every $\lambda > \aleph_0$ was constructed. The combinatorics we used here relay on [Sh 1], Ch. VIII 2.6 and it amounts to building a model of power $\lambda^{\aleph_0}$ omitting countable types along the way, the method is proved in *ZFC*, nevertheless it has features of the diamond. It has been used also in Gobel and Corner [CG] and Gobel and Shelah [GS1], [GS2]. See more on the method and on refinements of it in [Sh 4] and [Sh 3] and mainly [Sh 5].

§1 The combinatorial principle

Content: Let $\lambda > \kappa$ be fixed infinite cardinal.

We shall deal with the case $cf\ \lambda > \aleph_0$, $\lambda^{\aleph_0} = \lambda^{\kappa}$, and usually $\kappa = \aleph_0$. Let L be a set of function symbols, each with $\leq \kappa$ places, of power $\leq \lambda$. Let $\mathfrak{M}$ be the L-algebra freely generated by $\mathrm{T} \overset{def}{=} {}^{\kappa>}\lambda (= \{\eta : \eta$ a sequence $< \omega$ of ordinals $< \lambda)$ (We could have as well considered T as a set of urelements, and let $\mathfrak{M}$ be the family $H_{<\kappa^+}(\mathrm{T})$ of sets hereditarily of cardinality $\leq \kappa$ build from the urelements]. For $\eta \in \mathrm{T} \cup {}^{\kappa}\lambda$ let $orco(\eta) = \{\eta(i) : i < \ell(\eta)\}$, for a sequence $\bar{\eta} = \langle \eta_i : i < \beta \rangle$ let $orco(\bar{a}) = \bigcup_{i<\beta} orco(\eta_i)$, for $a = \tau(\bar{\eta}) \in \mathfrak{M}$ let $orco(\eta) = orco(\bar{\eta})$ and $orco(\langle a_i : i < \beta \rangle) = \bigcup_{i<\beta} orco(a_i)$, and similarly for a set.

1.2 Explanation: We shall let B_0 be the Boolean Algebra freely generated by $\{\eta : \eta \in \mathrm{T}\}$, B_0^c its completion and we can interprete B_0^c as a subset of $\mathfrak{M}$ (each $a \in B_0^c$ has the form $\bigcup_{n<\omega} \tau_n$ where τ_n is a Boolean combination of members of T, so as we have in L $\aleph_0$-place function symbols there is no problem). As the $\eta \in \mathrm{T}$ may be over-used we replaced them for this purpose by x_η (e.g. let $F \in L$ be a monadic function symbol, $x_\eta = F(\eta)$).

Our desired Boolean Algebra B will be a subalgebra of B_0^c containing B_0.

1.3 Definition :

1) Let L_n be fixed vocabularies (= signatures), $|L_n| \leq \kappa$, $L_n \subset L_{n+1}$, (with each predicate function symbol finitary for simplicity, let $P_n \in L_{n+1} - L_n$ be monadic predicates.

2) Let $\mathfrak{J}_n$ be the family of sets (or sequences) of the form $\{(f_\ell, N_\ell) : \ell \leq n\}$ satisfying

a) $f_\ell : {}^{\ell \geq} \kappa \to \mathrm{T}$ is a *tree embedding* i.e.

(i) f_ℓ is length preserving i.e. η, $f_\ell(\eta)$ have the same length.

(ii) f_ℓ is order preserving i.e. for $\eta, \nu \in {}^{\ell \geq \kappa}$, $\eta \lessdot \nu$ iff $f_\ell(\eta) \lessdot f_\ell(\nu)$.

b) $f_{\ell+1}$ extend f_ℓ (when $\ell + 1 \leq n$)

c) N_ℓ is an L'_ℓ-model of power $\leq \kappa$, $|N_\ell| \subseteq |\mathcal{M}|$, where $L'_\ell \subset L_\ell$.

d) $L'_{\ell+1} \cap L_\ell = L'_\ell$ and $N_{\ell+1} \restriction L'_\ell$ extends N_ℓ

e) if $P_m \in L'_{m+1}$, then $P_m^{N_\ell} = |N_m|$ when $m < \ell \leq n$ and

f) $\mathrm{Rang}\,(f_\ell) - \bigcup_{m<\ell} \mathrm{Rang}\,(f_m)$ is included in $|N_\ell| - \bigcup_{m<\ell} |N_m|$.

3) Let $\mathfrak{J}_\omega$ be the family of pairs (f, N) such that for some $(f_\ell, N_\ell)(\ell < \omega)$ the following holds:

(i) $\{(f_\ell, N_\ell) : \ell \leq n\}$ belongs to $\mathfrak{J}_n$ for $n < \omega$.

(ii) $f = \bigcup_{\ell<\omega} f_\ell$, $N = \bigcup_{n<\omega} N_n$, (i. e. $|N| = \bigcup_{n<\omega} |N_n|$, $L(N) = \bigcup_n L(N_n)$, and $N \restriction L(N_n) = \bigcup_{n<m<\omega} N_m \restriction L(N_n)$)

4) For any $(f, N) \in \mathfrak{J}_\omega$ let (f_n, N_n) be as above (it is easy to show that (f_n, N_n) is uniquely determined - notice d),e) in (2),) so for (f^α, N^α) we get (f_n^α, N_n^α)

5) Let $\mathfrak{J}_n = \{(f_n, N_n)$: for some $(f_\ell, N_\ell)(\ell < n), \{(f_\ell, N_\ell) : \ell \leq n\} \in \mathfrak{J}_n$ and

we adopt conventions of 4).

6) A branch of Rang (f) or of f (for f as in (3)) is just $\eta \in {}^{\omega}\lambda$ such that for every $n < \omega$, $\eta \restriction n \in$ Rang (f).

1.4 Explanation of our Intended Plan (of Constructing e.g the Boolean Algebra)

We will be given $W = \{(f^{\alpha}, N^{\alpha}) : \alpha < \alpha(*)\}$, so that every branch η of f^{α} converge to some $\zeta(\alpha)$, $\zeta(\alpha)$ non-decreasing (in α). We have a free object generated by $\mathbf{T}$ (B_0 in our case) and by induction on α we define B_{α}, increasing continuous, such that $B_{\alpha+1}$ is an extension of B_{α}, $a_{\alpha} \in B_{\alpha+1} - B_{\alpha}$ (usually $B_{\alpha+1}$ is generated by B_{α} and a_{α}, and a_{α} is in the completion of B_0). Every element will depend on few ($\leq \kappa$) members of $\mathbf{T}$, and a_{α} "depends" in a peculiar way: the set $Y_{\alpha} \subseteq \mathbf{T}$ on which it "depends" is $Y_{\alpha}^{0} \cup Y_{\alpha}^{1}$ where Y_{α}^{0} is bounded below $\zeta(\alpha)$ (i.e. $Y_{\alpha}^{0} \subseteq {}^{\omega >}\zeta$ for some $\zeta < \zeta(\alpha)$) and Y_{α}^{1} is a branch of f^{α} or something similar. See more in 1.8.

1.5 Definition of the Game: We define for $W \subseteq \mathcal{J}_{\omega}$ a game $Gm(W)$, which lasts ω-moves.

In the n-th move:

Player I: Choose f_n, a tree-embedding of ${}^{n\geq}\kappa$ into ${}^{n\geq}\lambda$, extending $\bigcup_{\ell<n} f_{\ell}$, such that Rang $(f_n) - \bigcup_{\ell<n}$ Rang (f_{ℓ}) is disjoint to $\bigcup_{\ell<n} |N_{\ell}|$; then

player II chooses N_n such that $\{(f_{\ell}, N_{\ell}) : \ell \leq n\} \in \mathcal{J}_n$.

In the end player I wins if $(\bigcup_{n<\omega} f_n, \bigcup_{n<\omega} N_n) \in W$.

1.6 Remark: We shall be interested in W such that player I wins (or at least does not lose) the game, but W is "thin". Sometimes we need a strengthening of the second player in two respects: he can force (in the n-th move) Rang (f_{n+1}) $-$ Rang (f_n) to be outside a "small" set, and in the zero move he can determine an arbitrary initial segment of the play.

1.7 Definition : We define, for $W \subseteq \mathcal{J}_{\omega}$, a game $Gm'(W)$ which lasts ω-

moves.

In the zero move

player I choose f_0, a tree embedding of ${}^{0\geq}\kappa$ to ${}^{0\geq}\lambda$ (but there is only one choice).

player II chooses $k < \omega$ and $\{(f_\ell, N_\ell) : \ell \leq k\} \in \mathcal{F}_k$, and $X_0 \subseteq \mathbf{T}$, $|X_0| < \lambda$.

In the n-th move, $n > 0$:

player I chooses f_{k+n}, a tree embedding of ${}^{(k+n)\geq}\kappa$ into ${}^{(k+n)\geq}\lambda$, with Range $f_{k+n} - \bigcup_{\ell<k+n}$ Rang f_ℓ disjoint to $\bigcup_{\ell<k+n} N_\ell \cup \bigcup_{\ell<n} X_\ell$.

player II choose N_{k+n} such that $\{(f_\ell, N_\ell) : \ell \leq k+n\} \in \mathcal{F}_{k+n}$ and $X_n \subseteq \mathbf{T}$, $|X_n| < \lambda$.

1.8 Remark: What do we want from W?: First that by adding an element (to B_0) for each (f, N) we can "kill" every undesirable endomorphism, for this it has to encounter every possible endomorphism, and this will be served by "W a barrier". For this $W = \mathcal{F}_\omega$ is O.K. but we also want W to be thin enough so that various demands will have small interaction, for this disjointness and more are demanded.

1.6 Definition : 1) We call $W \subseteq \mathcal{F}_\omega$ a ***strong barrier*** if player I wins in $\mathrm{Gm}(W)$ and even $\mathrm{Gm}'(W)$ (which just means he has a winning strategy.)

2) We call W a ***barrier*** if player II does not win in $\mathrm{Gm}(W)$ and even does not win in $\mathrm{Gm}'(W)$.

3) We call W disjoint *if* for any distinct $(f^\ell, N^\ell) \in W$ $(\ell = 1,2)$, f^1 and f^2 has no common branch.

1.7 The Existence Theorem : 1) If $\lambda^{\aleph_0} = \lambda^\kappa$, $cf\ \lambda > \aleph_0$ then there is a strong disjoint barrier.

2) Suppose $\lambda^{\aleph_0} = \lambda^\kappa$, $cf\ \lambda > \aleph_0$. Then there is $W = \{(f^\alpha, N^\alpha) : \alpha < \alpha^*\} \subseteq \mathcal{F}_\omega$ and a function $\zeta : \alpha^* \to \lambda$ such that:

(a) W is a strong disjoint barrier, moreover for every stationary $S \subseteq \{\delta < \lambda : cf\ \delta = \aleph_0\}$ $\{(f^\alpha, N^\alpha) : \alpha < \alpha^*, \zeta(\alpha) \in S\}$ is a disjoint barrier.

(c) $cf\ (\zeta(\alpha)) = \aleph_0$ for $\alpha < \alpha^*$.

(d) Every branch of f^α is an increasing sequence converging to $\zeta(\alpha)$.

(e) If $\bar{\eta}$ is a sequence from T (of any length $\gamma < \kappa^+$), $\tau(\bar{x})$ a term, $\ell(\bar{x}) = \gamma$ and $\tau(\bar{\eta}) \in N^\alpha$ then $\bar{\eta} \subseteq N^\alpha \cap \mathrm{T}$.

(f) If $\zeta(\alpha) = \zeta(\beta)$, $\alpha + \kappa^{\aleph_0} \leq \beta < \alpha^*$ and η is a branch of f^β then $\eta \restriction k \notin N^\alpha$ for some $k < \omega$.

(g) If $\lambda = \lambda^\kappa$ we can demand: if η is a branch of f^α and $\eta \restriction k \in N^\beta$ for all $k < \omega$ (where $\alpha, \beta < \alpha^*$) then $N^\alpha \subseteq N^\beta$ (and even $N_n^\alpha \in N^\beta$ if $\mathfrak{M} = H_{<\kappa^+}(\mathrm{T})$.)

§2 Preliminaries on Boolean Algebras

We review here some easy material from [Sh 2].

2.1 Definition : 1) For any endomorphism h of a Boolean Algebra B, let $Ex\ Ker(h) = \{x_1 \cup x_2 : h(x_1) = 0$, and $h(y) = y$ for every $y \leq x_2\}$.

$Ex\ Ker^*(h) = \{x \in B :$ in $B / Ex\ Ker(h)$, below $x / Ex\ Ker(h)$, there are only finitely many elements$\}$.

2) A Boolean Algebra is endo-rigid *if* for every endomorphism h of B, $B / Ex\ Ker(h)$ is finite (equivalently: $1_B \in Ex\ Ker^*(h)$).

3) A Boolean algebra is indecomposable *if* there are no two disjoint ideal I_0, I_1 of B, each with no maximal member which generate a maximal ideal $(\{a_0 \cup a_1 : a_0 \in I_0, a_1 \in I_1\})$.

4) A Boolean algebra B is $\aleph_1$-compact if for pairwise disjoint $d_n \in B (n < \omega)$ for some $x \in B$, $x \cap d_{2n+1} = 0$, $x \cap d_{2n} = d_{2n}$.

2.2 Lemma : 1) A Boolean algebra B is endo-rigid *iff* every endomorphism of B is the endomorphism of some scheme (see Definition 2.3 below).

2) A Boolean algebra B is endo-rigid and indecomposable *iff* every endomorphism of B is the endomorphism of some simple scheme (see Def 2.3 below).

2.3 Definition : (1) A scheme of an endomorphism of B consists of a partition $a_0, a_1, b_0, \ldots, b_{n-1}, c_0, \ldots, c_{m-1}$ of 1, maximal nonprincipal ideal I_ℓ below b_ℓ for $\ell < n$, nonprincipal disjoint ideals I_ℓ^0, I_ℓ^1 below c_ℓ which generates a maximal ideal below c_ℓ for $\ell < m$, a number $k < n$, and a partition $b_0^*, \ldots, b_{n-1}^*, c_0^*, \ldots, c_{m-1}^*$ of $a_0 \cup b_0 \cup \cdots \cup b_{k-1}$. We assume also that $[k+m > 0 \Longrightarrow a_0 = 0]$, $[(n-k)+m > 0 \Longrightarrow a_1 = 0]$ and except in those cases there are no zero elements in the partition.

(2) the scheme is simple if $m = 0$.

(3) The endomorphism of the scheme is the unique endomorphism $T:B \to B$ such that:

(i) $Tx = 0$ when $x < a_0$ or $x \in I_\ell, \ell < k$, or $x \in I_\ell^0, \ell < m$.

(ii) $Tx = x$ when $x \le a_1$ or $x \in I_\ell, k \le \ell < n$ or $x \in I_\ell^1, \ell < m$.

(iii) $T(b_\ell) = b_\ell^*$ when $\ell < k$.

(iv) $T(b_\ell) = b_\ell \cup b_\ell^*$ when $k \le \ell < n$.

(v) $T(c_\ell) = c_\ell \cup c_\ell^*$ when $\ell < m$.

2.4 Claim: If h is an endomorphism of a Boolean Algebra B,and $B/\,Ex\ Ker(h)$ is infinite *then* there are pairwise disjoint $d_n \in B(n < \omega)$ such that $h(d_n) \not\subseteq d_n$. By easy manipulation we can assume that $h(d_n) \cap d_{n+1} \ne 0$, and if B satisfies the c.c.c. then $\{d_n : n < \omega\}$ is a maximal antichain.

2.5 Lemma : 1) Every endo-rigid Boolean Algebra B is Hopfian and and dual Hopfian. Even $B + B$ is Hopfian (and dual Hopfian) but not rigid.

Proof : Easy to check using 2.2, 2.3.

§3 The Construction.

3.1 Main Theorem : Suppose $\lambda > \aleph_0$. Then there is a B.A. (Boolean Algebra) B such that:

1) B satisfies the c.c.c.

2) B has power $\lambda^{\aleph_0}$, and density character λ.

3) B is endo-rigid and indecomposable.

Proof: We concentrate on the case $cf(\lambda) \geq \aleph_1$ (on the case $cf\,\lambda = \aleph_0$ see [Sh 5, §2, §3]) we shall use Theorem 1.3, and let $W = \{(f^\alpha, N^\alpha) : \alpha < \alpha^*\}$, the function ζ, $\mathcal{M}$ and $T = {}^{\omega>}\lambda$ be as there.

Stage A: Let B_0 be the B.A. freely generated by $\{x_\eta : \eta \in T\}$, let $x_\eta = a_\eta$ and B_0^c be its completion. For $A \subseteq B_0^c$ let $\langle A \rangle_{B_0^c}$ be the Boolean subalgebra A generates. As B_0 satisfies the c.c.c every element of B_0^c can be represented as a countable union of members of B_0, so w.l.o.g. $B_0^c \subseteq \mathcal{M}$. We say $x \in B_0^c$ is based on $J \subseteq {}^{\omega>}\lambda$ if it is based on $\{x_\nu : \nu \in J\}$ [i.e. $X = \bigcup_n y_n$, each y_n is in the subalgebra generated by $\{x_\nu : \nu \in J\}$] and let $\underline{d}(x)$ be the minimal such I. We shall now define by induction on $\alpha < \alpha^*$, the truth value of "$\alpha \in J$ ", η_α, and members a_α, b_n^α, $c_m^\alpha, d_m^\alpha, \tau_m^\alpha$ of B_0^c such that , letting $B_\alpha = \langle B_0, a_i : i < \alpha, i \in J \rangle_{B_0^c}$:

1) η_α is a branch of Rang (f^α), $\eta_\alpha \neq \eta_\beta$ for $\beta < \alpha$.

2) if $\alpha \in J$, then for some $\xi < \zeta(\alpha)$:
$a_\alpha = \bigcup_m (\tau_m^\alpha \cap d_m^\alpha)$ where $\langle d_m^\alpha : m < \omega \rangle$ is a maximal antichain of non zero elements (of B_0^c) $\bigcup_m \underline{d}(d_m^\alpha) \subseteq {}^{\omega>}\xi$, $\tau_m^\alpha \in \langle x_\rho : \eta_\alpha \restriction m \leq \rho, \rho \in T \rangle_{B_0^c}$, and $\tau_m^\alpha \cap d_m^\alpha > 0$.

3) if $\alpha \in J$, $b_n^\alpha, d_n^\alpha \in N_0^\alpha$, $c_n^\alpha, \tau_m^\alpha \in N^\alpha$ (hence each is based on $\{x_\nu : \nu \in {}^{\omega>}\lambda, \nu \in N^\alpha\}$), and $b_n^\alpha \cap b_m^\alpha = 0$ for $n \neq m$.

4) for $\beta < \alpha$, $\beta \in J$, B_α omit $p_\beta = \{x \cap b_n^\alpha = c_n^\alpha : n < \omega\}$.

Remark: Many times we shall write $\beta < \alpha < \alpha^*$ or $w \subseteq \alpha < \alpha^*$ intead $\beta \in \alpha \cap J$, $w \subseteq \alpha \cap J$.

Before we carry the construction note:

3.2 Crucial Fact: For any $x \in B_\alpha$ there are $k, \xi < \zeta$, and $\alpha_0 < \cdots < \alpha_k$ such that $\zeta(\alpha_0) = \zeta(\alpha_1) = \zeta(\alpha_2) = \cdots = \zeta(\alpha_k) = \zeta$, x is based on $\{x_\nu : \nu \in {}^{\omega>}\xi$ *or* $\nu \in \underline{d}(\tau_m^{\alpha_\ell})$, for some $\ell \le k$, $m < \omega\}$, but for some $\xi_1 < \zeta$, and for no $m < \omega$ and $\ell \in \{0, \ldots, k\}$ is x based on $\{x_\nu : \eta_{\alpha_\ell} \restriction m \not\leq \nu \in {}^{\omega>}\lambda\}$.

Stage B. Let us carry the construction. For $\xi < \lambda, w \subseteq \alpha^*$ let

$$I_{\xi,w} = \{\nu : \nu \in {}^{w>}\xi \quad or \quad \nu \in \bigcup_{\substack{m<\omega \\ \gamma\in w}} \underline{d}(\tau_m^\gamma)\}$$

We let $\alpha \in J$ *iff* $|N^\alpha| \subseteq B_\alpha, N^\alpha = (B_0^c \restriction |N^\alpha|, h_\alpha)$ where h_α is an endomorphism of $B_0^c \restriction |N^\alpha|$ (hence maps N_n^α into N_n^α for $n < \omega$) and there are $d_m^\alpha \in N_0^\alpha$ for $m < \omega$, $d_m^\alpha \neq 0$, $d_m^\alpha \cap d_\ell^\alpha = 0$ for $m \neq \ell$, such that for some $\xi < \zeta(\alpha)$ each d_m^α is based on ${}^{\omega>}\xi$, and there are a branch η_α of Rang (f^α) and $\tau_m^\alpha \in N^\alpha(m < \omega)$ as in 1), 2) above, such that if we add $\bigcup_{n<\omega} (\tau_n^\alpha \cap d_\ell^\alpha)$ to B_α, each $p_\beta(\beta < \alpha)$ is still omitted as well as $\{x \cap h_\alpha(d_m^\alpha) = h_\alpha(d_m^\alpha \cap \tau_m^\alpha) : m < \omega\}$ and $\langle d_m^\alpha : m < \omega \rangle$ is a maximal antichain.

If $\alpha \in J$ we choose $\eta^\alpha, d_n^\alpha, \tau_m^\alpha$, satisfying the above and let $b_m^\alpha = h_\alpha(d_m^\alpha)$, $c_m^\alpha = h_\alpha(d_m^\alpha \cap \tau_m^\alpha)$.

The Boolean algebra B is B_{α^*}. We shall investigate it and eventually prove it is endo-rigid (in 3.11) and indecomposable (in 3.12) (3.1(1), 3.1(2) are trivial).

Note also

3.3 Fact: 1) For $\nu \in {}^{\omega>}\lambda$, x_ν is free over $\{x_\eta : \eta \in {}^{\omega>}\lambda, \eta \neq \nu\}$ hence also over the subalgebra of B_0^c of those elements based on $\{x_\eta : \eta \in {}^{\omega>}\lambda, \eta \neq \nu\}$.

2) For every branch η of f^α such that $\eta \neq \eta_\beta$ for $\beta < \alpha, \xi < \zeta(\alpha)$; and finite $w \subseteq \alpha$ *there is* k such that $\{\rho : \eta \restriction k \trianglelefteq \rho \in \mathrm{T}\}$ is disjoint to ${}^{\omega>}\xi \cup \bigcup\{N^\beta \cap \mathrm{T} : \beta \in w, \beta + 2^{\aleph_0} \le \alpha\} \cup \bigcup\{\underline{d}(\tau_n^\beta) : n < \omega, \beta \in w\}$.

From 3.2 we can conclude:

3.4 Fact: If $\xi < \zeta(\beta), \beta < \alpha$, $I \subseteq \mathrm{T}$ finite then every element of B_α, based on $I \cup {}^{\omega>}\xi$ is in B_β.

3.5 Notation: 1) Let B^ξ be the set of $a \in B_0^c$ supported by ${}^{\omega>}\xi$

2) For $a \in B_0^c$, $\xi < \lambda$ let $pr_\xi(x) = \cap \{a \in B^\xi : x \le a\}$.

3) For $\xi < \lambda$ let $\varepsilon(\xi) = Min\{\gamma : \zeta(\gamma) > \xi\}$.

4) For $\gamma < \alpha^*$ let $B_{<\gamma>} = \langle \{x_\eta : \eta \in {}^{\omega>}\zeta(\gamma)\} \cup \{x_\beta : \beta < \gamma\} \rangle_{B_0^c}$.

5) For $I \subseteq {}^{\omega>}\lambda$, $w \subseteq \alpha^*$, let $B(I,w) = \langle \{x_\eta : \eta \in I\} \cup \{x_\beta : \beta \in w \cap J\} \rangle$.

6) For $\xi < \lambda$ let $B_{[\xi]} = \langle \{x_\eta : \eta \in {}^{\omega>}\xi\} \cup \{x_\beta : \zeta(\beta) \le \xi\} \rangle_{B_0^c}$.

3.6 Fact: 1) B^ξ is a complete Boolean subalgebra of B_0^c.

2) $pr_\xi(x)$ is well defined for $x \in B_0^c$.

3) if $\xi_0 < \xi_1 < \lambda$, $x \in B_0^c$ then $pr_{\xi_0}(pr_{\xi_1}(x)) = pr_{\xi_0}(x)$.

4) If $\xi < \lambda$, $w \subseteq T$ is finite then for the function $pr_{\xi,w}(x) = \cap \{y \in \langle B^\xi \cup \{x_\nu : \nu \in w\} \rangle : x \le y\}$ is well defined.

3.7 Fact: 1) For $x \in B_{\alpha^*}$, $\xi < \lambda$, the element $pr_\xi(x)$ belong to $\langle B^\xi \cup \{x_\nu : \nu \in w\} \rangle$.

2) For $x \in B_{\alpha^*}$, $\xi < \lambda$, $w \subseteq {}^{\omega>}(\xi+1)$, the element $pr_{\xi,w}(x)$ belongs to $B({}^{\omega>}\xi, w)$.

Proof: 1) We prove this for $x \in B_\alpha$, by induction on α (for all ξ).

Note that $pr_\xi(\bigcup_{\ell<n} x_\ell) = \bigcup_{\ell<n} pr_\xi(x_\ell)$.

Case i: $\alpha = 0$, or even $(\forall \beta < \alpha)\ [\zeta(\beta) \le \xi]$.

Easy; if $x = \tau(a_0, \ldots, a_{n-1}, x_{\nu_0}, \ldots, x_{\nu_{m-1}})$ where τ is a Boolean term, $a_\ell \in B_{[\xi]}$, $\nu_\ell \in {}^{\omega>}\lambda - {}^{\omega>}\xi$; by the remarks above w.l.o.g. $x = \bigcap_{\ell<n+m} \tau_\ell$, $\tau_\ell \in \{a_\ell, 1-a_\ell\}$ when $\ell < n$, $\tau_\ell \in \{x_{\nu_{\ell-n}}, 1-x_{\nu_{\ell-n}}\}$ when $n \le \ell < n+m$, and the sequence $\langle x_{\nu_0}, \ldots, x_{\nu_{n-1}} \rangle$ is with no repetition, then clearly $pr_\xi(x) = \bigcap_{\ell<n} \tau_\ell \in B_{[\xi]}$;

Case ii: α limit.

Trivial as $B_\alpha = \bigcup_{\beta<\alpha} B_\beta$.

Case iii: $\alpha = \beta + 1$.

By the induction hypothesis w.l.o.g. $x \notin B_\beta$. As $x \in B_\alpha$ there are disjoint $e_0, e_1, e_2 \in B_\beta$ such that $x = e_0 \cup (e_1 \cap a_\beta) \cup (e_2 - a_\beta)$. It suffices to prove that $pr_\xi(e_0)$, $pr_\xi(e_1 \cap a_\beta)$, $pr_\xi(e_2 - a_\beta) \in B_{[\xi]}$, the first is trivial and w.l.o.g. we concentrate on the second. There are $\xi_0 < \zeta(\beta)$ and $k < \omega$ such that e_1 is based on $J \overset{def}{=} {}^{\omega>}\lambda - \{\rho : \eta_\beta \restriction k \trianglelefteq \rho \in {}^{\omega>}\lambda\}$ and each $d_n^\beta (n < \omega)$ is based on ${}^{\omega>}\xi_0$. By Case i, we can assume $\xi < \zeta(\beta)$ hence w.l.o.g. $\xi < \xi_0$, and by the induction hypothesis and 3.6(3) it suffices to prove $pr_{\xi_0}(e_1 \cap a_\beta) \in B_{[\xi]}$. W.l.o.g. $e_1 \cap d_m^\alpha = 0$ for $m < k$ and now clearly $pr_{\xi_0}(e_1 \cap a_\beta) = e_1$ as $pr_{\xi_0}(e_1 \cap d_m^\alpha \cap \tau_m^\alpha) = e_1 \cap d_m^\alpha$ for $m \geq k$, (because d_m^α, e_1 are based on J, ${}^{\omega>}\xi_0 \subseteq J$ and τ_m^α is based on ${}^{\omega>}\lambda - J$ and is > 0).

2) Same proof.

3.8 Lemma : Suppose I, w satisfies:

$(*)_{I,w}$ $I \in {}^{\omega>}\lambda$, $w \subseteq \alpha^*$, I is closed under initial segments, and for every $\alpha < \alpha^*$ *if* $\bigwedge_{m<\omega} (\eta_\alpha \restriction m \in I)$ then $\tau_m^\alpha, d_m^\alpha$ are based on I and belong to $B(I,w)$.

Then for any countable $C \subseteq B_{\alpha^*}$ there is a projection from $\langle B(I,w), C \rangle_{B_0^*}$ onto $B(I,w)$.

Proof : We can easily find $I(*), w(*)$ such that $w \subseteq w(*) \subseteq \alpha^*$, $|w(*) - w| \leq \aleph_0$, $I \subseteq I(*) \subseteq {}^{\omega>}\lambda$, $|I(*) - I| \leq \aleph_0$ and if $\alpha \in w(*) - w$, then $\tau_m^\alpha, d_m^\alpha \in B(I(*), w(*))$. Let $w(*) - w = \{\alpha_\ell : \ell < \omega\}$, and we define by induction on ℓ a natural number $k_\ell < \omega$, such that the sets $\{\nu \in {}^{\omega>}\lambda : \nu$ appears is $\tau_m^{\alpha_\ell}$ for some $m > k_\ell\}$ are pairwise disjoint and disjoint to I. Now we can extend the identity on $B(I,w)$ to a projection h_0 from $B(I(*),w)$ onto $B(I,w)$ such that if $\ell < \omega, m > k_\ell$, then $h_0(\tau_m^{\alpha_\ell} \cap d_m^{\alpha_\ell}) = 0$. Now we can define by induction on $\alpha \in (w(*) - w) \cup \{0, \lambda\}$ a projection h_α from $B(I(*), w \cup (w(*) \cap \alpha))$ onto

$B(I,w)$ extending h_β for $\beta < \alpha$ (and $\beta \in (w(*)-w) \cup \{0\}$). For $\alpha = 0$ we have defined, for $\alpha = \lambda$ we get the conclusion, and in limit stages takes the union. In successive stages there is no problem by the choice of h_0, and the k_ℓ's.)

3.9 Claim: If B' is an uncountable subalgebra of B_{α^*} *then* there is an antichain $\{d_n : n < \omega\} \subset B'$ and for no $x \in B$, $x \cap d_{2n} = 0$, $x \cap d_{2n+1} = d_n$ for every n provided that

(*) no one countable $I \subseteq {}^{\omega>}\lambda$ is a support for every $a \in B'$.

Proof : We now define by induction on $\alpha < \omega_1, d_\alpha, I_\alpha$, such that:

(i) $I_\alpha \subseteq {}^{\omega>}\lambda$ is countable.

(ii) $\bigcup_{\beta<\alpha} I_\beta \subseteq I_\alpha$ and for α limit, equality holds.

(iii) $d_\alpha \in B'$ is supported by $I_{\alpha+1}$ but not by I_α.

There is no problem in this.

By (iii) for each α there are $\tau_\alpha^0 \in \langle a_\eta : \eta \in I_\alpha \rangle_{B_0^c}$, $\tau_\alpha^1, \tau_\alpha^2 \in \langle a_\eta : \eta \in I_{\alpha+1} - I_\alpha \rangle_{B_0^c}$ such that $\tau_\alpha^1 \cap \tau_\alpha^2 = 0$, $\tau_\alpha^0 \cap \tau_\alpha^1 \le d_\alpha$, $\tau_\alpha^0 \cap \tau_\alpha^2 \le 1 - d_\alpha$.

By Fodour's lemma w.l.o.g. $\tau_\alpha^0 = \tau^0$ (i.e. does not depend on α). For each α there is $n(\alpha) < \omega$ such that

$$\tau_\alpha^0 \in \langle a_\eta : \eta \in I_\alpha \cap {}^{n(\alpha)\ge}\lambda \rangle_{B_0^c}, \tau_\alpha^1, \tau_\alpha^2 \in \langle a_\eta : \eta \in (I_{\alpha+1} - I_\alpha) \cap {}^{n(\alpha)\ge}\lambda \rangle_{B_0^c}$$

Again by renaming w.l.o.g. $n(\alpha) = n(*)$ for every α. Let for $n < \omega$, $d^n = d_n - \bigcup_{\ell<n} d_\ell$, $\tau^n = \tau^0 \cap \bigcap_{\ell<n} \tau_\ell^2 \cap \tau_n^1$, so easily $d^n \in B'$, $\langle d^n : n < \omega \rangle$ is an antichain, $\tau^n \le d^n$ and $\tau^n \in \langle a_\eta : \eta \in {}^{n(*)\ge}\lambda \rangle_{B_0^c}$. Suppose $x \in B$ $x \in B, x \cap d^{2n} = 0$, $x \cap d^{2n+1} = d^{2n+1}$. Hence for $n < \omega$, $x \cap \tau^{2n} = 0$, $x \cap \tau^{2n+1} = \tau^{2n+1}$. But by 3.8 (for $I = {}^{n(*)\ge}\lambda$), there is such x in $\langle a_\eta : \eta \in {}^{n(*)\ge}\lambda \rangle_{B_0^c}$, an easy contradiction.

So we have proven that for every $\aleph_1$-compact $B' \subseteq B_{\alpha^*}$, some countable

$I \subseteq {}^{\omega>}\lambda$ support every $x \in B'$.

3.10 Claim: No infinite subalgebra B' of B_{α^*} is $\aleph_1$-compact.

Proof : Suppose there is such B' , and let ξ be minimal such that there is such $B' \subseteq B_{[\xi]}$.

Part I: if (*) a) $B' \subseteq B_{\alpha^*}$ is $\aleph_1$-compact and infinite and

b) $B' \subseteq B_{[\xi]}$,

then

c) for every $\zeta < \xi$ and $x \in B' - \{y : \{z \in B' : z \le y\}$ is finite$\}$, there is $x_1 \in B', x_1 \le x$ such that for no $y \in B_{[\zeta]}$, $y \cap x = x_1$.

So assume B', satisfies a) and b) but they fail c) for $\zeta < \xi$ and $x \in B'$, where $\{y : y \le x, y \in B'\}$ is infinite. So for every $z \in B'$, there is $g(z) \in B_{[\zeta]}$ such that $g(z) \cap x = z \cap x$ (use $x_1 = z \cap x$). Let B^a be the subalgebra of $B_{[\zeta]}$ generated by $\{g(z) : z \in B'\}$. Clearly $\{y \in B' : y \le x\} = \{t \cap x : t \in B^a\}$. Let $x^* = pr_\zeta(x)$, (it is in $B_{[\zeta]}$ by 3.7(1)) and let $B^b = \{t \cap x^* : t \in B^a\} \cup \{t \cup (1 - x^*) : t \in B^a\}$. Clearly B^b is a subalgebra of $B_{[\zeta]}$, and $1 - x^*$ is an atom of B^b;B^b is infinite as there are in B' distinct $x_n \le x$, so $g(x_n) \in B^a$ hence $g(x_n) \cap x^* \in B^b$, as $x \le x^*$ and $[n \ne m \Longrightarrow g(x_n) \cap x \ne g(x_m) \cap x]$ clearly $[n \ne m \Longrightarrow g(x_n) \cap x^* \ne g(x_n) \cap x^*]$. We shall prove that B^b is $\aleph_1$-compact, thus contradicting the choice of ξ. Let $d_n \in B^b$ be pairwise disjoint, and we want to find $t \in B^b$, $t \cap d_{2n} = 0$, $t \cap d_{2n+1} = d_{2n+1}$ (for $n < \omega$). Clearly w.l.o.g. $d_n \le x^*$ (as $1 - x^*$ is an atom of B^b). So $d_n = t_n \cap x^*$ for some $t_n \in B^a$, hence easily $t_n \cap x \in B'$ so for some $x_n \in B'$, $x_n \le x$ and $t_n \cap x = x_n \cap x = x_n$. So $x_n = g(x_n) \cap x$. For $n \ne m$,

$$x_n \cap x_m = (t_n \cap x) \cap (t_m \cap x) \le (t_n \cap x^*) \cap (t_m \cap x^*) = d_n \cap d_m = 0$$

As B' is $\aleph_1$-compact there is $y \in B'$, $y \cap x_{2n} = 0$, $y \cap x_{2n+1} = x_{2n+1}$. Now $g(y), d_n, t_n$ belongs to $B_{[\zeta]}$ and (as $x_n \le x \le x^*$):

(i) $g(y) \cap d_{2n} \cap x = g(y) \cap t_{2n} \cap x =$
$g(y) \cap x_{2n} \cap x = y \cap x_{2n} \cap x = 0.$

(ii) $g(y) \cap d_{2n+1} \cap x = g(y) \cap t_{2n+1} \cap x = g(y) \cap x_{2n+1} \cap x = y \cap x_{2n+1} \cap x = x_{2n+1} \cap x = t_{2n+1} \cap x = d_{2n+1} \cap x$.

Now by the definition of $x^* = pr_\xi(x)$, $[\tau \in B_{[\xi]} \wedge \tau \cap x = 0 \Longrightarrow \tau \cap x^* = 0]$ (as $1-\tau \in B_{[\xi]}$, $x \leq 1-\tau$) hence by (i) (for $\tau = g(y) \cap d_{2n}$):

(iii) $g(y) \cap d_{2n} \cap x^* = 0$.

Also by the definition of $x^* = pr_\xi(x)$:

$$\tau_1, \tau_2 \in B_{[\xi]} \wedge \tau_1 \cap x = \tau_2 \cap x \Longrightarrow \tau_1 \cap x^* = \tau_2 \cap x^*$$

(as $\tau_1 - \tau_2 \in B_{[\xi]}, x \leq 1-(\tau_1-\tau_2)$) hence by (ii)

(iv) $g(y) \cap d_{2n+1} \cap x^* = d_{2n+1} \cap x^*$.

But $d_n \leq x^*$, so from (iii) and (iv) $(g(y) \cap x^*) \cap d_{2n} = 0$, $(g(y) \cap x^*) \cap d_{2n+1} = d_{2n+1}$, and $g(y) \in B^a$ hence $g(y) \cap x^* \in B^b$. So B^b is $\aleph_1$-compact this contradicts the minimality of ξ, so we finish Part I.

Part II: if B^1 is $\aleph_1$-compact $B^1 \subseteq B^2$, $B^2 = \langle B^1 \cup \{z\} \rangle$ then B^2 is $\aleph_1$-compact.

The proof is straightforward. [If $d_n \in B^2$ are pairwise disjoint, let $d_n = (d_n^1 \cap z) \cup (d_n^2 - z)$ for some $d_n^1, d_n^2 \in B^1$. Now w.l.o.g $d_n^1 \cap d_m^1 = 0$ for $n \neq m$ - otherwise replace then by $d_n^1 - \bigcup_{\ell<n} d_\ell^1$; Similarly $d_n^2 \cap d_m^2 = 0$, for $n \neq m$ So there are $y^\ell \in B^1$, $y^\ell \cap d_{2n}^\ell = 0$, $y^\ell \cap d_{2n+1}^\ell = d_{2n+1}^\ell = d_{2n+1}^\ell$, and $(y^1 \cap z) \cup (y^2 - z)$ is the solution.]

Part III. ξ cannot be a successor ordinal.

Proof: Let B' satisfy (*).

Suppose $\xi = \zeta+1$, and by 3.9 there is a countable $I \subseteq {}^{\omega>}\xi$ which support every $a \in B'$. w.l.o.g. I is closed under initial segments and $k = |I - {}^{\omega>}\zeta|$ is minimal. Now Part I can be applied with $\langle B_{[\zeta]}, \{a_\eta : \eta \in w\} \rangle_{B_0^*}$, for any finite $w \subseteq I$ of power $< k$ instead $B_{[\xi]}$ (using 3.7(2) instead 3.7(1)). So by applying Part I (to $\langle B_{[\xi]}, \{a_\eta : \eta \in w\} \rangle_{B_0^*}$) we can add to its conclusion:

d) *for every* finite $w \subseteq I$, $|w| < |I-^{\omega>}\zeta|$ and $x \in B'$ for which $\{y \in B' : y \leq x\}$ is infinite, *there is* $x_1 \in B', x_1 \leq x$ *such that for* no $y \in \langle B_{[\zeta]} \cup \{a_\eta : \eta \in w\}\rangle_{B_0^*}, y \cap x = x_1$.

Now $I-^{\omega>}\zeta$ is infinite [otherwise let $B'' = \langle B' \cup \{a_\eta : \eta \in I-^{\omega>}\zeta\}\rangle_{B_0^*}$, easily it is infinite and $\aleph_1$-compact by Part II and then we apply Part I : for $I-^{\omega>}\zeta = \{\eta_0, \ldots, \eta_{k-1}\}$ and for $u \subseteq \{0, \ldots, k-1\}$, let $x_u \overset{def}{=} \cap \{x_{\eta_\ell} : \ell \in u\} \cap \{1 - x_{\eta_\ell} : \ell < k, \ell \notin u\}$ so $x_u \in B''$, $1 = \bigcup\{x_u : u \subseteq \{0, \ldots, k-1\}\}$, hence for some u, $\{y \in B'' : y \leq x_u\}$ is infinite; ζ, x_u contradict the conclusion of Part I.

As B' is $\aleph_1$-compact, for any $x \in B'$ such that $\{y \in B' : y \leq x\}$ is infinite, x can be splitted in B' to two elements satisfying the same i.e. $x = x^1 \cup x^2$; $x^1 \cap x^2 = 0$, $\{y \in B' : y \leq x^\ell\}$ is infinite for the $\ell = 1,2$. Let $I-^{\omega>}\zeta = \{\eta_\ell : \ell < \omega\}$, so we can find pairwise disjoint $e_n \in B'$, $\{y \in B' : y \leq e_n\}$ is infinite; now by d) above for each n we can find d_{2n}, d_{2n+1}, such that $e_n = d_{2n} \cup d_{2n+1}$, $d_{2n} \cap d_{2n+1} = 0$ and that for no $y \in \langle B_{[\zeta]} \cup \{a_{\eta_\ell} : \ell < n\}\rangle$, $y \cap (d_{2n} \cup d_{2n+1}) = d_{2n+1}$. As B' is $\aleph_1$-compact there is $y \in B'$ such that $y \cap (d_{2n} \cup d_{2n+1}) = d_{2n+1}$ for every n. So for no n $y \in \langle B_{[\zeta]} \cup \{a_{\eta_\ell} : \ell < n\}\rangle_{B_0^*}$.

As $y \in B'$ clearly $y \in B_{[\zeta+1]}$, but y is based on ${}^{\omega>}\zeta \cup \{a_{\eta_\ell} : \ell < \omega\}$ so by 3.7(2) $y \in \langle B_{[\zeta]} \cup \{a_{\eta_\ell} : \ell < \omega\}\rangle_{B_0^*}$, hence by Stage B for some n, $y \in \langle B_{[\zeta]} \cup \{a_{\eta_\ell} : \ell < n\}\rangle_{B_0^*}$, contradiction to $y \cap (d_{2n} \cup d_{2n+1}) = d_{2n+1}$.

Part IV: Let B', satisfy (*) of Part I. By 3.9 for some countable $I \subseteq {}^{\omega>}\xi$, every $b \in B'$ is based on I. By Part III ξ is not a successor ordinal, so necessarily $cf(\xi) = \aleph_0$, let $Fi(B') = \{x \in B' : \{y \in B' : y \leq x\}$ is finite$\}$. Next we shall show:

(**) for some finite $w \subseteq \{\gamma : \zeta(\gamma) = \xi\}$ and $x^* \in B' - Fi(B')$ for every $y < x^*$ from B', for some $z \in \langle \bigcup_{\zeta<\xi} B_{[\zeta]} \cup \{a_\alpha : \alpha \in w\}\rangle_{B_0^*}$, $z \cap x^* = y$.

Suppose (**) fail, and we define by induction $n < \omega$, x_n, y_n, w_n such that :

(i) $x_n \in B'$,

(ii) $1 - \bigcup_{i<n} x_i \notin Fi(B')$

(iii) $w_n \subseteq \{\gamma : \zeta(\gamma) = \xi\}$ is finite.

(iv) $w_n \subseteq w_{n+1}$

(v) $y_n \leq x_n, y_n \in B'$.

(vi) for no $z \in \langle \bigcup_{\zeta<\xi} B_{[\zeta]} \cup \{a_\alpha : \alpha \in w_n\} \rangle_{B_0^*}$ is $z \cap x_n = y_n$.

For $n = 0$ $1 \notin Fi(B')$.

For every n let w_n be a finite subset of $\{\gamma : \zeta(\gamma) = \xi\}$ extending $\bigcup_{\ell<n} w_\ell$, such that for every $\ell < n$, $x_\ell, y_\ell \in \langle \bigcup_{\zeta<\xi} B_{[\zeta]} \cup \{a_\alpha : \alpha \in w_n\} \rangle_{B_0^*}$. Then as $1 - \bigcup_{\ell<n} x_i \notin Fi(B')$, and as B' is $\aleph_1$-compact, there is $x_n \leq 1 - \bigcup_{i<n} x_i$, $x_n \in B'$, $1 - \bigcup_{\ell \leq n} x_i \notin Fi(B')$ and $x_n \notin Fi(B')$. Now as (**) fails, w_n, x_n does not satisfy the requirements on w, x^* in (**), so there is $y_n \in B'$, $y_n \leq x_n$ such that for no $z \in \langle \bigcup_{\zeta<\xi} B_{[\zeta]} \cup \{a_\alpha : \alpha \in w_n\} \rangle_{B_0^*}$ is $z \cap x_n = y_n$.

As B' is $\aleph_1$-compact, for some $z^* \in B'$, $z^* \cap x_n = y_n$ for every n. As $z^* \in B'$ for some finite $w^* \subseteq \varepsilon(\xi)$, $z^* \in \langle \bigcup_{\zeta<\xi} B_{[\zeta]} \cup \{a_\alpha : \alpha \in w^*\} \rangle_{B_0^*}$. As w^* is finite, for some $n(*) < \omega$, $w^* \cap (\bigcup_{n<\omega} w_n) \subseteq w_{n(*)}$. Let $\zeta < \xi$ be such that: $\underline{d}(d_n^\alpha) \subseteq {}^{\omega >}\zeta$ for $\alpha \in w_{n(*)+1} \cup w^*$, $n < \omega$ and $x_n, y_n \in \langle B_{[\zeta]} \cup \{a_\alpha : \alpha \in w_{n(*)+1}\} \rangle_{B_0^*}$, for $n \leq n(*) + 1$ and $z^* \in \langle B_{[\zeta]} \cup \{a_\alpha : \alpha \in w^*\} \rangle_{B_0^*}$. By 3.8 we can easily get a contradiction to (vi). So (**) holds.

Let $t_0, \ldots, t_m \in B_{[\xi]}$ be such that $\bigcup_{\ell=1}^{m} t_\ell = 1$ and $(\forall \ell \leq m)(\forall \alpha \in w)$ $[t_\ell \leq a_\alpha \vee t_\ell \cap a_\alpha = 0]$. There is an $\ell \leq m$ such that $\{y \cap t_\ell : y \leq x^*$ and $y \in B'\}$ is infinite. It is clear (by Part II) that $B'' = \langle B', t_\ell \rangle_{B_0^*}$ is $\aleph_1$-compact: also $x^* \cap t_\ell \in B'' - Fi(B'')$. Now if $y \in B'', y \leq x^* \cap t_\ell$ then for some $y' \in B', y = y' \cap t_\ell$ and w.l.o.g. $y' \leq x^*$, so for some $z \in \langle \bigcup_{\zeta<\xi} B_{[\zeta]} \cup \{a_\alpha : \alpha \in w\} \rangle_{B_0^*}$ $z \cap x^* = y'$ hence $z \cap (x^* \cap t_\ell) = y$, and by the choice of t_ϱ, for some $z' \in \bigcup_{\zeta<\xi} B_{[\zeta]}$, the equation $z' \cap (x^* \cap t_\varrho) = z \cap (x^* \cap t_\varrho) = y$ holds.

So B'', $x^{**} \overset{def}{=} x^* \cap t_\varrho$ satisfy the requirements in (*). Now we use (c) of Part I. As $cf(\xi) = \aleph_0$, let $\xi = \bigcup_{n<\omega} \zeta_n$, and we define by induction on $n < \omega, x_n, y_n$ such that :

(i) $x_n \in B''$, $x_n \le x^{**}$

(ii) $x^{**} - \bigcup_{\ell<n} x_\ell \notin Fi(B'')$

(iii) $y_n \in B'$, $y_n \le x_n$

(iv) for no $z \in B_{[\zeta_n]}$, $z \cap x_n = y_n$.

As B'' is $\aleph_1$-compact, for some $z^* \in B''$, $z^* \cap x_n = y_n$ for each n.

Now as B'', x^{**} satisfy (**), for some $z^{**} \in \bigcup_{\zeta<\xi} B_{[\zeta]}$ $z^* \cap x^{**} = z^{**} \cap x^{**}$. So for some n $z^{**} \in B_{[\zeta_n]}$, contradicting (iv) above. Thus we have finished the proof of 3.9.

3.11 Claim: B_{α^*} is endo-rigid.

Proof: Suppose h is as counterexample, i.e. h is an endomorphism of B_{α^*} but $B_{\alpha^*} / Ex\ Ker(h)$ is infinite, and we shall get a contradiction.

Clearly if for some α, $N^\alpha = (|N^*|, h \restriction N^\alpha)$, h maps $N^* \cap B_{\alpha^*}$ into itself and $\alpha \in J$ (see Stage B) then $h(a_\alpha)$ realizes the type p_α, contradiction (by stage A, B_{α^*} omits p_α.) So we shall try to find such α which satisfy the requirements in Stage B for belonging to J. We assume $N^\alpha = (|N^\alpha|, h_\alpha)$, $|N^\alpha| \subseteq B_\alpha$, $h_\alpha = h \restriction N^\alpha$, h_α maps $N^\alpha \cap B_\alpha$ onto itself, and N_0^α contains some elements we need and somewhat more (see latter). As W is a barrier this is possible. We then will choose η_α, an ω-branch of f^α, distinct from η_β for $\beta < \alpha$ [if $\beta + 2^{\aleph_0} \le \alpha$ this follows, the rest exclude $< 2^{\aleph_0}$ branches of f^α but there are $2^{\aleph_0}$ such branches], a maximal antichain $\langle d_n : n < \omega \rangle$ of B_α, $d_n \in N_0^\alpha$, and $\tau_n \in N^\alpha$ in $\langle x_\nu : \eta_\alpha \restriction n \underline{\lessdot} \nu \in \mathrm{T} \rangle_{B_0^\alpha}$, and let $b_n = h(d_n)$, $c_n = h(d_n \cap \tau_n)$,

$p_\alpha = \{x \cap b_n = c_n : n < \omega\}$, and $a_\alpha = \bigcup_{n<\omega} (d_n \cap \tau_n) \in B_0^c$. All should have superscript $\bar{d},\bar{\tau}$ (where $\bar{d} = \langle d_n : n < \omega \rangle$, $\bar{\tau} = \langle \tau_n : n < \omega \rangle$) but we usually omit them or write $a_\alpha[\bar{\tau},\bar{d}]$, $p_\alpha[\bar{\tau},\bar{d}]$ etc.

The choice of $\bar{d},\bar{\tau}$ (and η_α which is determined by $\bar{\tau}$) is done by listing the demands on them (see Stage B) and showing a solution exists. The only problematic one is (a) (omitting p_β for $\beta \le \alpha$) and we partition it to three cases :

(I) $\zeta(\beta) < \zeta(\alpha)$ or $\zeta(\beta) = \zeta(\alpha)$, $\beta + 2^{\aleph_0} \le \alpha$,

(II) $\zeta(\beta) = \zeta(\alpha)$, $\beta < \alpha < \beta + 2^{\aleph_0}$.

(III) $\beta = \alpha$.

We shall prove that every $\bar{\tau},\bar{d}$ are O.K. for (I), that for any family $\{(\bar{d}^i,\eta^i,\bar{\tau}^i): i < 2^{\aleph_0}\}$ (η a branch of f^α, etc.) with pairwise distinct η^i's, all except $< 2^{\aleph_0}$ many are O.K. for instance of (II), and that there is a family of $2^{\aleph_0}$ triples $(\bar{d},\eta,\bar{\tau})$ satisfying (III) with pairwise distinct η^i's. This clearly suffices.

Case I: $\zeta(\beta) < \zeta(\alpha)$ or $\zeta(\beta) = \zeta(\alpha), \beta + 2^{\aleph_0} \le \alpha$.

Suppose some $x \in \langle B_\alpha, a_\alpha[\bar{\tau},\bar{d}] \rangle_{B_0^c}$ realizes p_β. Clearly there is a partition $\langle e_\ell : \ell < 4 \rangle$ of 1 (in B_α) such that $x = e_0 \cup (e_1 \cap a_\alpha[\bar{\tau},\bar{d}]) \cup (e_2 - a_\alpha[\bar{\tau},\bar{d}])$. Choose $\xi < \zeta(\alpha)$ large enough and finite $w \subseteq \alpha$ so that $[\zeta(\beta) < \zeta(\alpha) \Longrightarrow \zeta(\beta) < \xi]$, $d_n, h_\alpha(d_n)$ b_n^β, are based on $\{x_\nu : \nu \in {}^{\omega>}\xi\}$ (for $n < \omega$) and c_ℓ^β ($\ell < \omega$), e_0, e_1, e_2, e_3 are based on $J = \{\nu \in T : \eta_\alpha \restriction k \not\le \nu\}$, where $k < \omega$ also satisfies such that $\eta_\alpha(k) > \xi$, $\eta_\alpha \restriction k \notin N_\beta$.

We claim:

(*) there is $m < \omega$ such that $b_m^\beta \cap (e_1 \cup e_2) - \bigcup_{n \le k} d_n \ne 0$.

For suppose (*) fail, then as $a_\alpha[\bar{\tau},\bar{d}] \cap (\bigcup_{n \le k} d_n) \in B_\alpha$, w.l.o.g. $(e_1 \cup e_2) \cap \bigcup_{n \le k} d_n = 0$ (otherwise let

$$e_0' = e_0 \cup (e_1 \cap a_\alpha[\bar{\tau},\bar{d}] \cap \bigcup_{n \le k} d_n) \cup (e_2 \cap \bigcup_{n\le k} d_n - a_\alpha[\bar{\tau},\bar{d}])$$

$$e_1' = e_1 - \bigcup_{n\le k} d_n,$$

$$e_1' = e_2 - \bigcup_{n \le k} d_n).$$

so if x realizes p_β then so does e_0, but $e_0 \in B_\alpha$ contradicting an induction hypothesis . So (*) holds.

Now as $\langle d_n : n < \omega \rangle$ is a maximal antichain in B_α, for some $\ell < \omega$, $d_\ell \cap (b_m^\beta \cap (e_1 \cup e_2 - \bigcup_{n \le k} d_n)) \ne 0$. Necessarily $\ell > k$. So for some $\varepsilon \in \{1,2\}$, $d_\ell \cap b_m^\beta \cap e_\varepsilon \ne 0$. As x realizes p_β, $x \cap (d_\ell \cap b_m^\beta \cap e_\varepsilon) = d_\ell \cap c_n^\beta \cap e_\varepsilon$ which is based on J. But we know that $x \cap (d_\ell \cap b_\varepsilon^\beta \cap e_\varepsilon)$ is $d_\ell \cap b_m^\beta \cap e_1 \cap a_\alpha[\bar{\tau},\bar{d}] = d_\ell \cap b_m^\beta \cap e_1 \cap \tau_\ell$ (if $\varepsilon=1$) or $d_\ell \cap b_m^\beta \cap e_2 \cap (1 - a_\alpha[\bar{\tau},\bar{d}]) = d_\ell \cap b_m^\beta \cap e_2 \cap 1 - \tau_\ell$ (if $\varepsilon=2$). As $d_\ell \cap b_m^\beta \cap e_\varepsilon \ne 0$ is based on J, $\ell > k, \eta_\alpha(k) > \xi$, τ_ℓ is free over J, (see Fact 3.3(2)) necessarily $x \cap (d_\ell \cap b_m^\beta \cap e_\varepsilon)$ is not based on J, contradiction.

Case II: $\beta < \alpha < \beta + 2^{\aleph_0}$.

We shall prove that if η^ℓ, τ^ℓ are appropriate (for $\ell = 1$, 2) and $\eta^1 \ne \eta^2$ then p_β cannot be realized in both $\langle B_\alpha, a[\bar{\tau}_\ell,\bar{d}] \rangle_{B_\delta^*}$.

As there is a perfect set of appropriate η's it will suffice to prove that for each ω-branch η of Rang (f^α) for some appropriate $\bar{\tau}$ $\langle B_\alpha, a^{\bar{\tau}} \rangle_{B_\delta^*}$ omit $p_\alpha = p_\alpha[\bar{\tau},\bar{d}]$ which will be in done in Case III.

Note that $I_\beta^\alpha = \{e \in B_\alpha :$ for some $x \le e$ for every n $x \cap b_\beta^n \cap e = c_\beta^n \cap e\}$ is an ideal.

The details are easy.

Case III: $\beta = \alpha$.

This case is splitted into several subcases. Let η_α be any ω-branch of f^α, $\eta_\alpha \ne \eta_\beta$ whenever $\beta < \alpha < \beta + 2^{\aleph_0}$. Let $I^* = \bigcup \{\underline{d}(h(x)) : x \in B_\alpha\}$. We shall assume $|I^*| \leq \aleph_0 \Longrightarrow I^* \subseteq N_0^\alpha$, so in this case p_α is omitted by $B_{\alpha+1}$ or B_{α^*} iff it is omitted by B_α (by 3.7(1)). As accomplishing our aim is easier we shall ignore this case (work as in III 4 and use quite arbitrary p_β.

Subcase III 1.: For some $\rho^* \in T$, and $a^* \in B_\alpha - Ex\ Ker^*(h)$ for every $\rho, \rho^* \leq \rho \in T$ for some $\tau \in \langle x_\eta : \rho \leq \eta \in T \rangle_{B_0^\alpha}$, $\tau \cap a^* \neq 0 = h(\tau \cap a^*)$.

As we are interested not in (f^α, N^α) itself, but in h, by using $Gm'(W)$, w.l.o.g. $\rho^* \in$ Range (f^α). By 3.9 (for Rang (h), which by assumption, is infinite) and easy manipulations (see 2.4 and [Sh 2]) there is a maximal antichain $\langle d_n : n < \omega \rangle$ of B_{α^*} such that for no $x \in B_\alpha$, $x \cap h(d_{2n}) = h(d_{2n})$ and $x \cap h(d_{2n+1}) = 0$. W.l.o.g. $\{d_n : n < \omega\} \subseteq N_0^\alpha$.

It suffices to prove the conclusion for any ω-branch η_α of Range $(f^\alpha), \rho^* < \eta_\alpha \not\in \{\eta_\beta : \beta < \alpha\}$. We define by induction on n, $\tau_n \in N_n^\alpha$, $\tau_n \in \langle x_\eta : \eta_\alpha \restriction n \leq \eta \rangle_{B_0^\alpha}$, $\tau_n \neq 0, 1$ and $h(\tau_{2n}) = 1, h(\tau_{2n+1}) = 0$. (possible by the assumption of subcase III 1), so we finish this subcase.

Subcase III 2. For some $a^* \in B_\alpha$, $\{h(x) - a^* : x \in B_*, x \leq a^*\}$ is infinite.

Clearly $B^a = \{h(x) - a^* : x \in B_{\alpha^*}, x \leq a^*\} \cup \{1 - (h(x) - a^*) : x \in B_{\alpha^*}, x \leq a_\alpha\}$ is a subalgebra of B_{α^*} (with a^* an atom). By assumption (of this subcase) B^a is infinite. So by 3.9 there are $e_n \in B^a$, pairwise disjoint, and $\neg(\exists x \in B_\alpha) \bigwedge_n (x \geq e_{2n} \wedge x \cap e_{2n+1} = 0)$. As a^* is an atom of B^a w.l.o.g. $e_n \leq 1 - a^*$, hence there is $d_n \leq a^*$ (in B_{α^*}), such that $h(d_n) = e_n$. Clearly $h(d_n - \bigcup_{\ell < n} d_n) = e_n - \bigcup_{\ell < n} e_\ell = e_n$, so w.l.o.g. the d_n are pairwise disjoint. So by easy manipulation for some $\langle d_n : n < \omega \rangle$ the following holds:

(i) $d_0 = 1 - a^*$

(ii) $\langle d_n : n < \omega \rangle$ is a maximal antichain of B_{α^*}.

(iii) for no $x \leq 1 - a^*$, $x \cap h(d_{2n+2}) - a^* = h(d_{2n+2}) - a^*$,

$$x \cap h(d_{2n+1}) - a^* = 0$$

We can assume that $d_n, h(d_n) \in N_0^a$.

Let $\bar{\tau}^0 = \langle \tau_n^0 : n < \omega \rangle$ be a suitable sequence, (for our η_a) then so are $\bar{\tau}^\ell = \langle \bar{\tau}_n^\ell : n < \omega \rangle$, for $\ell < 4$ where

$$\tau_{2n}^1 = 1-\tau_{2n}^0,\ \tau_{2n+1}^1 = \tau_{2n+1}^0;$$

$$\tau_{2n}^2 = \tau_{2n}^0, \qquad \tau_{2n+1}^2 = 1-\tau_{2n+1}^0;$$

$$\tau_{2n}^3 = 1-\tau_{2n}^0, \quad \tau_{2n+1}^3 = 1-\tau_{2n+1}^0.$$

Suppose for each $\ell < 4$, in $\langle B_a, a_a[\bar{\tau}^\ell, \bar{d}] \rangle_{B_0^a}$ there is an element y^ℓ which satisfies $y^\ell \cap h(d^\ell) - a^* = h(\tau_n^\ell \cap d_n) - a^*$ for $1 \leq n < \omega$. W.l.o.g. $y^\ell \leq 1-a^* = d_0$ hence $y^\ell \in B_a$. Now $(y^0 \cup y^1) \cap (y^2 \cup y^3) \in B_a$ contradict (iii) above.

Subcase III 3. For some $a^* \in B_{a^*} - Ex\ Ker^*(h)$, and $\rho^* \in \mathrm{T}$, for every ρ, $\rho^* \underline{\lessdot} \rho \in \mathrm{T}$ there is $\tau \in \langle x_\nu : \rho \underline{\lessdot} \nu \in \mathrm{T} \rangle_{B_0^a}$ such that $h(\tau \cap a^*) \cap a^* = \tau \cap a^*$.

Clearly the function $h' : B_{a^*} \restriction a^* \to B_{a^*} \restriction a^*$ defined by $h'(x) = h(x) \cap a^*$ is an endomorphism; W.l.o.g. the assumption of subcase III 2 fail hence $\{h(x) - a^* : x \leq a^*\}$ is finite, hence the range of h' is infinite (as $a^* \notin Ex\ Ker^*(h)$, so by 2.4 there is $x \leq a^*$ such that $h(x) \cap a^* - x \neq 0$; we know that $\underline{d}(x)$ is countable, hence for some $\rho^{**}, \rho^* \underline{\lessdot} \rho^{**} \in T$ and $\{\nu : \rho^{**} \underline{\lessdot} \nu \in \mathrm{T}\}$ is disjoint to $\underline{d}(a^*) \cup \underline{d}(x) \cap \underline{d}(h(x))$. Now by the hypothesis of subcase III 3 we can easily find $\tau_n \in \langle x_\nu : \rho^{**} \underline{\lessdot} \nu \in \mathrm{T} \rangle_{B_0^a}$, with pairwise disjoint $\underline{d}(\tau_n)$ and $h(\tau_n \cap a^*) \cap a^* = \tau_n \cap a^*$. So

$$h(\tau_n \cap x) \cap (a^* - x) =$$

$$h((\tau_n \cap a^*) \cap x) \cap (a^* - x) = h(\tau_n \cap a^*) \cap h(x) \cap (a^* - x) =$$

$$(h(\tau_n \cap a^*) \cap a^*) \cap h(x) \cap (a^* - x) = (\tau_n \cap a^*) \cap h(x) \cap (a^* - x) =$$

$$= \tau_n \cap h(x) \cap (a^* - x) = \tau_n \cap (h(x) \cap a^* - x)$$

It is $\neq 0$ [as $\underline{d}(\tau_n) \cap (\underline{d}(x) \cup \underline{d}(h(x)) \cup \underline{d}(a^*)) = \phi$) and $h(x) \cap a^* - x \neq 0$, $\tau_n \neq 0$], and for different n we get different values. So

$\{h(y \cap x) \cap (a^*-x) : x \in B_{\alpha^*}\}$, is infinite. Hence $\{h(y \cap x)-x; y \in B_{\alpha^*}\}$ is infinite, leading to the assumption of Subcase III 2 (with x here for a^* there).

Subcase III. 4. For some $\rho^* \in T$, and $a^* \in B_{\alpha^*} - Ex\ Ker^*(h)$ for every $\tau \in \langle x_\nu : \rho^* \underline{<} \nu \in T\rangle_{B_0^\alpha}$ $h(\tau \cap a^*) \cap a^*$ is based on $\{\nu : \rho^* \underline{\not<} \nu \in T\}$.

W.l.o.g. the hypothesis of subcase III 1 fail, hence $\{h(\tau \cap a^*) : \tau \in \langle x_\nu : \rho^* \underline{<} \nu \in T\rangle_{B_0^\alpha}\}$ is infinite. As also w.l.o.g. the hypothesis of subcase III 2 fail we get $\{h(\tau \cap a^*) \cap a^* : \tau \in \langle x_\nu : \rho^* \underline{<} \nu \in T\rangle_{B_0^\alpha}\}$ is infinite. So by 3.9 we we can find $d_n \in \langle x_\nu : \rho^* \underline{<} \nu \in T\rangle_{B_0^\alpha}$ such that $\langle d_n : n < \omega\rangle$ is a maximal antichains in B_0^α, and there is no $x \in B_{\alpha^*}$, $x \cap h(d_{2n}) = h(d_{2n})$, $x \cap h(d_{2n+1}) = 0$, and $d_0 = 1-a^*$.

As before we can assume $\rho^* \in$ Rang (f^*) and $d_n \in N_0^\alpha$ for $n < \omega$. We suppose $\eta_\alpha \notin \{\eta_\beta : \beta < \alpha\}$ is an ω-branch of $f^\alpha, \rho^* < \eta_\alpha$.

For any suitable $\bar{\tau}$, if $y[\bar{\tau},\bar{d}] \in \langle B_\alpha, a_\alpha[\bar{\tau},\bar{d}]\rangle_{B_0^\alpha}$ satisfies $\tau_n \in \langle x_\nu : \rho^* \le \nu \in T\rangle_{B_0^\alpha}$ and $y[\bar{\tau},\bar{d}] \cap h(d_n) = h(\tau_n \cap d_n)$, (for every n) then by 3.3 we easily get $y[\bar{\tau},\bar{d}] \in B_\alpha$, and then get contradiction by trying four $\bar{\tau}$'s, as in subcase III2.

Subcase III. 5. There are $\rho^* \in T$ and atomless countable subalgebra $Y \subseteq B_{\alpha^*}$ and pairwise disjoint $c_\ell \in Y (\ell < \omega)$ such that for every ℓ and $\rho_\ell \in \{\rho : \rho^* \underline{<} \rho \in T\}$ for some $\tau_\ell \in \langle x_\nu : \rho_\ell \underline{<} \nu \in T\rangle_{B_0^\alpha}$, the following holds: for no $x \in B_0^\alpha$ is $\underline{d}(x) \subseteq \{\nu : \rho_\ell \underline{\not<} \nu \in T\}$ and $x \cap h(c_\ell) \cap c_\ell - \tau_\ell = h(c_\ell \cap \tau_\ell) \cap c_\ell - \tau_\ell$.

Let $\langle d_n : n < \omega\rangle$ be a maximal antichain of B_{α^*} such that $d_{2n} = c_{2n}$.

So w.l.o.g. $Y \cup \{d_n : n < \omega\} \subseteq N_0^\alpha, \rho^* \in$ Rang (f^α) (using $Gm'(W)$), and even $\rho^* < \eta_\alpha$, and each N_m^α is closed under the functions h and $\rho_\ell \to \tau_\ell$ (implicit in the assumption of the subcase).

We can now choose by induction on n, $\tau_n \in N_n^\alpha$,

$$\tau_n = \left\langle x_\nu : \eta_\alpha \restriction n \trianglelefteq \nu \in T \right\rangle_{B_\delta^c}$$

such that

(*) (a) for even n, for no $x \in B_0^c$ based on $\{\nu : \eta_\alpha \restriction n \not\trianglelefteq \nu \in T\}$ is $x \cap h(d_n) \cap d_n - \tau_n = h(d_n \cap \tau_n) \cap d_n - \tau_n$.

Why is this sufficient? We let $\bar{d} = \langle d_n : n < \omega \rangle$, and $\bar{\tau} = \langle \tau_n : n < \omega \rangle$. So assume some $y[\bar{\tau},\bar{d}] \in \langle B_\alpha, a_\alpha[\bar{\tau},\bar{d}] \rangle_{B_\delta^c}$ realizes $p_\alpha[\bar{\tau},\bar{d}]$, i.e. satisfies $y[\bar{\tau},\bar{d}] \cap h(d_n) = h(d_n \cap \tau_n)$ for every n. As $y[\bar{\tau},\bar{d}] \in \langle B_\alpha, a_\alpha[\bar{\tau},\bar{d}] \rangle_{B_\delta^c}$ for some pairwise disjoint $e_0[\bar{\tau},\bar{d}], e_1[\bar{\tau},\bar{d}], e_2[\bar{\tau},\bar{d}] \in B_\alpha$, $y[\bar{\tau},\bar{d}] = e_0[\bar{\tau},\bar{d}] \cup (e_1[\bar{\tau},\bar{d}] \cap a_\alpha[\bar{\tau},\bar{d}]) \cup (e_2[\bar{\tau},\bar{d}] - a_\alpha[\bar{\tau},\bar{d}])$.
For some $m(*) < \omega$, $\underline{d}(e_0[\bar{\tau},\bar{d}]) \cup \underline{d}(e_1[\bar{\tau},\bar{d}]) \cup \underline{d}(e_2[\bar{\tau},\bar{d}])$ is disjoint to $\{\nu : \eta_\alpha \restriction m(*) \trianglelefteq \nu \in T\}$ (see 3.3(2)).

Now we compute for n even $> m(*)$:

$$z \overset{\text{def}}{=} h(d_n \cap \tau_n) \cap d_n - \tau_n =$$

$$= y[\bar{\tau},\bar{d}] \cap h(d_n) \cap d_n - \tau_n \text{ (by the choice of } y[\bar{\tau},\bar{d}])$$

$$= (e_0[\bar{\tau},\bar{d}] \cup (e_1[\bar{\tau},\bar{d}] \cap a_\alpha[\bar{\tau},\bar{d}]) \cup (e_2[\bar{\tau},\bar{d}] - a[\bar{\tau},\bar{d}])) \cap h(d_n) \cap d_n - \tau_n =$$
$$= (e_0[\bar{\tau},\bar{d}] \cap h(d_n) \cap d_n - \tau_n) \cup ((e_1[\bar{\tau},\bar{d}] \cap a_\alpha[\bar{\tau},\bar{d}]) \cap h(d_n) \cap d_n - \tau_n) \cup$$
$$\cup ((e_2[\bar{\tau},\bar{d}] - a_\alpha[\bar{\tau},\bar{d}]) \cap h(d_n) \cap d_n - \tau_n)$$

But $a_\alpha[\bar{\tau},\bar{d}] \cap d_n = \tau_n \cap d_n$ hence

$$(e_1[\bar{\tau},\bar{d}] \cap a_\alpha[\bar{\tau},\bar{d}]) \cap d_n = (e_1[\bar{\tau},\bar{d}] \cap \tau_n) \cap d_n$$

$$(e_2[\bar{\tau},\bar{d}] - a_\alpha[\bar{\tau},\bar{d}]) \cap d_n = (e_2[\bar{\tau},\bar{d}] - \tau_n) \cap d_n$$

Hence

$$z = (e_0[\bar{\tau},\bar{d}] \cap h(d_n) \cap d_n - \tau_n) \cup ((e_1[\bar{\tau},\bar{d}] \cap \tau_n) \cap h(d_n) \cap d_n - \tau_n) \cup$$
$$((e_2[\bar{\tau},\bar{d}] - \tau_n) \cap h(d_n) \cap d_n - \tau_n)$$

But the second term is zero and in the third the first $-\tau_n$ is redundant, so

$$z = (e_0[\bar{\tau},\bar{d}] \cap h(d_n) \cap d_n - \tau_n) \cup (e_2 \cap h(d_n) \cap d_n - \tau_n) =$$

$$= (e_0[\bar\tau,\bar d] \cup e_2[\bar\tau,\bar d]) \cap h(d_n) \cap d_n - \tau_n$$

We can conclude

$$(e_0[\bar\tau,\bar d] \cup e_2[\bar\tau,\bar d]) \cap h(d_n) \cap d_n - \tau_n = h(d_n \cap \tau_n) \cap d_n - \tau_n$$

contradicting the choice of τ_n.

To finish Case III (hence the proof of 3(10) we need only

Why the five subcases exhaust all possibilities?

Suppose none of III 1-5 occurs. By not subcase III 1 for some $\rho^0 \in T$,

a) $h(\tau) \neq 0$ for every $\tau \in \langle x_\eta : \rho^0 \leq \eta \in T \rangle_{B_{\S}}$.

Let Y be the $\langle x_{\rho^0 \frown \langle i \rangle} : i < \omega \rangle_{B_{\S}}$. As Y is countable, for some $i(*) < \lambda$, $\{\nu : \rho^0 \frown \langle i(*) \rangle \leq \nu \in T\}$ is disjoint to $\bigcup\{\underline{d}(y) \cup \underline{d}(h(y)) : y \in Y\}$. As "not subcase III 5" for some ρ^1, $\rho^0 \frown \langle i(*) \rangle \leq \rho^1 \in T$, and

(b) there are no pairwise disjoint non zero $c_\ell \in Y (\ell < \omega)$, such that for every $\rho^1_\ell, \rho^1 \leq \rho^1_\ell \in T$ for some $\tau_\ell \in \langle x_\nu : \rho^1_\ell \leq \nu \in T \rangle_{B_{\S}}$, the following holds:

(*) for no $x \in B_{\S}$, $\underline{d}(x) \subseteq \{\nu : \rho^1_\ell \not\leq \nu \in T\}$ and

$$x \cap h(c_\ell) \cap c_\ell - \tau_\ell = h(c_\ell \cap \tau_\ell) \cap c_\ell - \tau_\ell$$

Clearly

c) $\bigcup\{\underline{d}(y) \cup \underline{d}(h(y)) : y \in Y\}$ is disjoint to $\{\nu : \rho^1 \leq \nu \in T\}$.

Let $Z = \{c \in Y$: for some $\rho^1_c, \rho^1 \leq \rho^1_c \in T$ for no $\tau \in \langle x_\nu : \rho^1 \leq \nu \in T \rangle_{B_{\S}}$ does (*) of (b) hold (with c, τ instead c_ℓ, τ_ℓ)$\}$.

By (b) among any $\aleph_0$ pairwise disjoint members of Y, at least one belong to Z.

It is quite easy to define $y_n \in Z((n < \omega)$ such that $[y_n \in Ex\ Ker^*(h) \Rightarrow y_n \in Ex\ Ker(h)]$, $[m < n \Rightarrow y_n \cap y_m = 0]$, and for every $y \in Y - \{0\}$ for some n, $y \cap (\bigcup_{\ell < n} y_\ell) \neq 0$ or $y_n \leq y$. So (by the choice of

Y) $\langle y_n : n < \omega \rangle$ is a maximal antichain of B_0^c. We shall show $y_n \in Ex\ Ker(h)$; fix n for a while, and suppose $y_n \notin Ex\ Ker(h)$, and let $\rho_n^1, \rho^1 \leq \rho_n^1 \in T$ be such that for no $\tau \in \langle x_\nu : \rho_n^1 \leq \nu \in T \rangle_{B_0^c}$ does (*) of (b) hold.

Now for each $\tau \in \langle x_\nu : \rho_n^1 \leq \nu \in T \rangle_{B_0^c}$ as $y_n \in Z$, clearly [as (*) of (b) fail for y_n, τ (and ρ_n^1)] for some $x_1 \in B_0^c$, $\underline{d}(x_1) \subset \{\nu : \rho_n^1 \not\leq \nu \in T\}$ and $x_1 \cap h(y_n) \cap y_n - \tau = h(y_n \cap \tau) \cap y_n - \tau$. Applying the failure of (*) of (b) for $y_n, 1-\tau, \rho_n^1$ we get $x_2 \in B_c^0$, $\underline{d}(x_2) \subset \{\nu : \rho_n^1 \not\leq \nu \in t\}$ and $x_2 \cap h(y_n) \cap y_n - (1-\tau) = h(y_n \cap (1-\tau)) \cap y_n - (1-\tau)$; note that $h(y_n \cap \tau) \leq h(y_n)$, and $h(y_n \cap (1-\tau)) = h(y_n) - h(y_n \cap \tau)$. By these equations and as $y_n\ h(y_n), x_1, x_2$ are based on $\{\nu : \rho_n^1 \not\leq \nu \in T\}$ (by (c) and their choice resp.) clearly for some partition of $1, e_0^\tau, e_1^\tau, e_2^\tau, e_3^\tau \in B_0^c$, based on $\{\nu : \rho^1 \not\leq \nu \in T\}$:

(i) $h(\tau \cap y_n) \cap y_n = e_0^\tau \cup (e_1^\tau \cap \tau) \cup (e_2^\tau - \tau)$.

Now for any $\tau, \sigma \in \langle x_\nu : \rho_n^1 \leq \nu \in T \rangle$, easily (as h is an endomorphism):

(ii) $h((\tau \cup \sigma) \cap y_n) \cap y_n = (h(\tau \cap y_n) \cap y_n) \cap (h(\sigma \cap y_n) \cap y_n)$.

(iii) $h((\tau \cup \sigma) \cap y_n) \cap y_n = (h(\tau \cap y_n) \cap y_n) \cup (h(\sigma \cap y_n) \cap y_n)$.

We can apply (i) to τ, σ and also to $\tau \cap \sigma, \tau \cup \sigma$, and substitute in (ii) (iii).

We get that

(α) $e_2^\tau \cap e_2^\sigma = 0$ if $\underline{d}(\tau) \cap \underline{d}(\sigma) = 0$, $\tau, \sigma \in \langle x_\nu : \rho^1 \leq \nu \in T \rangle_{B_0^c}$ (otherwise substitute (i) in (ii) and intersect with $e_2^\tau \cap e_2^\sigma$) and get $(h((\tau \cap \sigma) \cap y_m) \cap (e_2^\tau \cap e_2^\sigma) = (e_2^\tau - \tau) \cap (e_2^\sigma - \sigma) = e_2^\tau \cap \tau_2^\sigma \cap (\tau \cup \sigma)$, and by the assumptions on the $\underline{d}(e_2^\tau)$, $\underline{d}(e_2^\sigma)$, $\underline{d}(\tau)$, $\underline{d}(\sigma)$ we get $(h((\tau \cap \sigma) \cap y_n) \cap y_n) \cap (e_2^\tau \cap e_2^\sigma) \notin \langle \{x : \underline{d}(x) \subset \{\nu : \rho_m^1 \leq \nu \in T\} \cup (\tau \cap \sigma) \rangle_{B_0^c}$ contradiction to (i) for $\sigma \cap \tau$).

So let $\{\tau^i : i < \alpha\}$ be maximal such that $\underline{d}(\tau_i)$ are pairwise disjoint $e_2^{\tau^i} \neq 0$, and $\tau^i \in \langle x_\nu : \rho_n^1 \underline{<} \nu \in T \rangle_{B_0^c}$, then $\alpha < \omega_1$, and we can choose ρ_n^2 such that:
$\rho_n^1 \underline{<} \rho_n^2 \in T$, and $[\tau \in \langle x_\nu : \rho_n^2 \underline{<} \nu \in \tau \rangle_{B_0^c} \Longrightarrow e_2^\tau = 0]$.

Next we can get

(β) $e_1^\tau \cap e_0^\sigma = 0$ (if $\underline{d}(\tau) \cap \underline{d}(\sigma) = 0$, and $\tau, \sigma \in \langle x_\nu : \rho_n^2 \underline{<} \nu \in T \rangle_{B_0^c}$).

The proof is similar to that of (α), using $\tau \cap \sigma$.

As B_0^c satisfies the $\aleph_1$-c.c. we can find $\{\tau^i : i < \omega\} \subseteq \langle x_\nu : \rho_n^2 \underline{<} \nu \in T \rangle_{B_0^c}$, such that (in B_0^c) $e_\ell^* \overset{def}{=} \bigcup_{i<\omega} e_\ell^{\tau^i} = \bigcup \{e_\ell^\tau : \tau \in \langle x_\nu : \rho_n^2 \underline{<} \nu \in T \rangle_{B_0^c}\}$ for $\ell = 0, 1$. We can find $\rho_n^3, \rho_n^2 \underline{<} \rho_n^3 \in T$, such that $\bigcup_{\ell<\omega} \underline{d}(\tau^i)$ is disjoint to $\{\nu : \rho_n^3 \underline{<} \nu \in T\}$. So for every $\tau \in \langle x_\nu : \rho_n^3 \underline{<} \nu \in T \rangle_{B_0^c}$, $e_0^\tau \leq e_0^*$ (by the choice of e_0^*), and $e_0^\tau \cap e_1^{\tau^i} = 0$ for $i < \omega$ (by (β)) hence $e_0^\tau \cap e_1^* = 0$, hence

(γ) $e_0^\tau \leq e_0^* - e_1^*$.

Similarly

(δ) $e_1^\tau \leq e_1^* - e_0^*$.

Now we can prove that $e_1^\tau = e_1^\sigma$ when $\underline{d}(\tau) \cap \underline{d}(\sigma) = 0$, $\tau, \sigma \in \langle x_\nu : \rho_n^3 \underline{<} \nu \in T \rangle_{B_0^c}$, (repeat the proof of (α) intersecting with $e_1^\tau - e_1^\sigma$ or with $e_1^\sigma - e_1^\tau$). By the transitivity of equality $e_1^\tau = e_1^\sigma$ when $\tau, \sigma \in \langle x_\nu : \rho_n^3 \underline{<} \nu \in T \rangle_{B_0^c}$. So let $e_1 \in B_{\alpha^*}$ be the common value, so

$(*)$ $h(\tau \cap y_n) \cap y_n = e_0^\tau \cup (e_1 \cap \tau)$ for $\tau \in \langle x_\nu : \rho_n^3 \underline{<} \nu \in T \rangle_{B_0^c}$; and $e_0^\tau \leq y_n - e_1$.

Let $e_0 = y_n - e_1$, so $y_n = e_0 \cup e_1$, $e_0 \cap e_1 = 0$.

so $e_0^\tau \leq e_0$ for every $\tau \in \langle x_\nu : \rho_n^3 \underline{<} \nu \in T \rangle_{B_0^c}$.

As $y_n \notin Ex\ Ker^*(h)$, at least one of the elements, e_0, e_1 is not in $Ex\ Ker^*(h)$. As not subcase III 2, for $\ell = 1,2$ the homomorphism g_ℓ from $B_{\alpha^*} \restriction e_\ell$ to $B_{\alpha^*} \restriction (1-e_\ell)$, $g_\ell(x) = h(x) - e_\ell$ (for $x = e_1$) has a finite range. Hence for some ideal $\mathfrak{J}$ of B_0^c $y_n/\mathfrak{J}$ is a finite union of atoms and

for every $\tau \in \langle x_\nu : \rho_n^3 \trianglelefteq \nu \in T \rangle \cap \mathfrak{J}$

for $\ell = 0,1$ $h(\tau \cap y_n) \cap e_\ell = h(\tau \cap e_\ell) \cap e_\ell$

hence $h(\tau \cap e_\ell) \cap e_\ell = (e_0^\tau \cup (e_1^\tau \cap \tau)) \cap e_\ell$.

So (for $\tau \in \langle x_\nu : \rho_n^3 \trianglelefteq \nu \in T \rangle_{B_0^c} \cap \mathfrak{J}$):

$$h(\tau \cap e_0) \cap e_0 = e_0^\tau$$
$$h(\tau \cap e_1) \cap e_1 = \tau \cap e_1$$

If $e_1 \notin Ex\ Ker^*(h)$, we get contradiction to "not subcase III 3" [use ρ_n^3 for ρ^* there, now for any ρ, $\rho_n^3 \trianglelefteq \rho \in T$ choose pairwise disjoint $\tau_\ell \in \langle x_\nu : \rho \trianglelefteq \nu \in T \rangle_{B_0^c}$ for $\ell < \omega$ now by the choice of $\mathfrak{J}$ for at least one $\ell, \tau_\ell \in \mathfrak{J}$, so τ_ℓ is as required there]. So assume $e_0 \notin Ex\ Ker^*(h)$ and get contradiction to "not subcase III 4" [for some $\ell < m < \omega$ $x_{\rho_n^3 \frown \langle \ell \rangle} - x_{\rho_n^3 \frown \langle n \rangle}$ is in $\mathfrak{J}$, use $\rho_n^3 \frown \langle \alpha \rangle$, $e_0 \cap (x_{\rho_n^3 \frown \langle \ell \rangle} - x_{\rho_n^3 \frown \langle n \rangle})$ for ρ^*, a^* with α large enough].

So for each $n, y_n \in Ex\ Ker^*(h)$ (the y_n were chosen after (b)) hence $y_n \in Ex\ Ker(h)$, (by their choice) so let $y_n = y_n^0 \cup y_n^1$ (both in B_{α^*}), $h(y_n^0) = 0$, $h(x) = x$ for $x \le y_n^1$, $x \in B_{\alpha^*}$. Let $I \subseteq T$ be a countable set such that $\underline{d}(y_n^0), \underline{d}(y_n^1) \subseteq I$, and for $x \in B_{\alpha^*}$. $\underline{d}(h(x - y_n) \cap y_n) \subseteq I$ (by "not subcase III 2", for each n we have only finitely many elements of this form).

We can easily show that for every $x \in B_{\alpha^*}$ for some $a \in B_0^c$ based on I, $h(x) - x = a - x$, [as $\langle y_n : n < \omega \rangle$ is a maximal antichain in B_{α^*}, for this it suffices to show that for every $n < \omega$ there is $a_n \in B_{\alpha^*}^c$, $a_n \le y_n$ such that $(h(x) - x) \cap y_n = a_n - x$; But $(h(x) - x) \cap y_n$ is the union of $(h(x \cap y_n) - x) \cap y_n$ which is zero as $(\forall z \le y_n) h(z) \le z$ and of $(h(x - y_n) - x) \cap y_n$ which we know is based as wanted] So

$h(x) = e_0^x \cup (e_1^x \cap x) \cup (e_2^x - x)$

where each e_ℓ^* is based on I, $\langle e_\ell^* : \ell < 3 \rangle$ pairwise disjoint $e_\ell^* \in B_0^c$. As in the analysis above of $h(x \cap y_n) \cap y_n$, possibly increasing I, applied to $x \in B_{\underset{\alpha}{\bullet}}$ with $\underline{d}(x) \cap I = 0$, we get $e_2^x = 0, e_1^x = e_1$. If $e_1 \notin Ex\ Ker^*(h)$ we get contradiction to "not subcase III 3.". So $1 - e_1 \notin Ex\ Ker^*(h)$ and apply "not subcase III 4."

So we finish the proof of 3.11; so B_{α^*} is endo-rigid.

3.12 Lemma : B_{α^*} is indecomposable.

Proof : Suppose K_0, K_1 are disjoint ideals of B_{α^*}, each with no maximal members, which generate a maximal ideal of B_{α^*}. For $\ell = 1,2$ let $\{d_n^\ell : \ell < \omega\}$ be a maximal antichain $\subseteq K_\ell$ (they are countable as B_{α^*} satisfies the c.c.c., and may be chosen infinite as $K_\ell \neq \{0\}$, B_{α^*} is atomless). Let K be the ideal $K_0 \cup K_1$ generates.

Now, e.g. for some $\xi < \lambda$, $\{d_n^\ell : \ell < 2, n < \omega\} \subseteq B_{[\xi]}$. Clearly $a_{<\xi>} \in K$ or $1 - a_{<\xi>} \in K$. For notational simplicity assume $a_{<\xi>} \in K$. So $a_{<\xi>} = b^0 \cup b^1$, $b^\ell \in K_\ell$. Now $pr_\xi(b^\ell) \in B_{[\xi]}$ and is disjoint to each $d_n^{1-\ell}$, (as b^ℓ and is, $d_n^\ell \in B_{[\xi]}$,), so by the maximality of $\{d_n^{1-\ell} : n < \omega\}$, $pr_\xi(b^\ell)$ is disjoint to every member of $K_{1-\ell}$. As $K_0 \cup K_1$ generate a maximal ideal, clearly $pr_\xi(b^\ell) \in K_\ell$ [otherwise $pr_\xi(b^\ell) = 1 - c^1 \cup c^2$, for some $c^1 \in K_1$, $c^2 \in K_2$, and then $c^{1-\ell}$ is necessarily a maximal member of $K_{1-\ell}$, so $K_{1-\ell}$ is principal contradiction]. So $pr_\xi(b^0) \cup pr_\xi(b^2) < 1$ but $1 = pr_\xi(a_{<\xi>}) = \bigcup_{\ell=0}^{2} pr_\xi(b^\ell)$ contradiction.

3.13 Theorem : In 3.1 we can get $2^{\lambda^{\aleph_0}}$ such B. A. such that any homomorphism from one to the other has finite range.

Proof : Left to the reader (see [Sh 4, 3]).

References.

[CG] A. L. S. Corner and R. Gobel, Presenting endomorphism algbras - a uniform treatment, Proc. London Math. Soc. (to appear).

[DMR]

e. C. K. van Douwen, J. D. Monk and M. Rabin, Some questions about Boolean Algebras, *Algebra Universalis*, 11 (1980), 220-243.

[GS1]

R. Gobel and S. Shelah, Modules over arbitrary domains, I, *Math. Zeitschift*, accepted.

[GS2]

———, Modules over aribrary domain, II, *Fund. Math.*

[M] J. D. Monk, A very rigid Boolean algebra, *Israel J. Math.* 35 (1980), 135-150.

[Sh1]

S. Shelah, Classification Theory and the number of non isomorphic models, North Holland Publ. Co. 1978.

[Sh2]

———, Boolean algebras with few endomorphisms, *Proc. of A.M.S.* 14 (1979), 135-142.

[Sh3]

———, Existence of rigid like families of abelian p-groups, *Model theory and Algebra*. A memorial tribute to A. Robinson, ed. Arsano and Weisfenning. *Lecture Notes in Math.* 498 Springer Verlag, 1975, 385-402.

[Sh4]

———, A combinatorial principle and endomorphism rings of p-groups I. Proc. of the 1980/1 Jerusalem model theory year, *Israel J. Math. accepted*.

[Sh5]

———, A combinatorial principle and endomorphism rings of abelian groups II, Proc. of the Conference on Abelian groups, Iudine, Italy. April 1984.

[Sh6]

———, Why there are many non-isomorphic models for unsuperstable theoreies, *Proc. of the International Congress of Math.* Vancoucer (1974), 553-557.

[Sh7]

———, Abstract of A.M.S. (1984).

On the no(M) for M of singular power

Abstract: We prove that for λ singular of cofinality $\kappa > \aleph_0$, if $(\forall \mu < \lambda)\mu^\kappa < \lambda$ *then* for some model M, $M = (M, R^M)$, R a two place predicate, $\|M\| = \lambda$ and $no(M) = \{N/\approx : N \equiv_{\infty,\lambda} M, \|N\| = \lambda\}$ is quite arbitrary e.g. any $\mu < \lambda$ and λ^κ (hence 2^λ).

See [Sh 5] for the back ground: where the result were proved for M with relations with infinitely many places. By the present paper the only problem left, if we assume $V = L$, is whether $no(M) = \lambda$, may happen for M of cardinality λ for λ singular.

§1 On γ- systems of groups.

1.1 Definition : A γ-system will mean here a model of the form $\mathcal{A} = \langle G_\alpha, h_{i,j} \rangle_{\substack{i \le j < \gamma \\ \alpha < \gamma}}$ where

(i) G_i is a group with the unit $e_i = e^{G_i} = e_i^{\mathcal{A}}$, the G_i's are pairwise disjoint.

(ii) $h_{i,j}$ is a homomorphism from G_j into G_i when $i \le j$.

(iii) $h_{i_1,i_2} \circ h_{i_2,i_3} = h_{i_1,i_3}$ when $i_1 \le i_2 \le i_3 < \gamma$.

(iv) $h_{i,i}$ is the identity. (so we sometimes ignore them).

We denote γ-systems by $\mathcal{A}, \mathcal{B}$ and for a system $\mathcal{A}$, we write $G_i = G_i^{\mathcal{A}}, \gamma = \gamma^{\mathcal{A}}$ $h_{i,j} = h_{i,j}^{\mathcal{A}}$. Let $\|\mathcal{A}\| = \sum_{i<\kappa} \|G_i\|$. We omit the $\mathcal{A}$ when there is no danger of confusion.

Let $\gamma = \gamma^{\mathcal{A}}$, for $\beta \le \gamma$ let $\mathcal{A} \restriction \beta = \langle G_\alpha^{\mathcal{A}}, h_{i,j}^{\mathcal{A}} \rangle_{i \le j < \beta, \alpha < \beta}$. The really interesting case is $\gamma =$ limit.

1.2 Definition : For a γ-system $\mathcal{A}$ let $Gr(\mathcal{A}) = \{\mathbf{a} = \langle a_{i,j} : i \le j < \gamma \rangle : a_{i,j} \in G_i,\ a_{i,i} = e^{G_i}$ and if $\alpha \le \beta \le \varepsilon < \gamma$ then

$$a_{\alpha,\varepsilon} = h_{\alpha,\beta}(a_{\beta,\varepsilon})\ a_{\alpha,\beta}$$

Let $\mathbf{a} \restriction \beta = \langle a_{i,j} : i \le j < \beta \rangle$.

1.3 Definition : For $\mathbf{a} = \langle a_i : i < \gamma \rangle \in \prod_{i<\varepsilon} G_i$, let fact $(\mathbf{a}) = \langle a_{i,j} : i < j < \gamma \rangle$ where $a_{i,j} = h_{i,j}(a_j)^{-1}\ a_i$. Let Fact $(\mathcal{A}) = \}\{fact(\mathbf{a}) : \mathbf{a} \in \prod G_i^{\mathcal{A}}\}$.

1.4 Claim: The mapping $\mathbf{a} \to fact(\mathbf{a})$ is from $\prod_{i<\varepsilon} G_i$ into $Gr(\mathcal{A})$. So fact $(\mathcal{A})$ is a subset of $Gr(\mathcal{A})$.

Proof : Trivially $a_{i,j} \in G_i, a_{i,i} = e_i$, and if $\alpha \le \beta \le \varepsilon$;

$h_{\alpha,\beta}(a_{\beta,\varepsilon}) \circ a_{\alpha,\beta} = (h_{\alpha,\beta}(h_{\beta,\varepsilon}(a_\varepsilon)^{-1}) h_{\alpha,\beta}(a_\beta))(h_{\alpha,\beta}(a_\beta)^{-1} \circ a_\alpha) =$
$(h_{\alpha,\beta} h_{\beta,\varepsilon})(a_\varepsilon)^{-1} a_\alpha = h_{\alpha,\varepsilon}(a_\varepsilon)^{-1} a_\alpha = a_{\alpha,\varepsilon}$.

1.5 Definition : 1) $Gs(\mathcal{A}) = \{\bar{\mathbf{a}} \in Gr(\mathcal{A})$: for every $\beta < \gamma^{\mathcal{A}}$ $\langle a_{i,j} : i < j < \gamma \rangle \in Fact(\mathcal{A} \restriction \gamma)\}$. [2]

2) We define a relation $\approx_{\mathcal{A}}$ on $Gr(\mathcal{A})$ (let $\gamma = \gamma^{\mathcal{A}}$): $\mathbf{a} \approx_{\mathcal{A}} \mathbf{b}$ if for some $\langle g_i : i < \gamma \rangle \in \prod_{i<\gamma} G_i^{\mathcal{A}}$, for every $i < j < \gamma$ $b_{i,j} = h_{i,j}(g_j)^{-1} a_{i,j}\ g_i$.

We shall say that $\langle g_i : i < \gamma \rangle$ exemplify $\mathbf{a} \approx_{\mathcal{A}} \mathrm{B}$.

3) $\mathcal{A}$ is called *smooth* if for every limit $\beta < \gamma$, $Gr(\mathcal{A} \restriction \beta) = Fact(\mathcal{A} \restriction \beta)$.

1.6. Claim: For a γ-system $\mathcal{A}$:

1) $\approx_{\mathcal{A}}$ is an equivalence relation on $Gr(\mathcal{A})$ (hence also on $Gs(\mathcal{A})$).

2) If $\mathbf{a},\mathbf{b} \in Gr(\mathcal{A})$, $\beta < \gamma^{\mathcal{A}}$ and $\mathbf{a} \approx_{\mathcal{A}} \mathbf{b}$ then $\mathbf{b} \restriction \beta \approx_{\mathcal{A} \restriction \beta} \mathbf{b} \restriction \beta$.

3) For $\mathbf{a} \in Gr(\mathcal{A})$: $\mathbf{a} \in Fact(\mathcal{A})$ iff $\mathbf{a} \approx_{\mathcal{A}} \langle e_i^{\mathcal{A}} : i < j < \gamma^{\mathcal{A}} \rangle$ (where $e_i^{\mathcal{A}}$ is the unit of $G_i^{\mathcal{A}}$).

---

[2] Really $Gs(\mathcal{A}) = Gr(\mathcal{A})$, as if $\mathbf{a} = \langle a_{i,j} : i < j < \gamma \rangle \in Gr(\mathcal{A})$ then $\langle a_{i,\beta} : i < \beta \rangle$ witness $\mathbf{a} \restriction \beta \in Gs(\mathcal{A})$; but we shall not use this.

4) For $\mathbf{a},\mathbf{b} \in Gr(\mathcal{A})$, if $\mathbf{a} \approx_{\mathcal{A}} \mathbf{b}$ *then* $\mathbf{a} \in Gs(\mathcal{A}) \Longleftrightarrow \mathbf{b} \in Gs(\mathcal{A})$.

Proof: 1) Let us check the properties.

reflexivity for $\mathbf{a} \in Gr(\mathcal{A})$, $\mathbf{a} \approx_{\mathcal{A}} \mathbf{a}$: $\langle e_i^{\mathcal{A}}{:}i \quad j < \gamma \rangle$ exemplify this

symmetry: suppose $\bar{\mathbf{a}} \approx_{\mathcal{A}} \mathbf{b}$ and $\langle g_i{:}i < \gamma \rangle$ exemplify this, so for every $i \le j < \gamma$, $b_{i,j} = h_{i,j}(g_j)^{-1} a_{i,j} g_i$, hence $h_{i,j}(g_j) h_{i,j} g_i^{-1} = a_{i,j}$ but $h_{i,j}(g_j^{-1}) = (h_{i,j}(g_j))^{-1}$ (as $h_{i,j}$ is a homomorphism from G_j into G_i). So (for every $i \le j \le \gamma$)
$a_{i,j} = (h_{i,j}(g_j^{-1}))^{-1} b_{i,j}(g_i^{-1})$ so $\langle g_i^{-1}{:}i < \gamma \rangle$ exemplify $\mathbf{b} \approx_{\mathcal{A}} \mathbf{a}$.

transitivity: suppose $\mathbf{a} \approx_{\mathcal{A}} \mathbf{b}$, $\mathbf{b} \approx_{\mathcal{A}} \mathbf{c}$ and $\langle g_i^0{:}i < \gamma \rangle, \langle g_i^1{:}i < \gamma \rangle$ exemplify them (resp.) So for $i \le j < \gamma$, $b_{i,j} = h_{i,j}(g_j^0)^{-1} a_{i,j}\, g_i^0$ and $c_{i,j} = h_{i,j}(g_j^1)^{-1}$ $b_{i,j}\, g_i^1$, substituting we get

$$\begin{gathered} c_{i,j} = h_{i,j}(g_j^1)^{-1}(h_{i,j}(g_j^0)^{-1} a_{i,j} g_i^0) g_i^1 = \\ (h_{i,j}(g_j^0)\, h_{i,j}(g_j^1))^{-1} a_{i,j}(g_i^0 g_i^1) = \\ h_{i,j}(g_j^0 g_j^1)^{-1}\, a_{i,j}(g_i^0 g_i^1) \end{gathered}$$

So $\langle g_i^0 g_i^1{:}i < \gamma \rangle$ exemplify $\mathbf{a} \approx_{\mathcal{A}} \mathbf{c}$.

2) If $\langle g_i{:}i < \gamma \rangle$ exemplify $\mathbf{a} \approx_{\mathcal{A}} \mathbf{b}$ then $\langle g_i{:}i < \beta \rangle$ exemplify $\mathbf{a} \restriction \beta \approx_{\mathcal{A} \restriction \beta} \mathbf{b} \restriction \beta$.

3) Because $\langle g_i{:}i < \gamma \rangle$ exemplify $\langle e_i^{\mathcal{A}}{:}i < j < \gamma \rangle \approx_{\mathcal{A}} \mathbf{a}$ *iff* $a_{i,j} = h_{i,j}(g_j)^{-1} g_i$ (for every $i < j < \gamma$.) i.e. iff $\langle g_i : i < \gamma \rangle$ exemplify $\mathbf{a} \in Fact(\mathcal{A})$.

4) By 3) $\mathbf{c} \in Gs(\mathcal{A})$ iff for every $\beta < \gamma^{\mathcal{A}}$, $\mathbf{c} \restriction \beta \approx_{\mathcal{A}} \langle e_i^{\mathcal{A}}{:}i \le j < \beta \rangle$, and by 2) for $\beta < \gamma^{\mathcal{A}}$, $\mathbf{a} \restriction \beta \approx_{\mathcal{A}} \mathbf{b} \restriction \beta$, hence (as $\approx_{\mathcal{A} \restriction \beta}$ is as an equivalence relation)

$\mathbf{a} \restriction \beta \approx_{\mathcal{A}} \langle e_i^{\mathcal{A}}{:}i \le j < \beta \rangle$ iff $\mathbf{b} \restriction \beta \approx_{\mathcal{A}} \langle e_i^{\mathcal{A}}{:}i \le j < \beta \rangle$ and the result follows.

1.7 Definition : For a γ-system $\mathcal{A}$, let $no^*(\mathcal{A})$ be the cardinality of $Gs(\mathcal{A})/\approx_{\mathcal{A}}$ (i.e. the number of non $\approx_{\mathcal{A}}$-equivalent $\mathbf{a} \in Gs(\mathcal{A})$).

1.8 Lemma : Suppose $\mathcal{A}, \mathcal{B}$ are γ-systems,.

(i) H_i is a homomorphism from $G_i^{\mathcal{A}}$ onto $G_i^{\mathcal{B}}$.

(ii) for $i < j < \gamma$, $H_i \circ h^{\mathcal{A}}_{i,j} = h^{\mathcal{B}}_{i,j} \circ H_j$

(iii) for every $\beta < \gamma$, $\mathbf{a},\mathbf{b} \in Fact(\mathcal{A} \restriction \beta)$, satisfying $H_i(a_{i,j}) = H_i(b_{i,j})$ for $i < j < \beta$, a member $g^i_{\mathbf{a},\mathbf{b}} \in G^{\mathcal{A}}_i$ are defined for $i < \beta$ such that :

a) if $i < \alpha < \beta$ then $g^i_{\mathbf{a}\restriction\alpha,\mathbf{b}\restriction\alpha} = g^i_{\mathbf{a},\mathbf{b}}$.

b) $b_{i,j} = h^{\mathcal{A}}_{i,j}(g^j_{\mathbf{a},\mathbf{b}})^{-1} a_{i,j}\, g^i_{\mathbf{a},\mathbf{b}}$ for $i < j < \beta$.

Then $no^*(\mathcal{A}) \leq no^*(\mathcal{B})$.

Proof: We define a function H with domain $Gr(\mathcal{A}) : H(\mathbf{a}) = H(\langle a_{i,j}: i < j < \gamma \rangle) = \langle H_i(a_{i,j}): i < j < \gamma \rangle$. By (ii) we can check that H is into $Gr(\mathcal{B})$. We shall show later

(*) for $\mathbf{a},\mathbf{b} \in Gr(\mathcal{A})$, $\mathbf{a} \approx_{\mathcal{A}} \mathbf{b}$ iff $H(\mathbf{a}) \approx_{\mathcal{B}} H(\mathbf{b})$.

Applying this to $\mathcal{A} \restriction \beta$ (for $\beta < \gamma$) and noting that $H_i(e^{\mathcal{A}}_i) = e^{\mathcal{B}}_i$. $H(\langle e^{\mathcal{A}}_i : i < j < \beta \rangle) = \langle e^{\mathcal{B}}_i : i < j < \beta \rangle$ we see that for $\mathbf{a} \in Gr(\mathcal{A})$, $\beta < \gamma$.

[$\mathbf{a} \restriction \beta \in Fact(\mathcal{A} \restriction \beta)$ iff $H(\mathbf{a}) \restriction\ \in Gs(\mathcal{B})$]. So by (*) H induces a one to one map from $Gs(\mathcal{A})/\approx_{\mathcal{A}}$ into $Gs(\mathcal{B})/\approx_{\mathcal{B}}$, so $no^*(\mathcal{A}) \leq n^*(\mathcal{B})$.

Proof of (*): First suppose $\mathbf{a} \approx_{\mathcal{A}} \mathbf{b}$ and let $\langle g_i : i < \gamma \rangle$ exemplify this. So for every $i < j < \gamma$

$$b_{i,j} = h^{\mathcal{A}}_{i,j}(g_j)^{-1}\, a_{i,j} g_i$$

applying H_i we get $H_i(b_{i,j}) = H_i(h^{\mathcal{A}}_{i,j}(g_j)^{-1}) H_i(a_{i,j}) H_i(g_i)$

Now by (ii) $H_i(h^{\mathcal{A}}_{i,j}(g_j)^{-1}) = (H^{\mathcal{A}}_i(h_{i,j}(g_j))^{-1} = (h^{\mathcal{B}}_{i,j}(H_j(g_j)))^{-1}$, so

$$H_i(b_{i,j}) = h^{\mathcal{B}}_{i,j}(H_j(g_j))^{-1} H_i(a_{i,j}) H_i(g_i)$$

So $\langle H_i(g_i): i < \gamma \rangle$ exemplify that $H(\mathbf{a}) \approx_{\mathcal{B}} H(\mathbf{b})$.

Next suppose $H(\mathbf{a}) \approx_{\mathcal{B}} H(\mathbf{b})$ and let $\langle g^*_i : i < \gamma \rangle$ exemplify it. As H_i is a homomorphism from $G^{\mathcal{A}}_i$ onto $G^{\mathcal{B}}_i$, there are $g_i \in G^{\mathcal{A}}_i$, such that $H_i(g_i) = g^*_i$ (for $i < \gamma$). Now $H_i(b_{i,j}) = h^{\mathcal{B}}_{i,j}(g^*_j)^{-1} H_i(a_{i,j})\, g^*_i = h^{\mathcal{B}}_{i,j}(H_j(g_j))^{-1} H_i(a_{i,j}) H_i(g_i)$

$$= H_i(h^{\mathcal{A}}_{i,j}(g_j)^{-1}) H_i(a_{i,j}) H_i(g_i) = H_i(h^{\mathcal{A}}_{i,j}(g_j)^{-1} a_{i,j} g_i)$$

Let us define $\mathbf{c} \in Gr(\mathcal{A})$ by $c_{i,j} = b_{i,j}^{\mathcal{A}}(g_i)^{-1} a_{i,j} g_i$. It is easy to check that $\mathbf{c}$ really belongs to $Gr(\mathcal{A})$ and $\langle g_i : i < \gamma \rangle$ exemplify $\mathbf{a} \approx_{\mathcal{A}} \mathbf{c}$, and the above equation shows that $H(\mathbf{b}) = H(\mathbf{c})$, and by (iii) this implies $\mathbf{b} \approx_{\mathcal{A}} \mathbf{c}$ ($\langle g^i_{\mathbf{b},\mathbf{c}} : i < \gamma \rangle$ exemplify that). Together $\mathbf{a} \approx_{\mathcal{A}} \mathbf{b}$.

So we have proved (*) hence 1.8.

1.9 Claim: If in 1.8 in addition:

(iv) H_i^+ is a homomorphism from $G_i^{\mathcal{B}}$ into $G_i^{\mathcal{A}}$.

(v) $H_i \circ H_i^+$ is the identity (on $G_i^{\mathcal{B}}$)

(vi) $h_{i,j}^{\mathcal{B}} \circ H_j^+ = H_i^+ \circ h_{i,j}^{\mathcal{A}}$ for $i < j < \gamma$.

Then $no^*(\mathcal{A}) = n^*(\mathcal{B})$.

Proof : We define a function H^+ with domain $Gr(\mathcal{B})$: $H^+(\mathbf{a}) = \langle H_i^+(a_{i,j}) : i \le j < \gamma \rangle$. By (vi) $H^+(\mathbf{a})$ is always in $Gr(\mathcal{A})$. Clearly $H \circ H^+$ is the identity on $Gr(\mathcal{B})$, so let $\{\mathbf{c}^\zeta : \zeta < no^*(\mathcal{B})\}$ be pairwise non $\approx_{\mathcal{B}}$ equivalent members of $Gs(\mathcal{B})$, and let $\mathbf{a}^\zeta = H^+(\mathbf{c}^\zeta) \in Gr(\mathcal{A})$. So $H(\mathbf{a}^\zeta) = \mathbf{c}^\zeta$. From the proof of 1.8 we know that: $\mathbf{a}^\zeta \in Gs(\mathcal{A})$ because $\mathbf{c}^\zeta \in Gs(\mathcal{B})$, and for $\xi < \zeta < no^*(\mathcal{B})$—$\mathbf{a}^\xi, \mathbf{a}^\zeta$ are non $\approx_{\mathcal{A}}$ equivalent (because $\mathbf{c}^\zeta, \mathbf{c}^\xi$ are non $\approx_{\mathcal{B}}$ equivalent). So $no^*(\mathcal{A}) \ge no^*(\mathcal{B})$ hence we finish (by 1.8).

1.10 Claim: For a γ-system of abelian groups.

1) $Gr(\mathcal{A})$ here is the same as $Gr(\mathcal{A})$ from [Sh 5], Definition 3.4 (except that here we do not put the group structure.

2) $Fact(\mathcal{A})$ here is the same (set) as $Fact(\mathcal{A})$ from [Sh 5] Definition 3.5 .

3) For $\mathbf{a},\mathbf{b} \in Gr(\mathcal{A})$, $\mathbf{a} \approx_{\mathcal{A}} \mathbf{b}$, *iff* (in [Sh 5] notation), $\mathbf{a}-\mathbf{b} \in Fact(\mathcal{A})$.

4) $Gs(\mathcal{A})$ here is the same as $Gs(\mathcal{A})$ from [Sh 5] Definition 3.7(1).

5) $no^*(\mathcal{A})$ here is the same as the cardinality of $E'(\mathcal{A})$ (from [Sh 5] Definition 3.7(2)).

Proof : Straightforward.

1.11 Conclusion: For every regular $\kappa > \aleph_0$ and μ, for some κ-system, $\mathcal{A}$, $\|\mathcal{A}\| \leq \mu^\kappa$, and $no^*(\mathcal{A}) = \mu$.

1.12 Claim : Suppose $\mathcal{A}$ is a γ-system, γ limit and for $\ell = 1,2$ $\mathbf{a}^\ell = \langle a^\ell_{i,j} : i \leq j < \gamma \rangle$ belongs to $Gs(\mathcal{A})$.

Suppose further $S \subseteq \gamma$ is unbounded in γ and $a^1_{i,j} = a^2_{i,j}$ when $i,j \in S$. Then $\mathbf{a}^1 \approx_{\mathcal{A}} \mathbf{a}^2$.

Proof: For every $\beta < \gamma, \ell = 1,2$, $\mathbf{a}^\ell \restriction (\beta+1) \in Fact(\mathcal{A} \restriction (\beta+1))$ hence there is $g^\ell_\beta = \langle g^{\ell,\beta}_i : i \leq \beta \rangle \in \prod_{i \leq \beta} G^{\mathcal{A}}_i$ such that $a^\ell_{i,j} = (h^{\mathcal{A}}_{i,j}(g^{\ell,\beta}_j)^{-1})\ g^{\ell,\beta}_i$ when $i \leq j \leq \beta$. For $\alpha < \gamma$ let $\varepsilon(\alpha) = Min\{\beta : \alpha \leq \beta \in S\}$.

We want to find $g_i \in G^{\mathcal{A}}_i$ $(i < \gamma)$ such that $a^2_{i,j} = h_{i,j}(g_j)^{-1} a^1_{i,j}\ g_i$.

Now for $\ell = 1,2$, if $i \leq \varepsilon(i) \leq j$

$$
\begin{gathered}
a^\ell_{i,j} = h^{\mathcal{A}}_{i,\varepsilon(i)}(a^\ell_{\varepsilon(i),j})\ a^\ell_{i,\varepsilon(i)} = \\
h^{\mathcal{A}}_{i,\varepsilon(i)}(h^{\mathcal{A}}_{\varepsilon(i),j}(a^\ell_{j,\varepsilon(j)})^{-1}\ a^\ell_{\varepsilon(i),\varepsilon(j)})a^\ell_{i,\varepsilon(i)} = \\
h^{\mathcal{A}}_{i,j}(a^\ell_{j,\varepsilon(j)})^{-1}\ h^{\mathcal{A}}_{i,\varepsilon(i)}(a^\ell_{\varepsilon(i),\varepsilon(j)})\ a^\ell_{i,\varepsilon(i)}
\end{gathered}
$$

[apply twice Definition 1.2 first for $i, \varepsilon(i)$, j standing for α,β,ε, and second for $\varepsilon(i),\varepsilon(j)$ standing for α,β,ε].

Now if $i \leq j \leq \varepsilon(i)$, applying twice this equation (remembering $a^2_{\xi(i),\xi(j)} = a^2_{\xi(i),\xi(j)}$):

$$
\begin{gathered}
a^2_{i,j} = h_{i,j}(a^2_{j,\varepsilon(j)})^{-1} h^{\mathcal{A}}_{i,\varepsilon(i)}(a^2_{\varepsilon(i),\varepsilon(j)})\ a^2_{i,\varepsilon(i)} = \\
h_{i,j}(a^2_{j,\varepsilon(j)})^{-1} h^{\mathcal{A}}_{i,\varepsilon(i)}(a^1_{\varepsilon(i),\varepsilon(j)})\ a^2_{i,\varepsilon(i)} \\
= h_{i,j}(a^2_{j,\varepsilon(j)})^{-1}\ (h_{i,j}(a^1_{j,\varepsilon(j)})\ a^1_{i,j}\ (a^1_{i,\varepsilon(i)})^{-1})\ a^2_{i,\varepsilon(i)} = \\
h_{i,j}((a^2_{j,\varepsilon(j)})^{-1}\ a^1_{j,\varepsilon(j)})a^1_{i,j}((a^1_{i,\varepsilon(i)})^{-1}\ a^2_{i,\varepsilon(i)}) = \\
= h_{i,j}((a^1_{j,\varepsilon(j)})^{-1}\ a^2_{j,\varepsilon(j)})^{-1}\ a^1_{i,j}\ ((a^1_{i,\varepsilon(i)})^{-1}\ a^2_{i,\varepsilon(i)})
\end{gathered}
$$

This suggests to show that $\langle (a^1_{i,\varepsilon(i)})^{-1}\ a^2_{i,\varepsilon(i)} : i < \kappa \rangle$ exemplify $\mathbf{a}^1 \approx_{\mathcal{A}} \mathbf{a}^2$ as required. The missing case is $i < j < \gamma$ $j < \varepsilon(i)$; so $\varepsilon(i) = \varepsilon(j)$ and so we should prove $a^2_{i,j} = h_{i,j}((a^1_{j,\varepsilon(j)})^{-1} a^2_{j,\varepsilon(j)})^{-1} a^1_{i,j}((a^1_{i,\varepsilon(i)})^{-1}\ a^2_{i,\varepsilon(i)})$.

This is equivalent to $h_{i,j}(a^2_{j,\varepsilon(j)})\, a^2_{i,j}\, (a_{i,\varepsilon(i)})^{-1} = h_{i,j}(a^1_{j,\varepsilon(j)})\, a^1_{i,j}\, (a^1_{i,\varepsilon(i)})^{-1}$. Applying twice the equation from Definition 1.2 this is equivalent to $a^2_{i,\varepsilon(j)}(a^2_{i,\varepsilon(i)})^{-1} = a^1_{i,\varepsilon(j)}(a^1_{i,\varepsilon(i)})^{-1}$. As $\varepsilon(i) = \varepsilon(j)$ we finish.

§2 On γ- systems of automorphisms

For this section we make the assumption.

2.1 Assumption: M is an L-model, $P_i \in L$ monadic predicate, $P_i^M(i < \gamma)$ are pairwise disjoint and $|M| = \bigcup_{i<\gamma} P_i^M$. For such M let $M^{[\alpha]} = M \restriction \bigcup_{i\le\alpha} P_i^M$ for $\alpha < \gamma$.

2.2 Definition : 1) Let K^M be the class of L-models N such that $N = \bigcup_{i<\gamma} P_i^N$ and $N^{[\beta]} = N \restriction \bigcup_{i\le\beta} P_i^N$ is isomorphic to $M^{[\beta]}$ for every $\beta < \gamma$.

2) Let G^M_α be the group of automorphisms of $M^{[\alpha]}$.

3) Let $h^M_{i,j}$ (for $i \le j < \gamma$) be the following function with domain $G^M_j : h^M_{i,j}(g) = g \restriction M^{[i]}$.

4) Let $\mathcal{A} = \mathcal{A}^M = \langle G^M_\alpha, h^M_{i,j} : \alpha < \gamma, i < j < \gamma \rangle$. (i.e. as long as M is constant we can omit M).

2.3 Fact: 1) $h^M_{i,j}$ is a homomorphism from G^M_j into G^M_i.

2) $\mathcal{A}^M$ is a γ-system.

Proof : Immediate.

2.4 Definition: 1) We call $\mathbf{g} = \langle g_{i,j} : i \le j < \gamma \rangle$ a ***representation*** of $N \in K^M$ *if* there are isomorphism f_i from $M \restriction \bigcup_{\varepsilon\le i} P^M_\varepsilon$ onto $N \restriction \bigcup_{\varepsilon\le i} P_i^N$ (for $i < \gamma$) such that $g_{i,j} = (f_j^{-1} \restriction N^{(\alpha)}) \circ f_i$.

2) For $\mathbf{g}$, $f_i(i < \gamma)$ as above we say that $\langle f_i : i < \gamma \rangle$ ***exemplify*** $\mathbf{g}$ being a representation of N.

2.5 Fact: Every $N \in K^M$ has a representation.

Proof : By the definition of K^M (definition. 2.2(1)) there are f_i as required.

2.6 Fact: If $\mathbf{g}$ is a representation of N $(N \in K^M)$ then $\mathbf{g} \in Gr(\mathcal{A})$.

Proof : Let $\langle f_i : i < \gamma \rangle$ exemplify $\mathbf{g} \in Gr(\mathcal{A})$ is a representation of M. For each $i \leq j$, as f_j is an isomorphism from $M^{[j]}$ onto $N^{[j]}$ clearly f_j^{-1} is an isomorphism from $N^{[j]}$ onto $M^{[j]}$, hence $f_j^{-1} \restriction N^{[i]}$ is an isomorphism from $N^{[i]}$ onto $M^{[i]}$ clearly $(f_j^{-1} \restriction N^{[i]}) \circ f_i$ is an isomorphism from $M^{[i]}$ onto $M^{[i]}$ so it belongs to G_i^M. So $g_{i,j} \in G_i^M$.

Easily $g_{i,i}$ is the unit of G_i^M.

We can now check that for $i \leq j \leq \beta < \alpha$, $g_{i,\beta} = h_{i,j}^M(g_{j,\beta}) \circ g_{i,j}$; remembering the definition of $h_{i,j}^M$ this means that

$$(f_\beta^{-1} \restriction N^{[i]}) \circ f_i = ((f_\beta^{-1} \restriction N^{[j]}) \circ f_j) \restriction M^{[i]}) \circ (f_j^{-1} \restriction N^{[i]}) \circ f_i$$

or equivalent by, for every $x \in M^{[i]}$,

$$f_\beta^{-1} \circ f_i(x)) = f_\beta^{-1} \; f_j \;\; f_j^{-1} f_i(x)$$

which is obvious.

2.7 Fact: Let $\mathbf{g}^0$ be a representation of $N (\in K^M)$. Then $\mathbf{g} \in Gr(\mathcal{A})$ is also a representation of N *iff* $\mathbf{g} \approx_{\mathcal{A}} \mathbf{g}^0$.

Proof : First suppose that $\mathbf{g}^0 \approx_{\mathcal{A}} \mathbf{g}$, and let $\langle k_i : i < \gamma \rangle \in \prod_{i<\gamma} G_i^M$ exemplify this (see Definition. 1.2). So $g_{i,j} = h_{i,j}^M(k_j)^{-1} g_{i,j}^0 k_i$ (for $i \leq j \leq \gamma$). Let $\langle f_i : i < \gamma \rangle$ exemplify $\mathbf{g}^0$ being a representation of N (see Definition. 2.4(2)).

So $g_{i,j}^0 = (f_j^{-1} \restriction N^{[i]}) \circ f_i$, and we get

$$g_{i,j} = h_{i,j}^M(k_j)^{-1} \circ (f_j^{-1} \restriction N^{[i]}) \circ f_i \circ k_i =$$
$$(f_j \restriction M^{[i]} \circ h_{i,j}^M(k_j))^{-1} \circ (f_i \circ k_i)$$

[Note that $(f_j \restriction M^{[i]})^{-1} = f_j^{-1} \restriction N^{[i]}$]; we would like to show that $\langle f_i \circ k_i : i < \gamma \rangle$ exemplify $\mathbf{g}_{i,j}$ is a representation of N. Clearly $f_i \circ k_i$ is an isomorphism from $M^{[i]}$ onto $N^{[i]}$. The above equality will be the only missing information provided that we shall show that

$$f_j \restriction M^{[i]} \circ h_{i,j}(k_j) = (f_j \circ k_j) \restriction M^{[i]}$$

which is easy.

Second suppose $\mathbf{g} \in Gr(\mathcal{A})$ is a representation of N and we shall prove that $\mathbf{g} \approx_{\mathcal{A}} \mathbf{g}^0$.

Let $\langle f_i^0 : i < \gamma \rangle$ exemplify $\mathbf{g}^0$ being a representation of N and $\langle f_i : i < \gamma \rangle$ exemplify $\mathbf{g}$ being a representation of N (see Definition. 2.4(2)). So

$$g_{i,j}^0 = (f_j^0 \restriction N^{[i]})^{-1} \circ f_i^0,$$
$$g_{i,j} = (f_j \restriction N^{[i]})^{-1} \circ f_i$$

(for $i \leq j < \gamma$). Let $k_i \stackrel{def}{=} f_i^{-1} f_i^0$ (for $i < \gamma$). As f_i, f_i^0 are isomorphism from $M^{[i]}$ onto $N^{[i]}$ clearly k_i is an automorphism of $M^{[i]}$, i.e. it belongs to G_i^M. Now $f_i^0 = f_i k_i$ hence

$$g_{i,j}^0 = (f_j^0 \restriction M^{[i]})^{-1} \circ f_i^0 = ((f_j \circ k_j) \restriction M^{[i]})^{-1} \circ (f_i \circ h_i) =$$
$$= (k_i \restriction M^{[i]})^{-1} \circ (f_j \restriction M^{[i]})^{-1} \circ f_i \circ k_i =$$
$$(k_j \restriction M^{[i]})^{-1} \circ g_{i,j} \circ k_i$$

But easily $k_j \restriction M^{[i]} = h_{i,j}^M(k_i)$, so $\langle k_i : i < \gamma \rangle$ exemplify $\mathbf{g} \approx_{\mathcal{A}} \mathbf{g}^0$.

Fact 2.8: Suppose the models $N_1, N_2 \in K^M$ has representations $\mathbf{g}^1, \mathbf{g}^2$ respectively, then $N_1 \cong N_2$ iff $\mathbf{g}^1 \approx_{\mathcal{A}} \mathbf{g}^2$.

Proof : Let $\langle f_i^\ell : i < \gamma \rangle$ exemplify "$\mathbf{g}^\ell$ is a representation of N_ℓ" for $\ell = 1,2$. So $g_{i,j}^\ell = (f_j^\ell \restriction M^{[i]})^{-1} \circ f_i^\ell$ for $\ell = 1,2$, $i \leq j < \gamma$.

First assume N^1, N^2 are isomorphic, and let H be an isomorphism from N^1 onto N^2. For each $i < \gamma$, $H \restriction N_1^{[i]}$ is an isomorphism from $N_1^{[i]}$ onto $N_2^{[i]}$, hence $k_i \stackrel{def}{=} (f_i^2)^{-1}(H \restriction N_1^{[i]}) f_i^1$ is an isomorphism from $M^{[i]}$ onto $M^{[i]}$, i.e. $k_i \in G_i^M$. So for every $i, f_i^2 = (H \restriction N_1^{[i]}) \circ f_i^1 \circ k_i^{-1}$, and let $H_i \stackrel{def}{=} H \restriction N_1^{[i]}$ (so for $i < j, H_i = H_j \restriction N_1^{[i])]}$). Now for $i \leq j < \gamma$.

$$g_{i,j}^2 = (f_j^2 \restriction M^{[i]})^{-1} \circ f_i^2 =$$
$$= (H_j \circ f_j^1 \circ k_j^{-1} \restriction M^{[i]})^{-1} \circ (H_i \circ f_i^1 \circ k_i^{-1}) =$$
$$= (H_i \circ (f_j^1 \restriction M^{[i]}) \circ (k_j \restriction M^{[i]})^{-1})^{-1} \circ (H_i \circ f_i^1 \circ k_i^{-1}) =$$

$$= (k_j \restriction M^{[i]}) \circ (f_j^1 \restriction M^{[i]})^{-1} \circ H_i^{-1} \circ H_i \circ f_i^1 \circ k_i^{-1} =$$
$$= (k_j \restriction M^{[i]}) \circ (f_j^1 \restriction M^{[i]})^{-1} \circ f_i^1 \circ k_i^{-1} = (k_j \restriction M^{[i]}) \circ g_{i,j}^1 \circ k_i^{-2}$$

So $\langle k_i^{-1}: i < \gamma \rangle$ exemplify $\mathbf{g}^1 \approx_{\mathcal{A}} \mathbf{g}^2$.

Second, assume $\mathbf{g}^1 \approx_{\mathcal{A}} \mathbf{g}^2$ and let this be exemplified by $\langle k_i^{-1}: i < \gamma \rangle$. Define

$$H_i = f_i^2 \circ k_i \circ (f_i^1)^{-1}$$

It is easy to check that H_i is an isomorphism from $N_1^{[i]}$ for $i < \gamma$ and $H_i = H_j \restriction M^{[i]}$, for $i < j < \gamma$. So $\bigcup_{i<\gamma} H_i$ is an isomorphism from N_1 onto N_2.

2.9 Lemma : If $\mathbf{g}$ is a representation of $N \in K^M$ then $\mathbf{g} \in Gs(\mathcal{A})$.

Proof : Suppose not so for some $\beta < \gamma$, $\mathbf{g} \restriction \gamma \notin Fact(\mathcal{A} \restriction \gamma)$ so $\mathbf{g} \restriction \gamma, \langle e_i^{\mathcal{A}}: i \leq j < \beta \rangle$ are not $\approx_{(\mathcal{A} \restriction \beta)}$-equivalent. Apply 2.8 to $M^{[\beta]}$ instead M (and $\mathbf{g} \restriction \beta, \langle e_i^{\mathcal{A}}: i < j < \beta \rangle$, $N^{[\beta]}, M^{[\beta]}$), and get that $N^{[\beta]}, M^{[\beta]}$ are not isomorphic contradicting $N \in K^M$.

2.10 Lemma : Every $\mathbf{g} \in Gs(\mathcal{A})$ represents some $N \in K^M$.

Proof : We define by induction j

(a) an L-model N_j, such that $N_j \cong M^{[i]}$ and $N_i \subset N_j$ for $i \leq j$.

(b) an isomorphism f_j from $M^{[j]}$ onto N_j, such that for $i \leq j$; $g_{i,j} = (f_j \restriction M^{[i]}) \circ f_i$.

For $j = 0$, j successor there is no problem. For j limit $\bigcup_{i<j} N_i$ is isomorphic to $\bigcup_{i<j} M^{[i]} = M \restriction \bigcup_{i<j} P_i^M$ by 2.8, and multiplied by some $k \in Aut\ (M \restriction \bigcup_{i<j} P_i)$ it will be as required.

2.11 Conclusion: The numbers of non-isomorphic $N \in K^M$ is equal to $| Gs(\mathcal{A}) / \approx_{\mathcal{A}} |$.

Proof : By 2.5-2.10.

2.12 Lemma : If the following conditions hold, then every $N \in K^M$ is $L_{\infty,\lambda}$-equivalent to M.

a) Every function F of M are 1-place, amd for $x \in M^{[i]}$, $F_i^M(x) \in M^{[i]}$.

b) for any relation R of M for some $n < \omega$ and $i < \gamma$:

$$M \models (\forall x_1, \ldots, x_n)[R(x_1, \ldots, x_n) \to \bigwedge_{\ell=1}^{n} P_i(x_\ell)]$$

c) *if* $i < j < \gamma$, $g \in G_i^M$, g^* *a partial* automorphism of $M^{[j]}$,Dom (g^*) closed under the function of M, and $g \cup g^*$ is a partial automorphism of M and Dom (g^*) is in $\mathcal{I}_i$, (see below) *then* $g \cup g^*$ an be extended to an automorphism of $M^{[j]}$.

d) $\mathcal{I}_i$ is a family of subsets of M,$[i < j \Longrightarrow \mathcal{I}_i \subseteq \mathcal{I}_j]$ $\mathcal{I}_i$ closed under finite unions, and $[A \subseteq M,\ |A| < \lambda \Longrightarrow A \in \bigcup_{i<\gamma} \mathcal{I}_i]$.

Proof: Easy.

§3 Constructing the model.

3.1 Main Theorem : Suppose

(i) $\kappa = cf(\lambda) < \lambda$ and $(\forall \mu < \lambda)(\mu^{<\kappa} < \lambda)$.

(ii) $\mathcal{B}$ is a κ-system, and $|G_i^{\mathcal{B}}| < \lambda$ for $i < \kappa$.

Then there is a model M (with relations and functions of finitely many places only) of cardinality λ such that $no(M) = no^*(\mathcal{B})$.

3.1A Remarks: W.l.o.g. $M = (|M|, R^M)$ for some two- place relation R. (see [Sh 5], 1.4)

Notation: For $A \subseteq M$, let $cl_M(A)$ be the closure of A under the functions of M.

Proof : By 1.12 w.l.o.g. for $j < \kappa$ limit, $h_{j,j+1}^{\mathcal{B}}$ is onto $G_j^{\mathcal{B}}$, and if $x \in G_j^{\mathcal{B}}$, $x \neq e_j^{\mathcal{B}}$ then for some $i < j$, $h_{i,j}^{\mathcal{B}}(x) \neq e_i^{\mathcal{B}}$. By 1.12 w.l.o.g. $G_0^{\mathcal{B}}$ is trivial ($=\{e_0^{\mathcal{B}}\}$). Let $L = \{P_i, F_{i,j}, : i < j < \kappa\} \cup \{R_i : i < \kappa\}$, $P_i (i < \kappa)$ monadic predicates, $F_{i,j}$ one place function symbols, R_i three place predicate. Let $\lambda = \sum_{i<\kappa} \lambda_i$, $\lambda_i^{<\kappa} = \lambda_i < \lambda$, $\lambda_i > ((\sum_{j<i} \lambda_j^+ + |G_i^{\mathcal{B}}|)^{\kappa})^{+6}$. We shall now define by induction on

$j < \kappa$, M_j, G_j, H_j, H_j^+, $\mathcal{P}_i$ $(i < j)$ such that :

(A) (1) M_j is an L-model,

(A) (2) M_j is the disjoint union of $P_i^{M_j}(i < j)$ and $P_i^{M_j} = (\lambda_i, \lambda_i^{+2})$ when $i < j$, $P_i^{M_j} = \phi$ when $\kappa > i \geq j$

(A) (3) $F_{\alpha,\beta}^{M_j}$ is a 1-place function from $P_\beta^{M_j}$ into $P_\alpha^{M_j}$ (and not defined otherwise) for $\alpha < \beta < \kappa$.

(A) (4) for any R_i $R_i^{M_j}$ is a (three place) relation on $P_i^{M_j}$,

(A) (5) for $i < j$, $M_i = M_j \upharpoonright (\bigcup_{\varepsilon < i} P_\varepsilon^{M_j})$.

(B) (1) G_j is the group of automorphism of M_j if j is a successor ordinal, otherwise $G_j = \{k \in Aut(M_j)$: for some $a \in G_j^\beta$ for every $i < j$, $H_j(k \upharpoonright M_i) = h_{i,j}(a)\}$, (see below on H_j)

(B) (2) H_j is a homomorphism from G_j onto G_j^β.

(B) (3) for $i < j$, $k \in G_j$, $h_{i,j}^\beta(H_j(k)) = H_i(k \upharpoonright M_i)$.

(B) (4) G_j has cardinality $\leq \lambda_j^{+2}$.

(B) (5) H_j^+ is a homomorphism from G_j^β into G_j, $H_j \circ H_j^+$ is the identity (on G_j^β) and for $i < j, a \in G_j$, $H_j^+(a) \upharpoonright M_i = H_i^+(h_{i,j}^\beta(a))$.

(C) (1) $\mathcal{P}_i^j$ is a family of subsets of $(\lambda_j, \lambda_j^{+2})$ (when $i < j$).

(C) (2) if $A \in \mathcal{P}_i^j$, $i < \alpha < j$, then $cl_M(A) \cap (\lambda_\alpha, \lambda_\alpha^{+2}) \in \mathcal{P}_i^\alpha$.

(C) (3) for $i < \alpha < j$, $\mathcal{P}_i^j \subseteq \mathcal{P}_\alpha^j$.

(C) (5) every $g \in G_{j+1}$ maps any $A \in \mathcal{P}_i^j$ to a member of $\mathcal{P}_i^j$.

(C) (6) $\mathcal{P}_i^j$ is closed under union of $\leq \kappa$, (i.e if $A_\xi \in \mathcal{P}_i^j$ for $\xi < \zeta \leq \kappa$ then $\bigcup_{\xi < \zeta} A_\xi \in \mathcal{P}_i^j$).

(C) (7) every subset of $(\lambda_j, \lambda_j^{+2})$ of power $\leq \|M_i\|$ is included in some member of $\mathcal{P}_i^j$.

(D) (1) For $i < j$ let $Q_i^j = \{A \subseteq M_j$: for $\alpha < i$, $(\lambda_\alpha, \lambda_\alpha^{+2}) \subseteq A$ and for $\alpha \in [i,j)$, $A \cap (\lambda_\alpha, \lambda_\alpha^{+2}) \in \mathcal{P}_i^\alpha$ and $A = cl_{M_j}(A)\}$.

(D) (2) If $i < j$, $k_0, k_1 \in G_j$, $A \in Q_i^j$, k_0, k_1 are equal on $(\bigcup_{\alpha<i} P_\alpha^{M_j}) \cap A$ *then* $(k_0 \restriction A) \cup (k_1 \restriction \bigcup_{\alpha<i} P_\alpha^{M_j})$ can be extended to an automorphism k of M_j.

Moreover, if $a \in G_j^\beta$, $b_{i,j}^\beta(a) = H_i(k_1 \restriction M_i)$ then we can demand $H_j(k) = a$.

Clearly it suffices to carry the construction by induction, as then $M \stackrel{def}{=} \bigcup_{j<\kappa} M_j$ is as required by the previous Lemmas (i.e. by 2.12 every $N \in K_M$ is $L_{\infty,\lambda}$-equivalent to it (and clearly $[N \equiv_{\infty,\lambda} M \Longrightarrow N \in K_M]$) so $no(\mathfrak{A}) = \{N/\cong : N \in K_M\}$. But 2.11 this number is equal to $no^*(M) = |Gs(\mathfrak{A})/\approx_{\mathfrak{A}}|$ where $\mathfrak{A} = \mathfrak{A}^M$ (see Definition 2.2(4)). By 1.9 this number is $no^*(\mathcal{B})$. But $\mathcal{B}$ was chosen so that it is μ.)

Case I: $j = 0$.

Nothing to do.

Case II: j is limit.

In this case let $M_j = \bigcup_{i<j} M_i$, and there is no problem to check all the conditions. Note that in (D)(2) we can easily prove the second sentence.

Case III: $j + 1$ (assuming we have defined for j).

We shall define by induction on $\xi < \lambda_j^{+2}$, a group $G_{j,\xi}$, an ordinal $\alpha(\xi)$, an action of the group $G_{j,\xi}$ on $M_j \cup (\lambda_j, \alpha(\xi))$ and $H_{j,\xi}, P_{i,\xi}^j, F_{\alpha,j}^\xi, R^\xi$ such that

(i) for $\zeta < \xi$, $G_{j,\zeta}$ is a subgroup of $G_{j,\xi}$ and the action of $g \in G_{j,\xi}$ on $M_j \cup (\lambda_j, \alpha(\zeta))$ is extended too, and for $k \in G_{j,\xi}$, $k \restriction M_j \in G_j$.

(ii) $\alpha(\xi) \in (\lambda_j^{+1}, \lambda_j^{+2})$ and $\alpha(\xi)$ is increasing and continuous.

(iii) for ξ limit $G_{j,\xi} = \bigcup_{\zeta<\xi} G_{j,\zeta}$.

(iv) $H_{j,\xi}$ is a homomorphism from $G_{j,\xi}$ onto G_j^{β}.

(v) $F_{\alpha,j}^{\xi}$ is a one-place function from $(\lambda_j, \alpha(\xi))$ into $P_\alpha^{M_j}$ increasing and continuous in ξ.

(vi) $P_{i,\xi}^j$ is a family of subsets of $(\lambda_j, \alpha(\xi))$ such that $A^{[i]} \stackrel{def}{=} \{F_{\alpha,j}^{\xi}(x) : \alpha < j, x \in A\} \in Q_i^j$ for each $A \in P_{i,\xi}^j$, $i < j$.

(vii) if $A \in P_{i,\xi}^j$, $g \in G_{j,\xi}$ then $g(A) \in P_{i,\xi}^j$.

(viii) $P_{i,\xi}^j$ is closed under union of $\le \kappa$ members and it is increasing with ξ and if $cf\ \xi > \kappa$ then $P_{i,\xi}^j = \bigcup_{\zeta<\xi} P_{i,\zeta}^j$.

(ix) we can choose for every $\alpha(\xi)$ an increasing sequence $B_\varepsilon^\xi (\varepsilon < \lambda_j^+)$ such that $(\lambda_j, \alpha(\xi)) = \bigcup_{\varepsilon<\lambda_j^+} B_\varepsilon^\xi$, and B_ε^ξ has cardinality $\le \lambda_j$. We shall guarantee that for any $\xi < \lambda_j^{++}$, $\varepsilon < \lambda_j^+, i < j$ and $A \in Q_i^j$ for some ξ_1, $\xi < \xi_1 < \lambda_j^{+2}$, and $B \in P_{i,\xi_1}^j$, $B_\varepsilon^\xi \subseteq B$.

(x) if $k_0, k_1 \in G_{j,\xi}$, $A \in Q_i^j$ k_0, k_1 are equal on A, $a \in G_{j+1}^{\beta}$, $h_{j+1}^{\beta}(a) = H_j(k_1 \restriction M_i)$ *then* $(k_0 \restriction A) \cup (k_1 \restriction M_j)$ can be extended in some $G_{j,\zeta}(\xi \le \zeta < \lambda_j^{+2})$ to k, $H_{j,\zeta}(k) = a$.

(xi) R^ξ is a three place relation on $(\lambda_j, \alpha(\xi))$, increasing with ξ, but for $\zeta < \xi$, $R^\zeta = R^\xi \restriction (\lambda_j, \alpha(\zeta))$.

(xii) each $g \in G_{j,\xi}$ preserves R^ξ and $F_{\alpha,j}^{\xi}$.

(xiii) if $cf\ \xi = \lambda_j^+$, then $R(\alpha(\xi)-,-)$ define on $(\lambda_j, \alpha(\xi))$ a well-ordering [so *if* $g \in G_{j,\zeta}, \zeta > \xi$, g maps $(\lambda_j, \alpha(\xi))$ on itself *then*, $g \restriction (\lambda_j, \alpha(\xi))$ is determined by $g(\alpha(\xi))$].

(xiv) no $\alpha \ne \beta \in (\lambda_j, \alpha(\xi))$ realize the same quantifiers free, R_ξ-type over (λ_j, λ_j^+). (So together with (xiii) we have a strict control over the automorphism of M_{j+1}).

There is no problem to carry the induction on ξ hence on j, hence to finish the proof of 3.1.

REFERENCES

[C] C.C. Chang, Some remarks on the model theory of infinitary languages:, pp. 36-63 in: *The Syntax and Semantics of Infinitary Languages*, *Lecture Notes in Mathematics*, 72, ed., J. Barwise, Springer, 1968.

[HS] H.L. Hiller, and S. Shelah, "Singular cohomology in L", *Israel Journal of Mathematics*, vol. 26 (1977), pp. 313-319.

[NS] M. Nadel and J. Stavi, " $L_{\infty\lambda}$-equivalence, isomorphism and potential isomorphism ." *Transactions of the American Mathematical Society*, *vol. 236 (1978)*., *pp*. 51-74.

[P] E. A. Palyutin, "Number of models in L_{∞,ω_1} theories III," pp. 443-456 in *Algebra I Logika*, vol. 16, no. 4. (1977): English translation in *Algebra and Logic*, vol. 16, no. 4 (1977), pp. 299-309.

[S] D. Scott, "Logic with denumerably long formulas and finite strings of quantifiers," pp. 329-341 in *The Theory of Models*, North Holland, 1965.

[Sh1]
S. Shelah, "On the number of non-isomorphic models of cardinality $\lambda, L_{\infty\lambda}$-equivalent to a fixed model", *Notre Dame J. of Formal Logic* 22 (1981), 5-10.

[Sh2]
-------, "On the number of non-isomorphic models of power $\lambda \equiv_{\infty\lambda}$ to a model of power λ, when λ is weakly compact," *Notre Dame J. of Formal Logic*, 23 (1982), 21-26.

[Sh3]
------, "The consistency of $Ext(G,\mathbf{Z}) = \mathbf{Q}$" *Israel Journal of Mathematics*, 39 (1981), 283-288.

[Sn4]
-------, "A pair of non-isomorphic $\equiv_{\infty,\lambda}$, Models of Power λ for λ singular with $\lambda^\omega = \lambda$". *Notre Dame J. of Formal Logic*, *25 (1984)*, *97-104*.

[Sh5]
------, "On the possible number $no(M)$ = the numbers of non isomorphic models $L_{\infty,\lambda}$-equivalent of power λ for λ singular, *Notre Dame J. Formal Logic*, in press.

Non standard uniserial module over a uniserial domain exists

Our aim is to prove:

Theorem: (ZFC) There exist a non standard uniserial modules over some uniserial domain (see 12).

The paper is self contained. It uses forcing - this can be eliminated easily but for me this has no point. Our example is in $\aleph_1$ - we can replace it by any regular $\kappa > \aleph_0$. The problem appears in the version of a book of Fuchs and Salce on modules over uniserial domains in existence in April 1984.. An answer in the other direction would have simplified the subject, and I think, make unnecessary several proofs and distinctions.

I thank Silvana Bazzoni, Elizabetta Martinez and Claudia Mettel for going our of their way to tell me the problem during a dinner at the conference in Udine, to Fuch's for mentioning it and to Salce for impressing upon me the importance of solving it.

Subsequently Fuchs continues this work, investigating for which uniserial R there are such modules.

0. Definition and Notation: 1) Let R denote a uniserial domain, i.e., no zero divisors and $Id(R) = \{I: I$ an ideal of $R\}$ is linearly ordered by inclusion. Let $Q = Q_R$ be the field quotient. Let a,b,c,r,s denote member of R, x,y,z denote members of an R-module, M,N denote R-modules. Let $a \mid b$ mean a divides b.

2) An R-module is called standard if it is a homomorphic image of an R-submodule of Q (which is trivially an R-module) and $M \neq 0$.

3) An R-module is uniserial if its family of submodules is linearly ordered. (So we are assuming R itself is uniserial.)

0A Remark: Any standard R-module is uniserial.

This is well known.

1. Fact: Let M be a uniserial R-module; if $x \in M$, $ax \neq 0$ then for every $b \in R, (b \neq 0)$: $bx = 0$ if $(b/a)(ax) \neq 0$ and a divides b in R.

Proof: If in R $a \mid b$ let $b = ca$ so $bx = 0 \Longleftrightarrow cax = 0 \Longleftrightarrow (b/a)(ax) = 0$. So it suffices to prove $a \mid b$ assuming $bx = 0$, but if a does not divide b, b divides a so $a = db$, so $ax = dbx = d = 0$ contradicting, an assumption.

2. Definition : 1) We call $\langle a_{i,j} : i < j < \delta \rangle$ an I-representation of M (for M a uniserial module over a uniserial domain R) *if:*

(i) I is an ideal of $R, I \neq R$.

(ii) $a_{i,j} \in R$, $a_{i,j} \neq 0$.

(iii) for $\alpha < \beta < \gamma < \delta$, $a_{\alpha,\gamma} - a_{\alpha,\beta} a_{\beta,\gamma} \in a_{0,\gamma} I$.

(iv) there are $x_i \in M (i < \delta)$ such that M is generated by $\{x_i : i < \delta\}$, and:

$$I = \{r \in R : rx_0 = 0\}, \quad a_{ij} x_j = x_i$$

2) We call $\langle a_{i,j} : i < j < \delta \rangle$ an I- representation for R) if (i),(ii),(ii) above holds.

3. Claim: Every uniserial R-module M has an I-representation (for some ideal I of R).

Proof : Easy. Choose by induction on i, $x_i \in M (\neq 0)$ x_i not in the submodule generated by $\{x_j : j < i\}$. Say δ is the first for which x_δ is not defined. Clearly δ exists and is $< \|M\|^+$. For $i < j$, as $x_j \notin Rx_i$, by uniseriality $x_i \in Rx_j$ so for some $a_{i,j} \in R$, $x_i = a_{i,j} x_j$. Now for $\alpha < \beta < \gamma < \delta$, $a_{\alpha,\gamma} x_\gamma = x_\alpha = a_{\alpha,\beta} x_\beta = a_{\alpha,\beta}(a_{\beta,\gamma} x_\gamma)$. So $(a_{\alpha,\gamma} - a_{\alpha,\beta} a_{\beta,0}) x_\gamma = 0$. As $a_{0,\gamma} x_\gamma = x_0 \neq 0$, we finish by Fact 1.

Remark: Clearly $\delta > 0$ for $M \neq 0$, and if δ is a successor ordinal then M is standard.

4. Claim: 1) If $\langle a_{i,j} : i < j < \delta \rangle$ is an I- representation for R then some R-module M is I-represented by $\langle a_{i,j} : i < j < \delta \rangle$.

2) Moreover M is unique up to isomorphism and is uniserial

Proof: Let M be an R-module generated freely by $\{x_i : i < \delta\}$ except the relations:

(a) $rx_0 = 0$ (for $r \in I$)

(b) $x_i - a_{i,j}x_j = 0$ for $i < j < \delta$.

2) The uniqueness is trivial, so we shall prove that M constructed in (1) is uniserial. It is easy to see that (by the relations (b)).

(*) for every $y \in M$ for some $i < \delta$, $r \in R$: $y = rx_i$.

Now suppose K is a submodule of $M, K \neq M$, and we shall prove that for some $\xi < \delta$, $K \subseteq Rx_\xi$. This suffices [if K_1, K_2 are submodules of M, if $K_1 = M$ or $K_2 = M$ they are comparable so we finish; if $K_1, K_2 \neq M$ there are $\xi_1, \xi_2 < \delta$ such that $K_1 \subseteq Rx_{\chi_1}, K_2 \subseteq Rx_{\xi_2}$; let $\xi = Max\{\xi_1, \xi_2\}$, so K_1, K_2 are R- submodules of Rx_ξ, which is uniserial by OA, hence $K_1 \subseteq K_2$ or $K_2 \subseteq K_1$].

As $K \neq M$ for some ξ $x_\xi \notin K$. Assume $K \not\subseteq Rx_\xi$, so for some $y \in K$, $y \notin Rx_\xi$. By (*) above for some $\zeta < \delta$ and $r \in R$, $y = rx_\zeta$. Now $\xi < \zeta$ [otherwise $y = rx_\zeta \in Rx_\zeta \subseteq Rx_\xi$ contradiction to the choice of y]; As $y \neq 0, r \neq 0$, and $a_{\xi,\zeta} \neq 0$, in R r divides $a_{\xi,\zeta}$ or $a_{\xi,\zeta}$ divides r (or both).

If $a_{\xi,\zeta}$ divides r, then

$$y = rx_\zeta = (r / a_{\xi,\zeta})(a_{\xi,\zeta}x_\zeta) = (r / a_{\xi,\zeta})x_\xi \in Rx_\xi$$

contradiction to the choice of y.

If r divides $a_{\xi,\zeta}$ then

$$x_\xi = a_{\xi,\xi} x_\xi = (a_{\xi,\xi}/r) r x_\xi \in R(r x_\xi) = Ry \subseteq K$$

contradiction to the choice of ξ.

So $K \subseteq Rx_\xi$. We previously show that this (i.e. for every R-submodule K of M, $K \subseteq Rx_\xi$ for some ξ) suffice.

5. Lemma : A uniserial R-module with I-representation $\langle a_{i,j} : i < j < \delta \rangle$ is standard *iff* for some $c_i \in R (i < \delta)$ for every $i < j < \delta$:

(i) $\dfrac{c_i^{-1}}{a_{0,i}} - \dfrac{c_j^{-1}}{a_{0,j}/a_{i,j}} \in I$

(ii) $c_i^{-1} \in R$, i.e., each c_i is a unit.

5A. Remark: We can replace is (i),(iii), c_i^{-1} by c_i, c_j^{-1} by c_j.

Proof : First suppose that there are such $c_i (i < \delta)$. Let $J_i = (1/a_{0,i})R \subseteq Q$ and define a function from J_i into M by

$$f_i((1/a_{0,i})r) = r c_i x_i \quad \text{for } r \in R$$

Clearly f_i is a homomorphism from one R-module to another.

It is onto Rx_i as c_i is invertible in R.

We shall prove that

(*) for $i < j < \delta$, $f_i \subseteq f_j$.

This suffice as then $\bigcup_{i<\delta} f_i$ is a homomorphism from $\bigcup_{i<\delta} J_i$ onto M. For proving (*) it suffices to prove:

(**) $f_i(1/a_{0,i}) = f_j(1/a_{0,i})$

First $1/a_{0,i} \in \text{Dom}(f_j)$, [this is equivalent to $1/a_{0,i} \in R(1/a_{0,j})$ which is equivalent to $a_{0,j} \in Ra_{0,i}$, if this fails then by the uniseriality of R, for some $s \in R$ which is not a unit, $a_{0,i} = s\, a_{0,j}$ so

$$a_{0,j}\,(1 - s a_{i,j}) = a_{0,j} - a_{0,i} a_{i,j} \in a_{0,j} I$$

as R has no zero divisors, $1-sa_{i,j} \in I$; as s is not a unit sR is a proper ideal, but $1 = sa_{i,j} + (1-sa_{i,j}) \in sR + I$, but $sR \subseteq I$ or $I \subseteq sR$, so necessarily $I \subseteq sR, 1 \in I$ but then $x_0 = 0$ contradiction]. Second, we can confirm (**) remember we have shown above $a_{0,j} \in R\, a_{0,i}$ hence $a_{0,j}/a_{0,i} \in R$):

$$f_j(1/a_{0,i}) = f_j((a_{0,j}/a_{0,i})(1/a_{0,j})) = (a_{0,j}/a_{0,i})\, c_j x_j$$
$$f_i(1/a_{0,i}) = c_i x_i = c_i a_{i,j} x_j$$

So it is enough to show that

$$\left(\frac{a_{0,j}}{a_{0,i}} c_j - a_{i,j} c_i\right) x_j = 0$$

equivalently (see Fact 1):

$$\frac{a_{0,j}}{a_{0,i}} c_j - a_{i,j} c_i \in a_{0,j} I$$

equivalently

$$\frac{c_j}{a_{0,i}} - \frac{c_i}{a_{0,j}/a_{i,j}} \in I$$

Multiplying by $c_j^{-1} c_i^{-1}$ we get (i) of the hypothesis , i.e., the demand holds (Note that for a unit c ,$cI = I$).

We have proved the "if" part of Lemma 5.

For the only "if" part suppose J is an R-submodule of Q, $f; J \to M$ an onto homomorphism. W.l.o.g. $f(1) = x_0$ so $R \subseteq \mathrm{Dom}\, f$, $1 \notin \mathit{Ker}\, f = I$. For every i, let $x_i = f(y_i)$. $y_i \in J$. If $y_i \in R(1/a_{0,i})$ let for some $r \in R$, $y_i = r/a_{0,i}$, then $a_{0,i} y_i = r$ hence $f(r) = f(a_{0,i} y_i) = a_{0,i} f(y_i) = a_{0,i} x_i = x_0 = f(1)$, so $f(1-r) = 0$ hence $1-r \in I$, hence $r^{-1} \in R$ [otherwise $Rr \not\subseteq R$, so $Rr \cup R(1-r)$ is a proper ideal contradiction]. So $[y_i \in R(1/a_{0,i}) \Longrightarrow 1/a_{0,i} \in Ry_i]$. As $y_i, 1/a_{0,i} \in Q$, Q a uniserial R-module this implies $1/a_{0,i} \in Ry_i$, so for some $c_i \in R$, $1/a_{0,i} = c_i y_i$. As $y_i \in J$ clearly $1/a_{0,i} \in J$. Now

$$x_0 = f(1) = f(a_{0,i} \cdot (1/a_{0,i})) = a_{0,i} f(1/a_{0,i}) = a_{0,i} c_i x_i = c_i x_0$$
$$= a_{0,i} f(c_i y_i) = a_{0,i} c_i f(y_i)$$

so $(1-c_i)x_0 = 0$ hence $1-c_i \in I$, so as in an argument above c_i is a unit except when $I{=}R$ which is excluded.

So $1/a_{0,i} = c_i y_i$, $c_i \in R$ a unit. By (iii) of Definition 2 with G, i, j here standing for α, β, γ there, $1 - \frac{a_{0,i} a_{i,j}}{a_{0,j}} \in I$ so (when $I \neq R$) $a' \stackrel{def}{=} \frac{a_{0,i}}{a_{0,j}} a_{i,j}$ is a unit of R, as $a_{i,j} \in R$ this implies $\frac{a_{0,j}}{a_{0,i}} = a_{i,j} / a' \in R$. Now

$$0 = f(0) = f(1/a_{0,i} - 1/a_{0,i}) = f(1/a_{0,i}) - f((a_{0,j}/a_{0,i}) \cdot 1/a_{0,j}) =$$
$$f(c_i y_i) - (a_{0,j}/a_{0,i}) f(c_j y_j) =$$
$$c_i x_i - (a_{0,j}/a_{0,i}) c_j x_j = c_i a_{i,j} x_j - (a_{0,j}/a_{0,i}) c_j x_j =$$
$$(c_i a_{i,j} - (a_{0,j}/a_{0,i}) c_j) x_j$$

hence $[c_i a_{i,j} - [(a_{0,j}/a_{0,i}) c_j] / a_{0,j} \in I$ and we can finish.

For a while we make

6. Assumption: M is a non-standard model of $Th(\mathbf{Z})$ of power $\aleph_1$ not $\aleph_1$-like, $M = \bigcup_{i < \omega_1} M_i$, $M_i \prec M, M_i$ increasing continuous, each M_i countable, $p \in M$ a prime $R = R^p_M$ is $\{a/b ; a, b \in M, M \models$"$p$ does not divide b"$\}$./

Let $Q \supseteq R$ be the field of quotients of R.

Easily R is a uniserial domain. Let b be a member of M. let $\langle d(\alpha) : \alpha < \omega_1 \rangle$ be a sequence of members of M increasing, $d(\alpha) < b$, $b, p \in M_0$, $d(\alpha) \in M_{\alpha+1}$. Let Q_i be the field of quotients of M_i, $R_i = R \cap Q_i$.

Clearly we can find M as above, and then $b, d(\alpha)$.

7. Definition : Let $I = \{c \in R : p^b \mid c\}$, it is an ideal.

We define a set P; its members have the form:

$$\langle a_{i,j} : i < j, i \in u, j \in u \rangle$$

such that

(i) u a finite subset of $\omega_1, 0 \in u$.

(ii) for $\alpha < \beta < \gamma$ all in u,

$$(\frac{a_{\alpha,\gamma}-a_{\alpha,\beta}a_{\beta,\gamma}}{a_{0,\gamma}}) \in I.$$

(iii) $a_{\alpha,\beta}$ is divisible by $p^{d(\beta)-d(\alpha)}$ but not by $p^{d(\beta)-d(\alpha)+1}$ in R (exponentiation in M).

(iv) $a_{i,j} \in R_{j+1}$

[we write $a_{ij} = a^r_{ij}$, $u = u^r$ where $r = \langle a_{i,j}: i < j, i \in u, j \in u \rangle$].

We stipulate $a_{i,i} = 1$. The order of P is natural.

8. Fact: If $r = \langle a^r_{ij}: i < j \in u^r \rangle \in P$, $\xi < \omega_1$ *then* there is $q, r < q \in P, \xi \in u^q$.

Proof : If $\xi \in u^r$ let $q = p$, otherwise suppose $i_1 < \cdots < i_\ell < \xi < i_{\ell+1} < \cdots < i_m$, $u^r = \{i_1, \ldots, i_m\}$, (remember $i_0 = 0$) and let $a_{i,j} = a^r_{i,j}$.

We now define q:

$$u^q = u^r \cup \{\xi\}$$

$$a^q_{i,j} = \begin{cases} a_{i,j} & \text{if } i < j, i \in u^r, \ j \in u^r \\ a_{i,i_\ell} p^{d(\xi)-d(i_\ell)} & \text{if } i \in \{i_1, \ldots, i_\ell\}, \ j = \xi \\ \dfrac{a_{i_{\ell+1},j}\, a_{i_\ell, i_{\ell+1}}}{p^{d(\xi)-d(i_\ell)}} & \text{if } j \in \{i_{\ell+1}, \ldots, i_n\} \ \ i = \xi \end{cases}$$

We shall now check that $q \in P$.

Properties (i), (iii) and (iv) of Definition 7 are easy, so let us check (ii)).

So let $\alpha < \beta < \gamma$ be in u^r.

Case A: $\alpha = \xi$.

$$\frac{a^q_{\alpha,\gamma} - a^q_{\alpha,\beta} \ \ a^q_{\beta,\gamma}}{a^q_{0,\gamma}} = \text{(by the third case in the definition of } a^q_{\xi,j}).$$

$$\frac{a_{i_{\ell+1},\gamma}a_{i_\ell,i_{\ell+1}}p^{-(d(\alpha)-d(i_\ell))}-a_{i_{\ell+1},\beta}a_{i_\ell,i_{\ell+1}}p^{-(d(\alpha)-d(i_\ell))}a_{\beta,\gamma}}{a_{0,\gamma}}=$$

$$\frac{a_{i_\ell,i_{\ell+1}}}{p^{d(\alpha)-d(i_\ell)}}\quad\frac{a_{i_{\ell+1},\gamma}-a_{i_{\ell+1},\beta}a_{\beta,\gamma}}{a_{0,\gamma}}\in I$$

Because the left term is in R (by (iii) of Definition 7 for p) and the right term is in I (by (ii) of Definition 7 for p).

Case B: $\beta = \xi$.

$$\frac{a^q_{\alpha,\gamma}-a^q_{\alpha,\beta}a^q_{\beta,\gamma}}{a^q_{0,\gamma}}=$$

$$\frac{a_{\alpha,\gamma}-(a_{\alpha,i_\ell}p^{d(\beta)-d(i_\ell)})(a_{i_{\ell+1},\gamma}a_{i_\ell,i_{\ell+1}}p^{-(d(\beta)-d(i_\ell))})}{a_{0,\gamma}}=$$

$$\frac{a_{\alpha,\gamma}-a_{\alpha,i_\ell}a_{i_\ell,i_{\ell+1}}a_{i_{\ell+1},\gamma}}{a_{0,\gamma}}=$$

$$\frac{a_{\alpha,\gamma}}{a_{0,\gamma}}-a_{\alpha,i_\ell}\frac{a_{i_\ell,i_{\ell+1}}a_{i_{\ell+1},\gamma}}{a_{0,\gamma}}=$$

$$\frac{a_{\alpha,\gamma}}{a_{0,\gamma}}-a_{\alpha,i_\ell}\frac{a_{i_\ell,\gamma}}{a_{0,\gamma}}+a_{\alpha,i_\ell}\left(\frac{a_{i_\ell,\gamma}}{a_{0,\gamma}}-\frac{a_{i_\ell,i_{\ell+1}}a_{i_{\ell+1},\gamma}}{a_{0,\gamma}}\right)$$

$$=\frac{a_{\alpha,\gamma}-a_{\alpha,i_\ell}a_{i_\ell,\gamma}}{a_{0,\gamma}}+a_{\alpha,i_\ell}\frac{a_{i_\ell,\gamma}-a_{i_\ell,i_{\ell+1}}a_{i_{\ell+1},\gamma}}{a_{0,\gamma}}\in I$$

as the first term is in I (by (ii) of Definition 7 for p) and the second term is in I as a members of I times $a_{\alpha,i_\ell}\in R$ so as I is an ideal it belongs to I.

Case C: $\gamma = \xi$

$$\frac{a^q_{\alpha,\gamma}-a^q_{\alpha,\beta}a^q_{\beta,\gamma}}{a^q_{0,\gamma}}=$$

$$=\frac{a_{\alpha,i_\ell}p^{(d(\gamma)-d(i_\ell)}-a_{\alpha,\beta}a_{\beta,i_\ell}p^{d(\gamma)-d(i_\ell)}}{a_{0,i_\ell}p^{d(\gamma)-d(i_\ell)}}$$

$$=\frac{a_{\alpha,i_\ell}-a_{\alpha,\beta}a_{\beta,i_\ell}}{a_{0,i_\ell}}\in I$$

Case D: $\alpha,\beta,\gamma \neq \xi$.

Trivial.

So we have proved $q \in P$. Easily $p \leq q, \xi \in u^q$, so we finish.

9. Main Fact: Suppose $u_0 < u_1 < u_2$ (all finite subsets of ω_1, not empty for simplicity, $u < v$ means $\forall \alpha \in u \ \ \forall \beta \in v \ \ \alpha < \beta$) non empty, and

$$r^\ell \in P \quad \text{for} \quad \ell = 0,1,2,$$
$$u^{r^0} = u_0 \,,\; u^{r^1} = u_0 \cup u_1,\; u^{r^1} = u_0 \cup u_2$$
$$r^0 \leq r^1,\; r^0 \geq r^1$$

Let $\xi_\ell = Min\ u_\ell$ for $\ell = 1,2$, and $c_1, c_2 \in R$ are units of R.

Then we can find $r \in P$, $r^1 \leq r$, $r^2 \leq r$, such that

$$\frac{c_1}{a^r_{0,\xi_1}} - \frac{c_2}{a^r_{0,\xi_2} / a^r_{\xi_1,\xi_2}} \notin I$$

Let $\zeta_\ell = Max\ u_\ell$.

10. Subfact: We can find an element a of R such that

(α) $p^{d(\xi_2)-d(\zeta_1)}$ divides a but $p^{d(\xi_2)-d(\zeta_1)+1}$ does not divides a (in R).

(β) $$\frac{a^{r^2}_{\zeta_0,\xi_2} - a^{r_1}_{\zeta_0,\zeta_1} a}{a^{r^2}_{0,\xi_2}} \in I$$

(γ) $$\frac{c_1}{a^{r^1}_{0,\xi_1}} - \frac{c_2}{a^{r^2}_{0,\xi_2} / (a^{r^1}_{\xi_1,\zeta_1} a)} \notin I$$

(δ) $a \in M_{\xi_2+1}$

Proof: We shall choose some $t \in I \cap M_{\xi_2+1}$ and let

$$a = \frac{a^{r^2}_{\zeta_0,\xi_2} - a^{r^2}_{0,\xi_2} t}{a^{r_1}_{\zeta_0,\zeta_1}}$$

Now $t \in I$ guarantees (β) (just substitute and compute, and you shall get t) and $t \in M_{\xi_2+1}$ guarantee (δ) (as $\zeta_1, \zeta_0 \leq \xi_2$ and use (iv) from 7). Also (α) is immediate: $a^{r^2}_{0,\xi_2}$ is divisible by $p^{d(\xi_2)}$ hence $a^{r_2}_{0,\xi_2} t$ is divisible by

$p^{d(\xi_2)-d(\zeta_0)+1}$, but $a^{r^2}_{\zeta_0,\xi_2}$ is not; so $a^{r^2}_{\zeta_0,\xi_2} - a^{r^2}_{0,\xi_2}t$ is divisible by $p^{d(\xi_2)-d(\zeta_0)}$ but not by $p^{d(\xi_2)-d(\zeta_0)+1}$. Using (iii) of Definition 7 on $a^{r_1}_{\zeta_0,\zeta_1}$ we finish.

We are left with (γ), it means now

$$\frac{c_1}{a^{r^1}_{0,\xi_1}} - \frac{c_2}{a^{r^2}_{0,\xi_2}} \cdot a^{r^1}_{\xi_1,\zeta_1}\left(\frac{a^{r^2}_{\zeta_0,\xi_2} - a^{r^2}_{0,\xi_2}t}{a^{r_1}_{\zeta_0,\zeta_1}}\right) \notin I$$

this is equivalent to:

$$(*)\ \frac{c_1}{a^{r^1}_{0,\xi_1}} - \frac{c_2 a^{r^1}_{\xi_1,\zeta_1} a^{r^2}_{\zeta_0,\xi_2}}{a^{r^2}_{0,\xi_2} a^{r^1}_{\zeta_0,\zeta_1}} + \frac{c_2 a^{r^1}_{\xi_1,\zeta_1} a^{r^2}_{0,\xi_2} t}{a^{r^2}_{0,\xi_2} a^{r^1}_{\zeta_0,\zeta_1}} \notin I$$

If for $t = 0$ (*) holds, we finish, so we can assume

$s \overset{def}{=} \dfrac{c_1}{a_{0,\xi_1}} - \dfrac{c_1 a^{r^1}_{\xi_1,\zeta_1} a^{r_2}_{\zeta_0,\xi_2}}{a^{r_1}_{0,\xi_2} a^{r^1}_{\zeta_0,\zeta_1}} \in I$, so (*) is then equivalent to

$$(*)'\ \frac{c_2 a^{r^1}_{\xi_1,\zeta_1} a^{r^2}_{0,\xi_2}}{a^{r^2}_{0,\xi_2} a^{r^1}_{\zeta_0,\zeta_1}} t \notin I \qquad \text{i.e.,}\ \frac{c_2 a^{r^1}_{\xi_1,\zeta_1}}{a^{r^1}_{\zeta_0,\zeta_1}}\ t \notin I$$

By applying (iii) of Definition 7 to all $a_{i,j}$'s appearing in (*)' and remembering that for a unit c of R $cI = I$ and $c \in R$ is a unit iff p does not divide c for R, (*)' is equivalent to

$$(*)''\ t \in I \text{ but } \frac{p^{d(\zeta_1)-d(\xi_1)} p^{d(\xi_2)}}{p^{d(\xi_2)} p^{d(\zeta_1)-d(\zeta_0)}} t \notin I$$

which means $t \in I$ but $t / p^{d(\xi_1)-d(\zeta_0)} \notin I$, which is easily accomplished by choosing $t = p^b \in M_0$.

Now we define r:

$$u^r = u^{r^1} \cup u^{r^2} = u_0 \cup u_1 \cup u_2$$

$$a^{r}_{i,j} = \begin{cases} a^{r^1}_{i,j} & \text{if } i,j \in u^{r^1} \quad (a) \\ a^{r^2}_{i,j} & \text{if } i,j \in u^{r^2} \quad (b) \\ a_{i,\zeta_1}\, a\, a_{\xi_2,j} & \text{if } i = u_1,\ j \in u_2 \quad (c) \end{cases}$$

(remember $a_{\xi_2,\xi_2} = 1$)

Again condition (i) + (iii) + (iv) are easy. Let us try (ii).

So $\alpha < \beta < \gamma$.

Case A: $\alpha \in u_0, \beta \in u_1, \gamma \in u_2$.

$$\frac{a^{r}_{\alpha,\gamma} - a^{r}_{\alpha,\beta} a^{r}_{\beta,\gamma}}{a_{0,\gamma}} = \frac{a^{r^2}_{\alpha,\gamma} - a^{r^1}_{\alpha,\beta} a^{r^1}_{\beta,\zeta_1} a\, a^{r^2}_{\xi_2,\gamma}}{a_{0,\gamma}} \equiv mod\ I$$

$$\frac{a^{r^2}_{\alpha,\xi_2} a^{r^2}_{\xi_2,\gamma} - a^{r^1}_{\alpha,\beta} a^{r^1}_{\beta,\zeta_1} a\, a^{r^2}_{\xi_2,\gamma}}{a^{r^2}_{0,\gamma}} \equiv$$

$$a^{r^2}_{\xi_2,\gamma} \frac{a^{r^2}_{\alpha,\xi_2} - a^{r^1}_{\alpha,\beta} a^{r^1}_{\beta,\zeta_1} a}{a^{r^2}_{0,\gamma}} \equiv$$

$$\equiv \frac{a^{r^2}_{\xi_2,\gamma} a^{r^2}_{0,\xi_2}}{a^{r^2}_{0,\gamma}} \quad \frac{a^{r^2}_{\alpha,\xi_2} - a^{r^1}_{\alpha,\beta}\ a^{r^1}_{\beta,\zeta_1} a}{a^{r^2}_{0,\xi_2}}$$

Now $\dfrac{a^{r^2}_{\xi_2,\gamma} a^{r^2}_{0,\xi_2}}{a^{r^2}_{0,\gamma}}$ is a unit, so we can forget it

$$\frac{a^{r^2}_{\alpha,\xi_2} - a^{r^1}_{\alpha,\beta} a^{r^1}_{\beta,\zeta_1} a}{a^{r^2}_{0,\xi_2}} \equiv \frac{a^{r^0}_{\alpha,\zeta_0} a^{r^2}_{\zeta_0,\xi_2} - a^{r^1}_{\alpha,\beta} a^{r^1}_{\beta,\zeta_1} a}{a^{r^2}_{0,\xi_2}}$$

Now $(a^{r^1}_{\alpha,\beta} a^{r^1}_{\beta,\zeta_1}) \dfrac{a}{a^{r^2}_{0,\xi_2}} \equiv (a^{r^0}_{\alpha,\zeta_0} a^{r^1}_{\zeta_0,\zeta_1}) \dfrac{a}{a^{r^2}_{0,\xi_2}}\ mod\ I$ holds

[as $\dfrac{a^{r^1}_{\alpha,\beta} a^{r^1}_{\beta,\zeta_1} - a^{r^1}_{\alpha,\zeta_0} a^{r^1}_{\zeta_0,\zeta_1}}{a_{0,\zeta_1}} \equiv 0\ mod\ \dfrac{a_{0,\xi_2}}{a_{0,\zeta_1}, a} I$ holds, which hold by using twice (ii) of Definition 7, and computing power of p in the left side].

So

$$\frac{a^{r0}_{\alpha,\zeta_0}a^{r2}_{\zeta_0,\xi_2}-a^{r1}_{\alpha,\beta}a^{r1}_{\beta,\zeta_1}a}{a^{r}_{0,\xi_2}} = \frac{a^{r0}_{\alpha,\zeta_0}a^{r2}_{\zeta_0,\xi_2}-a^{r1}_{\alpha,\zeta_0}a^{r1}_{\zeta_0,\zeta_1}a}{a^{r}_{0,\xi_2}}$$

$$= a^{r0}_{\alpha,\zeta_0}\frac{a^{r2}_{\zeta_0,\xi_2}-a^{r1}_{\zeta_0,\zeta_1}a}{a_{0,\xi_2}} \in I$$

the "$\in$" holds by (β) above. So we finish Case A.

Case B: $\alpha,\beta \in u_1$, $\gamma \in u_2$

$$\frac{a^{r}_{\alpha,\gamma}-a^{r}_{\alpha,\beta}a^{r}_{\beta,\gamma}}{a^{r}_{0,\gamma}} = \frac{a^{r1}_{\alpha,\zeta_1}a\ a^{r2}_{\xi_2,\gamma}-a^{r1}_{\alpha,\beta}a^{r1}_{\beta,\zeta_1}a\ a^{r2}_{\xi_2,\gamma}}{a^{r2}_{0,\gamma}} =$$

$$a^{r2}_{\xi_2,\gamma}\ a\ [\frac{a^{r1}_{\alpha,\zeta_1}-a^{r1}_{\alpha,\beta}a^{r1}_{\beta,\zeta_1}}{a_{0,\gamma}}]$$

by computing power of p this term belongs to I iff

$$\frac{a^{r1}_{\alpha,\zeta_1}-a^{r1}_{\alpha,\beta}a^{r1}_{\beta,\zeta_1}}{a^{r1}_{0,\zeta_1}} \in I$$

which holds.

Case C: $\alpha \in u_1$ $\beta,\gamma \in u_2$.

$$\frac{a^{r}_{\alpha,\gamma}-a^{r}_{\alpha,\beta}a^{r}_{\beta,\gamma}}{a^{r}_{0,\gamma}} = \frac{a^{r1}_{\alpha,\zeta_1}a\ a^{r2}_{\xi_2,\gamma}-a^{r1}_{\alpha,\zeta_1}a\ a^{r2}_{\xi_2,\beta}a^{r1}_{\beta,\gamma}}{a_{0,\gamma}}$$

$$= a^{r1}_{\alpha,\zeta_1}a[\frac{a^{r2}_{\xi_2,\gamma}-a^{r2}_{\xi_2,\beta}a^{r2}_{\beta,\gamma}}{a_{0,\gamma}}] \in I$$

Case D: $\{\alpha,\beta,\gamma\} \subset u_0 \cup u_1$ or $\{\alpha,\beta,\gamma\} \subset u_0 \cup u_2$.

Trivial.

11. Conclusion: If $G \subset P$ is generic over V then in the new universal $V[G]$ over R there is a non standard uniserial R-module.

Proof : We can deal with I-representation. Let for $i < j < \omega_1$ $a_{i,j}$ be $a^{r}_{i,j}$ when $r \in G, \{i,j\} \subset u^r$, this is well defined as:

(A) $a_{i,j}$ has at most one value as G is directed.

(B) $a_{i,j}$ has at least one value [as by Fact B the sets $\{r \in P: i \in u^r\}$, $\{r \in P: j \in u^r\}$ are dense subsets of P, hence their intersection is. As G is generic, G is not disjoint to this intersection.] Now easily $\langle a_{i,j}: i < j < \omega_1 \rangle$ is an I- representation (over R). Why it represents a non standard uniserial module? Otherwise (letting $\underset{\sim}{a}_{i,j}$ be the name for $a_{i,j}$ defines above) there are P-name $\underset{\sim}{c}_i$ and $r \in P$ such that

(C) $r \Vdash_P$" $\underset{\sim}{c}_i$ is a unit of P, and $\dfrac{\underset{\sim}{c}_i}{\underset{\sim}{a}_{0,i}} - \dfrac{\underset{\sim}{c}_j}{\underset{\sim}{a}_{0,j} / \underset{\sim}{a}_{i,j}} \in I$ for every $i < j < \omega_1$".

As R consists of members of V, there are for $i < \omega_1$, $r_i \in P$, $r \leq r_i$ and $c_i^1 \in P$ $r_i \Vdash_P \underset{\sim}{c}_i = c_i^1$. Now using Fodor Lemma and Fact 9 we get a contradiction.

Originally we have then replaced forcing by $\diamondsuit_{\aleph_1}$, but it is better to have:

12. Theorem : (ZFC): There is a uniserial non standard module over some uniserial domain.

Proof : If we look carefully at the proof of this we can see that we have proved (and we shall prove):

(a) in $V[G]$, for every limit ordinal $\delta < \omega_1$ and unit $c \in R$, for every large enough $i < \delta$. $\dfrac{c}{a_{0,\delta} / a_{i,\delta}}$ is not I-equivalent to any member of R_δ.

13. Observation: If $\dfrac{c}{a_{0,\delta} / a_{i,\delta}} + I \notin \{x + I : x \in M_\delta\}$ and $i < j < \delta$ then $\dfrac{c}{a_{0,\delta} / a_{j,\delta}} + I \notin \{x + I : x \in M_\delta\}$.

Proof : Suppose $\dfrac{c}{a_{0,\delta} / a_{j,\delta}} = x + t$, $t \in I$, $x \in M_\delta$. Then

$$\frac{c}{a_{0,\delta} / a_{i,\delta}} = c \frac{a_{i,\delta}}{a_{0,\delta}} \equiv c \frac{a_{i,j} a_{j,\delta}}{a_{0,\delta}} = \quad mod\ I$$

$$a_{i,j} \left(\frac{c}{a_{0,\delta} / a_{j,\delta}} \right) = a_{i,j}(x + t) = a_{i,j} x + a_{i,j} t$$

Now $a_{i,j}x \in M_\delta$ (as $a_{i,j} \in M_{j+1} \subseteq M_\delta$, $x \in R_\delta$), and $a_{i,j}\, t \in I$ (as $a_{i,j} \in R$, $t \in I$).

Proof of (a): Suppose $r \in P$,

$r \Vdash_P$" $\delta < \omega_1$ is a limit ordinal, c a unit of R and δ, c, contradict (a)".

By Fact 8 w.l.o.g. $\delta \in u^r$. Now let $u_0 = u^r \cap \delta$, $u_2 = u^r - \delta$ $r^0 = r \restriction u_0$, $r^2 = r$, and find u_1, r_1 so that the assumptions of 9 holds ($u_2 \neq \phi$ as $\delta \in u_2$, $u_0 \neq \phi$ as $0 \in u_0$). Let $c_i = c$. We repeat the proof of 9 but in (γ) of 10 replace c_2 by c and $\notin I$ by $\notin I + M_{\xi_2}$, and drop $\dfrac{c_1}{a_{0,\xi_1}}$ i.e. we use

(γ)' $$\frac{c}{a^{r^2}_{0,\xi_2} / (a^{r^1}_{\xi_1,\zeta_1} a)} \notin I + M_\delta.$$

As we demand $a \in M_{\xi_2+1}$, and can assume M_{ξ_2+1} is quite large compared to M_{ξ_2} (though countable) there is no problem. [Let $e_i \in R$ ($i < \omega_1$) be distinct units, $e_i - e_j$ not divisible by p then for $i \neq j$: $\dfrac{c\, a^{r_1}_{\xi_1,\zeta_1}}{a^{r_1}_{\zeta_0,\zeta_1}} (p^b e_i) - \dfrac{c\, a^{r_1}_{\xi_1,\zeta_1}}{a^{r_1}_{\zeta_0,\zeta_1}} (p^b e_j) \notin I$; as M_δ is countable, for some i $\dfrac{c\, a^{r_1}_{\xi_1,\zeta_1}}{a^{r_1}_{\zeta_0,\zeta_1}} (p^b e_i) \notin I + M_\delta$. For being able to repeat the argument in M_{ξ_2+1} it is enough that in M_{ξ_2+1} there is a "finite" set to which every $x \in M_{\xi_2}$ "belongs", which is easy. Alternatively change the forcing as to allow us to choose $a \in M$, so that the forcing fail the $\aleph_1$-c.c. but is still proper see [Sh 2], Ch. III.] So we find r^1.

$$r \leq r^1 \in P, \quad \frac{c}{a^{r^1}_{0,\delta} / a^{r^1}_{i,\delta}} \notin I + M_\delta$$

Contradiction, so (a) holds. Note also

14. Observation: If M_α: ($\alpha < \omega_1$), $b, d(\alpha)(\alpha < \omega_1)$ are as in 6, $a_{i,j}$ satisfies (a) above, then $\langle a_{i,j} : i < j < \omega_1 \rangle$ is an I- representation of a non standard uniserial module.

Proof: Suppose $\langle c_i : i < \omega_1 \rangle$ exemplify the contrary. For a closed unbounded subset C of ω_1 for every $\delta \in C$

$$i < \delta \Longrightarrow c_i \in M_\delta$$

So $\frac{c_i}{a_{0,i}} \in M_\delta$ for $i < \delta$, hence $\frac{c_i}{a_{0,\delta}/a_{i,\delta}} + I \in \{x + I : x \in M_\delta\}$. Contradicting (a). So 14 holds.

Now the statement: there are $M_i (i < \omega_1)$ $b, d(\alpha)$ as in 6 and $a_{i,j}$ satisfying (a), can be expressed by a countable theory T in $L(aa)$ (note that we do not mind to replace ω_1 by a linear order K of power $\aleph_1$ such that $K = \bigcup_{i<\omega_1} K_i$, K_i increasing continuous each K_i countable $(\forall x \in K_i)(\forall y \in K_{i+1} - K_i)$ $(x < y)$ and K_i has a least upper bound). $L(aa)$ was introduced in Shelah [Sh 1], and thoroughly investigated in Barwise Kaufman and Makkai [BKM]. By the completeness theorem for $L(aa)$ (see [BKM]) the answer to "does T has a model" is absolute. As it has a model in $V[G]$ it has one in V.

15 Remark: We can replace $\aleph_1$ by any uncountable regular uncountable κ. Let $H(\aleph_2)$ be the family of sets of hereditary power $< \aleph_2$, and $\mathfrak{B}$ be $(H(\aleph_1), \in)$ expanded by (individual constants for) M, R, Q, I, $\langle M_i : i < \omega_1 \rangle$, $\langle d(i) : i < \omega_1 \rangle$, b and $\langle a_{i,j} : i < j < \omega_1 \rangle$. Now we can define by induction on $\alpha < \kappa^2$ $\mathfrak{B}_\alpha$ such that:

1) $\mathfrak{B}_\alpha$ is a model of power κ elementarily equivalent to $\mathfrak{B}$.

2) $\mathfrak{B}_\alpha$ $(\alpha < \kappa^2)$ is a continuous elementarily chain.

3) For every α there is $y_\alpha \in \mathfrak{B}_{\alpha+1}$ such that:

(a) $\mathfrak{B}_{\alpha+1} \models$ "y_α is a countable set".

(b) for every $x \in \mathfrak{B}_\alpha$, $\mathfrak{B}_{\alpha+1} \models$ "$x \in y_\alpha$".

(c) if α has cofinality κ and $\alpha < \beta \leq \kappa^2$ then $\mathfrak{B}_\beta \models$ "$x \in y_\alpha$", implies $x \in \mathfrak{B}_\alpha$.

Let $\mathbb{B}^* = \bigcup_{\alpha<\kappa^2} \mathbb{B}_\alpha$, $z_\alpha \in \mathbb{B}_{\alpha+1}$ be such that $\mathbb{B}_{\alpha+1} \models$ "z_α is $\sup(y_\alpha \cap \omega_1)$".

There is no problem to do this (e.g. use saturated models, possible as we can construct the models say in L), see Mekler and Shelah [M Sh]. Now use M,R,I $\langle a_{i,j} : \mathbb{B}^* \models$ "$i < j < \omega_1^{\mathbb{B}^*}$"$\rangle$ or equivalently $\langle a_\beta : \alpha < \beta < \kappa \rangle$ with $M_\alpha = M^{\mathbb{B}_\alpha}$. Note that we could replace κ^2 by $\kappa\mu$ if $cf\ \mu \geq \aleph_0$.

References

[CK] C. C. Chang and H. J. Keisler, Model Theory, North Holland Publ. Co. 1973.

[F] L. Fuchs, Infinite Abelian groups I,II. Academic Press, N.Y. 1970,1973.

[J] T. Jech. Set Theory. Academic Press.

[BKM]
K. J. Barwise, M. Kaufman, M. Makkai. Stationary Logic, Annals of Math. Logic, 13 (1978), 171-229.

[MSh]
A. Mekler and S. Shelah, Stationary Logic and its Friends, Proc. of the Jerusalem Model Theory Year 1984, *Notre Dame J. of Formal Logic*, to appear.

[Sh1]
S. Shelah, Generalized quantifiers and compact logic, Trans. of A.M.S. 204 (1975), 342-364.

[Sh2]
S. Shelah, Proper Forcing, Springer Lecture Notes, 940 (1982).

Remarks on the numbers of ideals of Boolean algebra and open sets of a topology

Abstract: We prove that the cardinals μ which may be the number of ideals of an infinite Boolean algebras are restricted: $\mu = \mu^{\aleph_0}$ and if $\kappa \leq \mu$ is strong limit then $\mu^{<\kappa} = \mu$. Similar results hold for the number of open sets of a compact space (we need $w(x)^{<s(x)} = 2^{<s(x)}$). We also prove that if $\mu \geq \beth_2$ is the number of open subsets of a Hausdorff space $X, \mu < \mu^{\aleph_0}$ *then* $0^\#$ exists, (in fact, the consequences of the covering lemma on cardinal arithmetic are violated). We also prove that if the spread μ of a Hausdorff space X satisfies $\mu > \beth_2(cf\ \mu)$ that the sup is obtained. For regular spaces $\mu > 2^{cf\ \mu}$ is enough.

Similarly for $3(X)$ and $h(X)$.

§0 Introduction.

We deal with some problems on Boolean algebras and their parallel on topological spaces. The problems are: what can be the number of ideals [open sets], and is the spread (and related cardinals) necessarily obtained (remember it is defined as a supremum.) Compare with the well known result that the cellularity (= first κ for which the κ- chain condition holds) is regular. We shall use freely the duality between a Boolean algebra and its space of ultrafilters. Recall

0.1 Definition : For a topological space X:

1) $s(X) = \sup\{|A| : A$ is a discrete subspace$\} + \aleph_0$ (note that A is a discrete subspace if $A = \{y_i : i < \alpha\}$ and for some open subsets $u_i (i < \alpha)$, $y_i \in u_j \Leftrightarrow i = j$).

2) $z(X) = \sup\{|A|: A = \{y_i : i < \alpha\}$, and for some open u_i $(i < \alpha)$, $i = j \Rightarrow y_i \in u_j \Rightarrow i \geq j\} + \aleph_0$.

3) $h(X) = \sup\{|A|: A = \{y_i : i < \alpha\}$ for some open $u_i (i < \alpha)$, $i = j \Rightarrow y_i \in u_j \Rightarrow i \leq j\} + \aleph_0$.

4) $\hat{s}(X), \hat{z}(X), \hat{h}(X)$ are defined similarly with $|A|^+$ instead $|A|$.

5) For a Boolean algebra B ,$\varphi(B)$ is $\varphi(X)$ where X is the space of ultrafilters of B.

On the problem of the attainment of the supremum when the cofinality is $\aleph_0$ see Hajnal and Juhasz [HJ 1], Juhasz [J1], Shelah [Sh 3] 1.1 (p. 252) and then Kunen and Roitman [KR].

On a counterexample for higher cofinalities see Roitman [R] and lately Juhaz and Shelah [JSh]. On the number of open subsets see Hajnal and Juhasz [HJ2] and Juhasz [J2]; the author observed in fall 1977 (see [Sh 6] for the main consequence) that by having a specific cardinal exponentiation function we can get from counterexample to the attainment of the spread when the cofinality is κ, a Hausdorff space X with $o(X)^\kappa > o(X)$ (this extra demand on the set theory has caused no trouble). This connected our two problems. The author had withdrawn another announcement of [Sh 6]: $o(X) = o(X)^{\aleph_0}$ for X a Lindelof space.

This work is written in the order it was conceived.

§1 The numbers of ideals of a Boolean Algebra

1.1 Theorem: Let B be an infinite Boolean Algebra, $Id(B)$ the set of ideals of B, $id(B)$ its power. Then $id(B) = id(B)^{\aleph_0}$.

Proof: Suppose not, $\lambda = Min\{\kappa : \kappa^{\aleph_0} \geq id(B)\}$, so $cf\ \lambda = \aleph_0$, $\lambda \leq id(B) < \lambda^{\aleph_0}$. Now $\lambda > 2^{\aleph_0}$ as $id(B) \geq 2^{\aleph_0}$, so $\lambda = \sum_n \lambda_n$, $\lambda_n < \lambda_{n+1} < \lambda$, $\lambda_n = \lambda_n^{\aleph_0}$. We define by induction on $n, a_n \in B$, $a_n \cap a_\ell = 0$ for $\ell < n$, $id(B \restriction a_n) \geq \lambda_n$, $id(B \restriction (1 - \bigcup_{\ell < n} a_\ell) \geq \lambda$. We should fail for some n, so w.l.o.g. for no $a \in B$, $id(B \upharpoonright a) \geq \lambda_n, id(B \upharpoonright (1-a)) \geq \lambda$. W.l.o.g. $n = 0$, so $\mathcal{J} = \{a \in B : id(B \upharpoonright a) \leq \lambda_0\}$ is a maximal ideal. Now $|B| < \lambda$ (otherwise $|\mathcal{J}| \geq \lambda$, each countable subset of $\mathcal{J}$ generates an ideal, there are $\geq \lambda^{\aleph_0} > id(B)$ such countable subsets, and each ideal of B of this form has power $\leq \lambda_0$ hence has at most $\lambda_0^{\aleph_0} = \lambda_0 < \lambda$ countable subsets. Contradiction). So W.l.o.g. $|B| < \lambda_0$. Now $Id^0(B) = \{I \in Id(B) : I \not\subset \mathcal{J}\} \subseteq \bigcup_{a \in \mathcal{J}} \{I \in Id(B) : 1-a \in I\}$ has power $\leq \sum_{a \in \mathcal{J}} id(B \upharpoonright a) \leq |B| + \lambda_0$

So $Id^0(B)$ has power $\leq \lambda_0$. Also $Id^1(B) = \{I \in Id(B) : I \subseteq \mathcal{J}$ but for some $a \in B-I$ there is no $b < a, b \in \mathcal{J}-I\}$ has power $\leq \lambda_0$ (for each such a, I $I \cap (B \upharpoonright a) = \mathcal{J} \cap (B \upharpoonright a)$, and for $I \cap (B \upharpoonright (1-a))$ we have $\leq id(B \upharpoonright (1-a)) \leq \lambda_0$ possibilities. So $Id^2(B) \stackrel{def}{=} Id(B) - Id^0(B) - Id^1(B)$ has cardinality $id(B)$. For each $I \in Id^2(B)$ choose by induction on $i, a_i \in \mathcal{J}-I$ such that $a_i \cap a_j \in I$ for $j < \alpha$, and let $\bar{a}^I = \langle a_i : i < \alpha \rangle$ be the resulting maximal sequence. Note that:

$\hat{s}(B) = Min\{\mu$: there are no $a_i \in B (i < \mu)$, a_i not in the ideal generated by $\{a_j : j \neq i\}\}$,

and let

$\kappa = Min\{\mu$: there are no μ pairwise disjoint non zero elements of $B\}$.

Clearly $\kappa \leq \hat{s}(B)$, and for $\mu < \hat{s}(B)$, $2^\mu \leq id(B)$ so $2^{<\hat{s}(B)} \leq id(B)$. It is known that $cf\ \hat{s}(B) > \aleph_0$, so $(2^{<\hat{s}(B)})^{\aleph_0} = 2^{<\hat{s}(B)}$ hence $2^{<\hat{s}(B)} < \lambda$ and w.l.o.g. $2^{<\hat{s}(B)} < \lambda_0$. Now easily if $\bar{a}^I = \bar{a}^J = \langle a_i : i < \alpha \rangle$, $I \cap (B \upharpoonright a_i) = J \cap (B \upharpoonright a_i)$ for $i < \alpha$, then $I = J$ (if e.g. $I \not\subset J$, choose $x \in I-J$, then x is a good candidate as a_α for J). We shall prove for each $\bar{a}$ that $\{I : I \in Id^2(B), \bar{a}^I = \bar{a}\} \leq \lambda^*$ for fixed $\lambda^* < \lambda$. By the argument above this is equal to $|\{\langle I \upharpoonright (B \upharpoonright a_i) : i \rangle : I \in Id^{(2)}, \bar{a}^I = \bar{a}\}|$ which is $\leq |\{\langle J_i : i < \alpha \rangle : J_i \subseteq B \upharpoonright a_i$ an ideal, $a_j \cap a_i \in J_i$ for $j \neq i\}|$. Let $\mu_i = |\{J : J \subseteq B \upharpoonright a_i$ an ideal (so $a_i \not\in J$) and for $j \neq i$, $a_j \cap a_i \in J\}|$. So the number is $\leq \prod_{i<\alpha} \mu_i$. Easily $\prod_{i<\alpha} \mu_i \leq id(B)$, and $\mu_i < \lambda_0$ but by cardinal arithmetic $(\prod_{i<\alpha} \mu_i)^{\aleph_0} = \prod_{i<\alpha} \mu_i$ (or $\prod_{i<\alpha} \mu_i \leq \lambda_0$) [you can see in 2.11], so $\prod_{i<\alpha} \mu_i < \lambda$. By more cardinal arithmetic (see 2.11) there is a bound λ^* as required.

So necessarily $|\{\bar{a}^I : I \in Id^2(B)\}| \geq \lambda$. Now each $\bar{a}^I$ has length $< \hat{s}(B)$ so $\lambda \leq |B|^{<\hat{s}(B)}$, and as $cf\ \hat{s}(B) > \aleph_0$, $cf\ \lambda = \aleph_0$ clearly there is $\mu < \hat{s}(B)$, $|B|^{\mu} \geq \lambda$. Let $\vartheta = Max\{\kappa, \mu^+\}$. So ϑ is regular $\vartheta \leq \hat{s}(B)$, B satisfies the ϑ-c.c. and $|B|^{<\vartheta} \geq \lambda$ and $2^{<\vartheta} \leq 2^{<\hat{s}(B)} \leq \lambda_0$. So $|B|^{<\vartheta} > 2^{<\vartheta}$. Let $\chi = Min\{\chi : \chi^{<\vartheta} \geq |B|\}$, then $\chi > 2^{<\vartheta}$, $\chi^{<\vartheta} = |B|^{<\vartheta} \geq \lambda$ and $(\forall \mu < \chi)\mu^{<\vartheta} < \chi$. By [Sh 1] 4.4 B has a subset of power χ no one in the ideal generated by the others. So $\chi < \hat{s}(B)$ so $2^{\chi} \leq id(B)$, but $2^{\chi} \geq \chi^{<\vartheta} \geq \lambda$ so $2^{\chi} \geq \lambda^{\aleph_0} > id(B)$ contradiction.

§2 On the number of open sets

2.1 Notation: 1) X is an infinite Hausdorff space, τ the family of open subsets of X, any $Y \subseteq X$ is equipped with the induced topology i.e $\tau^Y = \tau(Y) = \{U \cap Y : U \in X\}$. $\underset{\sim}{B}$ will denote a base of X.

2) Let $o(X) = |\tau|$, (and for $Y \subseteq X$, $o(Y) = |\{U \cap Y : U \in \tau\}|$.

3) $\hat{s}(X) = \{|A|^+ :$ A a discrete subspace of X, (i.e. (A, τ^A) is a discrete space $\}$.

4) $\underset{\sim}{B}$ is a strong base of X if for every $y \in X$, there is v, such that $y \in v \in \tau$, and $[y \in u \subseteq v, u \in \tau \Longrightarrow v \in \underset{\sim}{B}]$.

We shall assume in 2.3, 2.4:

2.2 Hypothesis: We assume λ is an infinite cardinal, $cf\ \lambda = \aleph_0$,

$(\forall \mu)(\aleph_0 \leq \mu < \lambda \rightarrow \mu^{\aleph_0} < \lambda)$ and at least one of the following holds:

(I) $\chi \leq o(X) < \lambda^{\aleph_0}$, $\chi = \lambda$

(II) $\chi \leq o(X) < \lambda^{\aleph_0}$, $\chi = \lambda^+$,

(III) $\chi \leq o(X) < \lambda^{\aleph_0}$, $\chi = \lambda$, and X is strongly Hausdorff (which means: for every infinite $A \subseteq X$ there are $p_n \in A$ and pairwise disjoint

$u_n \in \tau, p_n \in U_n$).

2.2A Explanation: We shall want to get a contradiction or at least get information on how an example like that looks like.

So we allow to replace X by X^* if $\chi \leq o(X^*) < \lambda^{\aleph_0}$ is still satisfied; but we shall use this for open X^* only.

2.3 Claim: Assume 2.2.

1) $\lambda > 2^{\aleph_0}$ and we can find λ_n, $\lambda_n = \lambda_n^{\aleph_0} < \lambda_{n+1} < \lambda$, $\lambda = \sum_{n<\omega} \lambda_n$.

2) W.l.o.g. there are no disjoint open sets $u, v (\in \tau)$ such that $o(u) \geq \chi$, $o(v) \geq \lambda$. (and even no open disjoint u, v such that $o(u) \geq \chi$, $o(v) \geq \lambda_0$) [and even no open u, v such that $o(u - v) \geq \chi$, $o(v - u) \geq \lambda_0$, but then we pass to a non-open subspace.]

3) W.l.o.g. every point y has an open number u_y (so $y \in u_y \in \tau$) such that $o(u_y) < \lambda$.

4) $o(X) \geq 2^{<\mathfrak{s}(X)}$; hence if $cf\ \mathfrak{s}(X) > \aleph_0$ then $\lambda > 2^{<\mathfrak{s}(X)}$ and w.l.o.g. $\lambda_0 > 2^{<\mathfrak{s}(X)}$.

5) if $|X| \geq \beth_2$ then $|X| < \lambda$ (and w.l.o.g. $|X| < \lambda_0$; similarly $|X| \geq 2^{2^\kappa} \Longrightarrow |X|^\kappa \leq o(X)$).

Proof : 1) If every $y \in X$ is isolated, X has $2^{|X|}$ open subsets, but X is infinite so $o(X) \geq 2^{\aleph_0}$. If $y^* \in X$ is not isolated we define by induction on $n, u_n, v_n \in \tau$ and y_n such that : $y^* \in u_n$, $y_n \in v_n$, $u_n \cap v_n = \phi$, and $v_{n+1} \subseteq u_n$, $u_{n+1} \subseteq u_n$. (choose $y_0 \in X, y_0 \neq y^*$ then choose v_0, u_0; if u_n is defined, choose $y_n \in u_n - \{y^*\}$ and then u_{n+1}, v_{n+1} using "X is Hausdorff".) So $\{u_n : n < \omega\}$ are open non empty pairwise disjoint hence $o(X) \geq |\{\bigcup_{n \in S} u_n : S \subseteq w\}| = 2^{\aleph_0}$.

In any case $o(X) \geq 2^{\aleph_0}$ but $\lambda \leq o(X)^{\aleph_0} > o(X)$ hence $o(X) > 2^{\aleph_0}$, but $o(X) < \lambda^{\aleph_0}$, so $2^{\aleph_0} < \lambda < \lambda^{\aleph_0}$.

so $(\forall\mu<\lambda)(\mu+\aleph_0<\lambda)$ hence $(\forall\mu<\lambda)\mu^{\aleph_0}<\lambda$ hence we can find λ_n as required.

2) Let $u_0=X$, define by induction on n, $1\le n<\omega$, u_n, v_n such that

(i) $u_n\in\tau, v_n\in\tau$; usually we demand they are disjoint.

(ii) $v_{n+1}\subseteq u_n, u_{n+1}\subseteq u_n$

(iii) $o(v_n - u_n - \bigcup_{\ell<n} v_\ell)\ge\lambda_n$

(iv) $o(u_n - v_n - \bigcup_{\ell<n} v_\ell)\ge\chi$

If we succeed, then v_ℓ are open, $v_n - \bigcup_{\ell\ne n} v_\ell \subseteq (v_n - u_n - \bigcup_{\ell<n} v_\ell)$ hence $o(v_n - \bigcup_{\ell\ne n} v_\ell)\ge\lambda_n$, so by Fact 2.3A below $o(X)\ge\prod_{n<\omega}\lambda_n=\lambda^{\aleph_0}>o(X)$ contradiction.

2.3A Fact: i) If $v_n\in\tau$ then $o(X)\ge\prod_{n<\omega} o(v_n-\bigcup_{\ell\ne n} v_\ell)$.

ii) If $v_i\in\tau(i<\alpha)$ then $o(X)\ge\prod_{i<\alpha} o(v_i-\bigcup_{j\ne i} v_j)$.

Proof : i) Let $\mu_n=o(v_n-\bigcup_{\ell\ne n} v_\ell)$ and let $v_i^n\in\tau(i<\mu_n)$ be such that $\{v_i^n\cap(v_n-\bigcup_{\ell\ne n} v_\ell): i<\mu_n\}$ are pairwise distinct. If $\rho\in\prod_{n<\omega}\mu_n$ let $v_\rho=\bigcup_{n<\omega}(v^n_{\rho(n)}\cap v_\ell)$. Clearly $v_\rho\in\tau$ and if $\rho\ne\nu\in\prod_{n<\omega}\mu_n$, then for some k, $\rho(k)\ne\nu(k)$, hence $v_\rho\cap(v_k-\bigcup_{\ell\ne k} v_\ell)=v^k_{\rho(k)}\cap(v_k-\bigcup_{\ell\ne k} v_\ell)$ $\ne v^k_{\nu(k)}\cap(v_k-\bigcup_{\ell\ne k} v_\ell)=v_\nu\cap(v_k-\bigcup_{\ell\ne k} v_\ell)$ hence $v_\rho\ne v_\nu$. So $o(X)=|\tau|\ge\prod_{n<\omega}\mu_n$ as required.

(ii) Similarly.

We return to the proof of 2.3.

(3) Let $Y=\bigcup\{v\in\tau: o(v)<\lambda\}$. If in $X-Y$ there is a non isolated point y^*, then the proof is as in 1) (with $y_n\in X-Y$). If every point of $X-Y$ is

isolated then: $o(X-Y) = 2^{|X-Y|}$. As $o(X)$ is infinite easily $o(X) = o(Y)$ or $o(X) = o(X-Y)$. The latter is impossible as $(2^{|X-Y|})^{\aleph_0} = 2^{|X-Y|}$ because it is infinite.

(4) If $y_i \in v_i \in \tau$, $y_i \notin v_j$, for $i < \alpha$, $i \neq j < \alpha$, then $\{\bigcup_{i \in S} v_i : S \subseteq \alpha\}$ is a family of $2^{|\alpha|}$ distinct open subsets of X, so $o(X) \geq 2^{|\alpha|}$. By the definition of $\hat{s}(X)$, $o(X) \geq 2^{<\hat{s}(X)}$. The second phrase is by cardinal arithmetic.

(5) Assume $|X| \geq \lambda$. For any countable $A \subseteq X$, the closure of A is a closed subset of X of power $\leq \beth_2$. The number of A is $|X|^{\aleph_0} > |X| \geq \beth_2$, and for any such A; $|\{B : B \subseteq X$ countable, the closure of B is the closure of $A\}|$ has power $\leq \beth_2$, so we finish.

2.4 Claim: Assume 2.2. 1) W.l.o.g.

(*) for every $y \in X$ for some $v_y \in \tau, y \in v_y$, $o(v_y) \leq \lambda_0$,

except possibly when: Hypothesis (I), holds (and not II or III) and $(\exists n)\ \lambda_n^{\beth_1} > \lambda$ (hence $\lambda_n^{\beth_1} \geq o(X)$).

2) $|X| < \lambda$ so w.l.o.g. $|X| < \lambda_0$ so X has strong base of power $< \lambda_0$.

Remark: So if $\lambda \geq \beth_2$, then (w.l.o.g.) $\lambda_0 > \beth_2$, $\lambda_0 = \lambda_0^{\aleph_0}$, $\lambda_0^{\beth_1} > \lambda_0^{+\omega}$, so $o^{\#}$ exist so the conclusion of [J2, 4.7, p. 97] holds.

Proof : 1) Let $Y_n = \bigcup\{v \in \tau : o(v) \leq \lambda_n\}$. By 2.3(3) $X = \bigcup Y_n$. If for some n $o(Y_n) \geq \chi$ we can replace X by Y_n. So assume $o(Y_n) < \chi$. Hence $Y_n \neq X$. If X is strongly Hausdorff choose $y_n \in X - Y_n$. As $X = \bigcup_{n<\omega} Y_n$, $Y_n \subseteq Y_{n+1}$, $\{y_n : n < \omega\}$ is infinite. By the definition of strongly Hausdorff applied to $\{y_n : n < \omega\}$ there are distinct $n(k) < \omega$, and $u_k \in \tau, y_{n(k)} \in u_k$, $\langle u_k : k < \omega \rangle$ pairwise disjoint. So $o(u_k) \geq \lambda_{n(k)}$, (as $y_{n(k)} \in u_k$) and $o(X) \geq \prod_k o(u_k) \geq \prod_{k<\omega} \lambda_{n(k)} = \lambda^{\aleph_0} > o(X)$ contr.

So we have dealt with Hypothesis III.

Next assume Hypothesis II, so

$$\sum_{n<\omega} o(Y_n) \le \sum_{n<\omega} \lambda = \lambda < \chi$$

So the following fact is sufficient.

2.4A Fact: If $Z_n \subset X$ is open (for $n < \omega$) $\sum_n o(Z_n) + \aleph_0 < o(\bigcup_n Z_n)$ then

$$o(\bigcup_n Z_n)^{\aleph_0} = o(\bigcup_n Z_n) = (\aleph_0 + \sum o(Z_n))^{\aleph_0}$$

Proof : Let $\vartheta = \aleph_0 + \sum o(Z_n)$.

We define a tree T with ω levels. Now T_n, the n'th level, will be $\{(u,n): u \subset \bigcup_{\ell<n} Z_\ell : u \in \tau\}$; the order will be: $(u,n) \le (v,m)$ iff $n \le m, u = v \cap (\bigcup_{\ell<n} Z_\ell)$. As $\bigcup_{\ell<n} Z_\ell$ is open (as well as $\bigcup_{\ell<\omega} Z_\ell$). $|T_\ell| = o(\bigcup_{\ell<n} Z_\ell) \le \sum_{\ell<n} o(Z_\ell) \le \vartheta$, and $o(\bigcup_{\ell<\omega} Z_\ell)$ is the number of ω-branches of T, so it is $> \vartheta \ge \sum |T_n|$. But in that case it is well known that the number of ω-branches of T is $\vartheta^{\aleph_0}$, as required. So we have proved 2.4A.

We are left with case I, and assume that for each n, $\lambda_n^{\beth_1} < \lambda$; let $C = \{(\vartheta^{\beth_1})^+ : \vartheta < \lambda\}$, $\varphi(Y) = o(Y)$, and apply 2.5A below, we get a contradiction.

Proof of 2.4(2): Let for $y \in X$ $v_y \in \tau, y \in v_y, o(v_y) < \lambda_0$. Suppose $|X| > \lambda_0^+$. Clearly $o(v_\tau) \ge |v_\tau|$ so $|v_\tau| < \lambda_0$. By Hajnal free subset theorem (see [J1]) there is $Y \subseteq X, |Y| = |X|$ such that $(\forall y \ne z \in Y)(y \notin v_z)$. So $|Y| < \hat{s}(X)$, so $o(X) \ge 2^{|Y|} = 2^{|X|}$ contradiction. So $|X| \le \lambda_0^+$, then $\{u \cap v_y : u \in \tau, y \in X\}$ is a strong basis of X of power $< \lambda_0^+ + \lambda_0$. Renaming we finish.

We can abstract from the proof of Kunen and Roitman [KJ] (or see [J2]), the following theorem. See 4.4(2), or 3.2A(2) for a simpler proof of 2.5(1)) even weakening (iv) to: $X \ne \bigcup_{\varphi(u)<\lambda_n} u$ for each n.

2.5 Lemma : 1) Suppose $cf\ \lambda = \aleph_0 < \lambda$, $\lambda = \sum_{n<\omega} \lambda_n, \lambda_n < \lambda$, X a topological space, and φ is a function from subsets of X to cardinals, satisfying:

(i) $\varphi(A) \leq \varphi(A \cup B) \leq \varphi(A) + \varphi(B)$.

(ii) $\varphi(X) \geq \lambda$

(iii) for an unbounded family C of cardinals $< \lambda$:

$\oplus$ if $\vartheta \in C$, $A_i \subseteq X (i < 2^{\aleph_0})$ and $\varphi(A_i) < \vartheta$ then $\varphi(\bigcup A_i) < \vartheta$.

(iv) $\varphi(\bigcup_{\varphi(u)<\lambda_n} u) < \lambda$.

Then there are open sets $u_n \subseteq X$ such that $\varphi(u_n - \bigcup_{\substack{\ell \neq n \\ \ell < \omega}} u_\ell) \geq \lambda_n$ for $n < \omega$.

2) We can replace $\oplus$ by $\oplus_a + \oplus_b$ where:

$\oplus_a$ if $\langle A_\eta : \eta \in {}^\omega 2 \rangle$ is a partition of X, and $\bigcup \{A_\eta : \eta(k) = 0\}$ is open for each $k < \omega, \vartheta \in C$ and $B \subseteq X, \varphi(B) \geq \vartheta$, *then* for some no-where dense set $K \subseteq {}^\omega 2$, $\varphi(B \cap \bigcup_{\eta \in K} A_\eta) \geq \vartheta$

and

$\oplus_b$ if $A_n \subseteq X$, $\vartheta \in C, \varphi(A_n) < \vartheta$ then $\varphi(\bigcup_{n<\omega} A_n) < \vartheta$.

3) If X is strongly Hausdorff, (i), (ii) suffice.

Proof : 1) We shall use (i) freely.

Case I: $\varphi(Y) < \lambda$ where $Y = \bigcup \{v : v \in \tau, \varphi(v) < \lambda\}$.

So $\varphi(X-Y) = \lambda$: if $X-Y$ has a non isolated point y^*, then we can define distinct $y_n \in X - Y_n - \{y^*\}$ and pairwise disjoint $u_n, y_n \in u_n \in \tau_n$, y^* not in the closure of u_n. So as $y_n \in Y$, $\varphi(u_n) \geq \lambda > \lambda_n$ and $u_n = u_n - \bigcup_{\ell < n} u_\ell$. So the u_n's are as required. So $X-Y$ is a discrete space hence $o(X-Y) = 2^{|X-Y|}$, but $o(X-Y) = \lambda$, contradiction.

So we can assume $\varphi(Y) \geq \lambda$, so w.l.o.g. $X = Y$ i.e.,

(*) for each $y \in X$ for some $v, y \in v \in \tau, \varphi(v) < \lambda$.

Case II: For every open $Y \subseteq X, \varphi(Y) \geq \lambda$ and $\vartheta < \lambda$, and $\langle v_y : y \in Y \rangle$ satisfying $y \in v_y \in \tau$ *there are* $p \in Y$, open $u, p \in u \subseteq v_p$ and open $Z \subset Y$, $\varphi(Z) \geq \lambda$ and v_z^0, a neighborhood of z, for $z \in Z$ such that: for every $z_m \in Z$, $\varphi(u - \bigcup_{n<\omega} v_{z_n}^0) \geq \vartheta$.

We define by induction on $n, 1 \leq n < \omega$, $p_n, u_n, y_n, \vartheta_n$ and $\langle v_y^n : y \in Y_n \rangle$ such that

(1) $Y_n \subset X, \varphi(Y_n) \geq \lambda$, $Y_{n+1} \subseteq Y_n$, Y is open.

(2) for $y \in Y_n$, v_y^n is an open neighborhood of y, $v_y^{n+1} \subseteq v_y^n$.

(3) $\vartheta_n \geq \lambda_n$ $\vartheta_{n+1} > \vartheta_n$.

(4) $p_n \in u_n \in \tau$, $\vartheta_n \leq \varphi(u_n) < \vartheta_{n+1}$, $u_n \subseteq v_{p_n}^n$

(5) for every $z_\ell \in Y_n (\ell < \omega)$ $\varphi(u_n - \bigcup_{\ell<\omega} v_{z_\ell}^n) \geq \vartheta_n$.

For $n = 0$ we stipulate $Y_0 = X$, $v_y^n (y \in Y_0)$ an open number of y with $\varphi(v_y^n)$ minimal and $\vartheta_0 = \beth_1 + \lambda_0$.

Suppose $Y_n, \langle v_y^n : y \in Y_n \rangle$ as defined. Choose $\vartheta_{n+1} < \lambda$ such that $\vartheta_{n+1} > \lambda_n$, $\vartheta_{n+1} > \vartheta_\ell$, $\varphi(u_\ell)$ when $0 < \ell < n+1$. Next apply the hypothesis of the case to Y_n, and ϑ_n and $\langle v_y^n : y \in Y_n \rangle$, so there are $p = p_{n+1} \in Y_n$, $u = u_{n+1}$, $Y = Y_{n+1}$, and $\langle v_z^{n,0} : z \in Y_{n+1} \rangle$ such that:

$Y_{n+1} \subseteq Y_n$, $\varphi(Y_{n+1}) \geq \lambda$, $p_{n+1} \in u_{n+1} \subseteq v_{p_{n+1}}^n$, $z \in v_z^{n,0} \in \tau$, and for $z_\ell \in Y_{n+1} (\ell < \omega)$, $\varphi(u_{n+1} - \bigcup_{\ell<\omega} v_{z_\ell}^{n,0}) \geq \vartheta_{n+1}$.

We let $v_z^{n+1} = v_z^{n,0} \cap v_z^n$.

Easily everything is o.k. Now in the end, as $u_\ell \subseteq v_{p_\ell}^n$ for $\ell < n$, and by (5) for n

$$\varphi(u_n - \bigcup_{\ell>n} u_\ell) \geq \varphi(u_n - \bigcup_{\ell>n} v_{p_\ell}^n) \geq \vartheta_n$$

As for $\ell < n, \varphi(u_\ell) < \vartheta_n$ clearly

$\varphi(u_n - \bigcup_{\ell \neq n} u_\ell) \geq \vartheta_n$, as required.

Case III: Not Cases I,II.

So (*) holds, and there are open $Y \subseteq X, \varphi(Y) \geq \lambda$, $\vartheta < \lambda$ and $\langle v_y : y \in Y \rangle, y \in v_y \in \tau$, witnessing the failure of Case II. W.l.o.g. $X = Y$, $\vartheta \in C$, $y \in u \subseteq v_y (u \in \tau) \Longrightarrow \varphi(u) \geq \varphi(v_y)$. If $\varphi(v_p) \geq \vartheta$, by (iii) $\oplus$:

(**) if $p \in u \in \tau$, $u \subseteq v_p$, *then* $\varphi(\{z \in Y$: for some $v \in T, z \in v, \varphi(v \cap u) < \vartheta\}) < \lambda$ [if this fails $p, u, Z = \{v: \varphi(v \cap u) < \lambda\}$ and $\langle v_z^0 : z \in Z \rangle$ where $z \in v_z^0$, $\varphi(v_z^0 \cap u) < \lambda$, exemplify Z, ϑ do not witness the failure the assumption of Case II].

Define by induction on n, $p_n^\ell \in Y, u_n^\ell \in \tau$, for $\ell = 1,2$ and ϑ_n such that:

(1) $p_n^\ell \in u_n^\ell, u_n^1 \cap u_n^2 = \phi$, $u_n^\ell \subseteq v_{p_n^\ell}$.

(2) $\vartheta < \vartheta_n \in C, \vartheta_n \geq \lambda_n, \vartheta_{n+1} > \vartheta_n, \vartheta$.

(3) $\vartheta_n \leq \varphi(u_n^1), \varphi(u_n^2) < \vartheta_{n+1}$.

(4) for every open neighborhood v of p_n^k, if $m < n$:

$$\varphi(u_m^\ell \cap v) \geq \vartheta_\ell$$

For $n = 0$ choose $\vartheta_0 \in C, \vartheta_0 > \lambda_0 + \vartheta$ then choose $p^1 \neq p^2$ in Y such that $\varphi(v_{p_0^\ell}) \geq \vartheta_0$ (possible by assumption (iv)) and then choose $u_0^\ell \in \tau, p_0^\ell \in u_0^\ell \subseteq v_{p_0^\ell}, u_0^1 \cap u_0^2 = \phi$. For $n+1$, choose first $\vartheta_{n+1} \in C, \vartheta_{n+1}$ larger than $\vartheta_n, \lambda_{n+1}, \varphi(u_0^\ell), \ldots, \varphi(u_n^\ell)$ for $\ell = 1,2$ (remember (*)). Now we should choose p_{n+1}^1, p_{n+1}^2, such that $\varphi(v_{p_{n+1}^\ell}) \geq \vartheta_{n+1}$, and for each $\ell \leq n$, (4) holds. Each demand excludes a set in $\{A : \varphi(A) < \lambda\}$, (note that $\bigcup\{v_p : o(v_p) < \vartheta\}$ satisfies this by assumption (iv)) so there are distinct p_{n+1}^1, p_{n+1}^2 as required, and now choose disjoint u_{n+1}^1, u_{n+1}^2, such that $p_{n+1}^\ell \in u_{n+1}^\ell \subseteq v_{p_{n+1}^\ell}$.

Define for $\eta \in {}^\omega 2$, $A_\eta = \bigcap_{\eta(n)=0} u_n^1 \cap \bigcap_{\eta(n)=1} (X - u_\eta^1)$.

We define by induction on $n < \omega$, η_n, k_n, m_n, such that

(a) $\eta_n \in {}^{\omega}2$

(b) $n \leq k_n < m_n < k_{n+1} < m_{n+1}$

(c) for $\ell < n, \eta_\ell(k_n) = \eta_\ell(m_n)$

(d) $\varphi(u^1_{k_n} \cap u^2_{m_n} \cap A_{\eta_n}) \geq \vartheta_{k_n}$.

For $n=0$ let $k_n = 0, m_n = 1$, now $\varphi(u^1_{k_n} \cap u^2_{m_n}) \geq \vartheta_{k_n}$ by condition (4) above. Then there is η_0 is required in (4) by $\oplus$. For $n > 0$ we first can find k_n m_n as required in (b),(c) and then η_n as above.

Now let $u_n = u^1_{k_n} \cap u^2_{m_n}$. So now by (c) $u_\ell \cap A_{\eta_m} = \phi$ for $\ell > m$, so $u_n - \bigcup_{\ell > n} u^1_{k_n} \cap u^2_{m_n} \supseteq A_{\eta_n}$ hence

$\varphi(u_n - \bigcup_{\ell > n} u_\ell) \geq \vartheta_n$; as $\varphi(u_\ell) < \vartheta_n$ for $\ell < n$, $\varphi(u_n - \bigcup_{\ell \neq n} u_\ell) \geq \vartheta_n$ so we finish.

2) Similar proof - instead $u^1_{k_n} \cap u^2_{m_n}$ we use finite such intersection and strengthen (4) accordingly (and $\{\eta_n\}$ is replaced by a no where dense set.)

Remark: If in 2.5(1) we weaken (iv) to $\varphi(X - \bigcup\{u: \varphi(u) < \lambda_n\}) \geq \lambda$, by changing φ so to satisfy (iv).

2.6 Lemma: 1) Suppose X is a Hausdorff space, $\underset{\sim}{B}$ a basis for X and $o(v) \leq \lambda_0$ for $v \in \underset{\sim}{B}$. Suppose further that $2^{<\hat{s}(X)} < o(X)$, $\lambda_0 < o(X)$ and for no $\kappa < \hat{s}(X)$, $(\lambda_0)^\kappa = o(X)$. Then $|\underset{\sim}{B}|^{<\hat{s}(X)} \geq o(X)$.

2) Under Hypothesis 2.2, if (*) of 2.4 holds, cf $\hat{s}(X) > \aleph_0$, and $\underset{\sim}{B}$ is a basis for X *then* $|\underset{\sim}{B}|^{<\hat{s}(X)} \geq o(X)$ (so for some χ and ϑ : $\chi^{<\vartheta} > o(X) \geq (\chi + 2^{<\vartheta})^{+\omega}$).

3) If X is a Hausdorff space $\beth_2 \leq o(X) < o(X)^{\aleph_0}$ *then* for some χ, ϑ : $(\chi + 2^{<\vartheta})^{+\omega} \leq o(X) < \chi^{<\vartheta}$.

2.6A Remark: The conclusion in 2.6(3) implies $0^\#$ exists by the covering lemma, and similarly much more.

* * *

We first prove some facts, where $\underset{\sim}{B}$ is a base of a Hausdorff space X and $o(v) \leq \lambda_0$ for $v \in B_0$.

2.7 Definition : 1) We say $\bar{v} = \langle v_i : i < \alpha \rangle$ is good for u (where $u, v_i \in \tau$) if

(i) $v_i - u \neq \phi$

(ii) $v_i \in \underset{\sim}{B}$ (hence $v_i \in \tau$)

(iii) for $i \neq j < \alpha, v_i \cap v_j \subseteq u$.

2) We say $\bar{v}$ is maximally good for u if $\bar{v}$ is good for u but for no $v \in \underset{\sim}{B}$ is $\bar{v}\char`^\langle v \rangle$ good for u.

2.8 Observation: 1) For every $u \in \tau$ there $\bar{v}$ maximally good for it.

2) If $\langle v_i : i < \alpha \rangle$ is good for u, then $\alpha < \hat{s}(X)$.

Proof : 1) Immediate.

2) By (i) of Definition 2.7(1)) there is $y_i \in v_i - u$. Now $y_i \in v_i - u \in \tau$, and $i \neq j \Longrightarrow y_i \notin v_j$ (as then $y_i \in v_j \cap v_i - u$.)

2.9 Fact: Let $G = \{\langle v_i : i < \alpha \rangle : v_i \in \underset{\sim}{B},\ v_i \not\subseteq \bigcup \{v_j : j < \alpha, j \neq i\}\ \}$.

1) If $\bar{v}$ is good for some u then $\bar{v} \in G$.

2) For each $\bar{v} = \langle v_i : i < \alpha \rangle \in G$ the following two sets has the same power:

$P_{\bar{v}} = \{u : \bar{v}$ is maximaly good for $u\}$.

$Q_{\bar{v}} = \{\langle J_i : i < \alpha \rangle : \bigcup_{j \neq i} (v_i \cap v_j) \subseteq J_i \subset v_i$, (so $J_i \neq v_i$) and J_i is open $\}$.

Proof : 1) Immediate.

2) We define H, a function with domain $P_{\bar{v}} : H(u) = \langle v_i \cap u : i < \alpha \rangle$

Clearly $H(u) \in Q_{\bar v}$. Now H is one to one: if $H(u_1) = H(u_2)$ but $u_1 \neq u_2$ then w.l.o.g. $u_1 \not\subseteq u_2$, choose $y \in u_1 - u_2$, then choose $v \in \underset{\sim}{B}$, $y \in v \subseteq u_1$. So y witness $v \not\subseteq u_2$; and for $i < \alpha, v \cap v_i \subseteq u_1$ (as $v \subseteq u_1$) but $v \cap v_i \subseteq v_i$, $u_1 \cap v_i = u_2 \cap v_i$ so also $v \cap v_i \subseteq u_2$. We conclude that v contradicts the maximality of $\bar v$ (as good for u_2). So H is one to one.

Now for any $\langle J_i : i < \alpha \rangle \in Q_{\bar v}$, $J \overset{def}{=} \bigcup_{i<\alpha} J_i$ is an open set and easily $v_i \cap v_j \subseteq J_i \subseteq J$ for $i \neq j, J \cap v_i = J_i$ and $v_i \not\subseteq J$. So $\bar v$ is good for J. Let $u^* = \bigcup \{u : \bar v$ is good for $u, u \cap v_i = J_i\}$. Easily $\bar v$ is maximally good for u^* and $H(u^*) = \langle J_i : i < \alpha \rangle$.

2.10 Fact: For $\bar v \in G$, for some $\mu^i_{\bar v}$, $|Q_{\bar v}| = \prod_{i<\ell(\bar v)} \mu^i_{\bar v}$, and $\mu^i_v \leq \lambda_0$.

Proof: Let $\mu^i_{\bar v} = |\{J \in \tau : \bigcup_{j \neq i} (v_j \cap v_i) \subseteq J \subseteq v_i\}|$. Clearly $\mu^i_{\bar v} \leq o(v_i)$, but $v_i \in \underset{\sim}{B}$ so $\mu^i_{\bar v} \leq \lambda_0$. By the definition of $Q_{\bar v}$, $|Q_{\bar v}| = \prod_i \mu^i_{\bar v}$.

2.11 Observation: By cardinal arithmetic:

1) If $\mu = \prod_{i<\alpha} \mu_i$ then $\mu = \prod_{\ell=1}^{n} (\chi_\ell)^{\kappa(\ell)}$, where $n < \omega$,, $\chi_\ell \leq \sup\{\mu_i : i < \alpha\}$, $\sum_{\ell=1}^{n} \kappa(\ell) = |\alpha|$. Also $(\forall i < \alpha)[\chi_\ell > \mu_i > \chi_{\ell+1} \rightarrow \kappa(\ell) \geq cf\ \chi_\ell]$ and $\kappa(\ell) = |\{i : \mu_i \leq \chi_\ell$, and $(\forall m)[\chi_m < \chi_\ell \Longrightarrow \chi_m < \mu]\}|$

2) In 1) if $\mu > \mu_i$ for each i, μ infinite then $\mu^{\aleph_0} = \mu$; in fact $\mu = \chi^{\kappa}$ for some $\chi \leq \sum_{i<\alpha} \mu_i, \aleph_0 \leq \kappa \leq |\alpha|$.

3) Suppose $\chi \geq 2^{<s}$, then $\{\prod_{i<\alpha} \mu_i : \alpha < s, \mu_i \leq \chi$ for each $i < \alpha$ but $\prod_{i<\alpha} \mu_i > \chi\}$ is finite.

4) If $\chi \geq 2^{<s} (s \geq \aleph_0)$ then for some $\vartheta < s$: $\chi^{\vartheta} = \chi^{<s}$.

Remark: In particular, in 3) $\{\lambda^\sigma : 2^\sigma < \lambda\}$ is finite. When I visited Budapest (in April 84) I learned that this already appeared explicitly in the Hungarian book of Hajnal on Set Theory.

Proof: 1) We define χ_ℓ by induction on ℓ, $\chi_1 > \chi_2 > \cdots$. Let $\chi_1 = \sup_{i<\alpha} \mu_i$.

If χ_ℓ is defined and is a successor cardinal, let $\chi_\ell = (\chi_{\ell+1})^+$. If χ_ℓ is defined, $\chi_\ell = 1$ let $n = \ell$.
If $\chi_\ell > 0$ is a limit cardinal, let $\chi_{\ell+1}$ be the minimal $\chi < \chi_\ell, \chi \geq 1$ such that for every χ^*, if $\chi < \chi^* < \chi_\ell$ then

(*) $|\{i<\alpha: \chi<\mu_i \leq \chi_\ell\}| = |\{i < \alpha: \chi^* < \mu_i \leq \chi_\ell\}|$.

Now χ exists as $\langle |\{i < \alpha: \chi < \mu_i \leq \chi_i\}|: \chi < \chi_\ell \rangle$ is a decreasing sequence.

Clearly for some ℓ $\chi_\ell = 1$, so $\ell = n$. Now $\prod_{i<\alpha} \mu_i = \prod_{\ell=1}^{n} \Pi\{\mu_i: \chi_{\ell+1} < \mu_i \leq \chi_\ell\}$ (remember $\mu_i \neq 0$, and we can ignore $\mu_i = 1$).

By (*), $\Pi\{\mu_i: \chi_{\ell+1} < \mu_i \leq \chi_\ell\} = \chi_\ell^{\kappa(\ell)}$, where $\kappa(\ell) = |\{i: \chi_{\ell+1} < \mu_i \leq \chi_\ell\}|$.

The last phrase is easy too.

2) Easy.

3) By 2) if $\prod_{i<\alpha} \mu_i \geq \chi$, $\mu_i \leq \chi, \alpha < s$ then for some $\vartheta \leq \chi, \kappa \leq |\alpha|$, $\vartheta^\kappa = \prod_{i<\alpha} \mu_i$, so $\vartheta^\kappa \leq \chi^\kappa \leq (\prod_{i<\alpha} \mu_i)^\kappa = (\vartheta^\kappa)^\kappa = \vartheta^\kappa$, hence $\prod_{i<\alpha} \mu_i = \chi^\kappa$ where $\kappa \leq |\alpha|$. So it suffices to prove $\{\chi^\kappa: \kappa < s\}$ is finite. Suppose $\chi^{\kappa(n)}$ are distinct for $n < \omega$, where for each n $\kappa(n) < s$. W.l.o.g. $\kappa(n) < \kappa(n+1)$. Let $\chi_n = Min\{\mu: \mu^{\kappa(n)} \geq \chi\}$, so easily:

(i) for each n, $\chi_n \geq \chi_{n+1}$.

(ii) $\chi_n^{\kappa(n)} = \chi^{\kappa(n)}$.

By (i) w.l.o.g. $\langle \chi_n: n < \omega \rangle$ is constant; as we have assumed $\{\chi^{\kappa(n)}: n < \omega\}$ are distinct, by (ii) $\{\chi_n^{\kappa(n)}: n < \omega\}$ are distinct.

But $(\forall \sigma < \chi_n) \sigma^{\kappa(n)} < \chi_n$, hence $(\forall \sigma < \chi_0)(\forall n < \omega)$ $(\sigma^{\kappa(n)} < \chi_0)$, and clearly $cf(\chi_n) \leq \kappa(n)$, so $\chi_n^{\kappa(n)} = \chi_n^{cf(\chi_n)} = \chi_0^{cf(\chi_0)}$. But $\chi_n^{\kappa(n)} = \chi^{\kappa(n)}$ are distinct, contradiction.

4) Follows from 3).

Proof of 2.6(1): Suppose $|\underset{\sim}{B}|^{<\mathfrak{s}(X)} < o(X)$. By 2.8(1), 2.9(1), $\tau = \bigcup\{P_{\bar{v}} : \bar{v} \in G\}$, hence $o(X) = |\tau| \leq \sum_{\bar{v}\in G} |P_{\bar{v}}|$. By 2.9(2) $o(X) \leq \sum_{\bar{v}\in G} |Q_{\bar{v}}|$, and by 2.8(2), $|G| \leq |\underset{\sim}{B}|^{<\mathfrak{s}(X)}$. So to get a contradiction it suffices to prove that $\sup\{|Q_{\bar{v}}| : \bar{v} \in G\} < o(X)$. By 2.10, $|Q_{\bar{v}}| = \prod_{i<\ell(\bar{v})} \mu^i_{\bar{v}}$ where $\mu^i_{\bar{v}} \leq \lambda_0$ (as $v_i \in \underset{\sim}{B}$ by an assumption) and $\ell(\bar{v}) < \mathfrak{s}(X)$ (by 2.8(2).) W.l.o.g. $(\forall i)(\mu^i_{\bar{v}} > 1)$.

Now by 2.11, for some natural number of $n(\bar{v})$ and cardinals $\mu_{\bar{v},\ell} \leq \lambda_0$ and $\kappa(\bar{v},\ell) \leq \ell(\bar{v}) < \mathfrak{s}(X)$, for $(\ell < n)$:

$$|Q_{\bar{v}}| = \prod_{\ell=1}^{n(\bar{v})} (\mu_{\bar{v},\ell})^{\kappa(\bar{v},\ell)}$$

so if $Q_{\bar{v}}$ is infinite, $Q_{\bar{v}} = \underset{\ell = 1,n}{Max}\,(\mu^{\kappa(\bar{v},\ell)}_{\bar{v},\ell})$.

But $(\mu_{\bar{v},\ell})^{\kappa(\bar{v},\ell)} \geq \lambda_0$ implies $(\mu_{\bar{v},\ell})^{\kappa(\bar{v},\ell)} = \lambda_0^{\kappa(\bar{v},\ell)}$ so $|Q_{\bar{v}}| \geq \lambda_0$, implies that for some $\kappa(\bar{v}) \leq \ell(\bar{v})$, $|Q_{\bar{v}}| = \lambda_0^{\kappa(\bar{v})}$. But $\ell(\bar{v}) < \mathfrak{s}(X)$.

So we have proved: if $|Q_{\bar{v}}| \geq \lambda_0$ then $|Q_{\bar{v}}| = \lambda_0^{\kappa(\bar{v})}$ where $\kappa(\bar{v}) < \mathfrak{s}(X)$. But we have assumed $(\lambda_0)^{\kappa(\bar{v})} \neq o(X)$ and we know $|Q_{\bar{v}}| = |P_{\bar{v}}| \leq o(X)$, so necessarily $|Q_{\bar{v}}| \geq \lambda_0 \Longrightarrow |Q_{\bar{v}}| < o(X)$. But $\lambda_0 < o(X)$ so $|Q_{\bar{v}}| < o(X)$. The same argument gives, $\sup\{|Q_{\bar{v}}| : \bar{v} \in G\} \leq \sup[\{\lambda_0\} \cup \{\lambda_0^{\kappa} : \kappa < \mathfrak{s}(X),\ \lambda_0^{\kappa} < o(X)\}]$ but by 2.11 this is $\lambda_0^{\kappa(0)}$, for some $\kappa(0) < \mathfrak{s}(X)$ hence this supremum is $< o(X)$, which we have shown is enough for 2.6(2).

Proof of 2.6(2): We use freely 2.3, 2.4. So (w.l.o.g.) $|X| < \lambda_0, X$ has a strong base $\underset{\sim}{B}$, $|\underset{\sim}{B}| < \lambda_0$, $o(v) < \lambda_0$ for $v \in \underset{\sim}{B}$, and $2^{<\mathfrak{s}(X)} \leq o(X)$. As $cf\ \mathfrak{s}(X) > \aleph_0$, $(2^{<\mathfrak{s}(X)})^{\aleph_0} = 2^{<\mathfrak{s}(X)}$ hence $2^{<\mathfrak{s}(X)} < \lambda$ hence w.l.o.g. $2^{<\mathfrak{s}(X)} < \lambda_0$. So all the assumptions of 2.6(1) hold, hence $|\underset{\sim}{B}|^{<\mathfrak{s}(X)} \geq o(X)$ as required. The last phrase holds if we choose $\chi = |\underset{\sim}{B}|$, $\vartheta = \mathfrak{s}(X)$. Note $(\chi + 2^{<\vartheta})^{+\omega} = (|\underset{\sim}{B}| + 2^{<\mathfrak{s}(X)})^{+\omega} \leq \lambda_0^{+\omega} \leq o(X)$ (as $\lambda_0^{\aleph_0} = \lambda_0$ also $(\lambda_0^{+n})^{\aleph_0} = \lambda_0^{+n}$) and $o(X) \leq |\underset{\sim}{B}|^{<\mathfrak{s}(X)}$.

Proof of 2.6(3): Now X satisfy I from Hypothesis 2.2. If (*) of 2.4 holds we finish by 2.6(2). Otherwise by 2.4 for some n $\lambda_n^{\beth_1} > \lambda$, hence $\lambda_n^{\beth_1} > o(X)$, hence $\lambda_n > \beth_2$ (as $o(X) \geq \beth_2$). Remember $\lambda_n^{\aleph_0} = \lambda_n$. Let $\chi = \lambda_n$, $\vartheta = \beth_1^+$, they satisfy the required conclusion.

A corollary of [Sh 1] 4.4 is

2.12 Observation: If B is in infinite Boolean algebra then $|B|^{<\mathfrak{s}(X)} \leq 2^{<\mathfrak{s}(B)}$.

Proof: Let κ be the cellularity of B, so κ is regular, $> \aleph_0, \kappa \leq \hat{\mathfrak{s}}(B)$, and let $\lambda = Min\{\lambda : \lambda^{<\kappa} \geq |B|\}$; as κ is regular $(\lambda^{<\kappa})^{<\kappa} = \lambda^{<\kappa}$. If $\lambda > 2^{<\kappa}$ then $(\forall \mu < \lambda)\ \mu^{\kappa} < \lambda$, and by [Sh 1] 4.4, $\lambda < \hat{\mathfrak{s}}(B)$ so $2^{\lambda} \geq \lambda^{<\kappa} \geq |B|$, hence

$$|B|^{<\mathfrak{s}(X)} \leq (|B|^{\lambda})^{<\mathfrak{s}(B)} \leq ((2^{\lambda})^{\lambda})^{<\mathfrak{s}(B)} = 2^{<\mathfrak{s}(B)}$$

If $\lambda \leq 2^{<\kappa}$, then $|B| \leq 2^{<\kappa}$; remember $\kappa \leq \hat{\mathfrak{s}}(B)$ now if $\kappa = \hat{\mathfrak{s}}(B)$, then $|B|^{<\mathfrak{s}(B)} = 2^{<\mathfrak{s}(B)}$ as $\kappa = \hat{\mathfrak{s}}(B)$ is regular; and if $\kappa < \hat{\mathfrak{s}}(B)$, $|B|^{<\mathfrak{s}(B)} \leq (2^{\kappa})^{<\mathfrak{s}(B)} = 2^{<\mathfrak{s}(B)}$.

2.13 Conclusion: 1) If B is a Boolean algebra, $id(B)^{\aleph_0} = id(B)$.

2) If X is locally compact Hausdorff space then $o(X)^{\aleph_0} = o(X)$.

Proof : 1) Let X be the space of ultrafilters of B, considering B as a basis. So $id(B) = o(X)$. By 2.6(2) (note X is strongly Hausdorff) $o(X) < o(X)^{\aleph_0}$ implies $|B|^{<\mathfrak{s}(B)} > o(X)$, but $o(X) \geq 2^{<\mathfrak{s}(X)} = 2^{<\mathfrak{s}(B)}$ contradicting 2.12.

2) We need the parallel of 2.12, which is proved by translating the proof of [Sh] 4.2, 4.4 to topology, which is done in 2.14 below.

2.14 Lemma : Let X be a locally compact Hausdorff compact space with cellularity κ.

1) If $(\forall \vartheta < \mu)(\vartheta^{<\kappa} < \mu)$ (so $2^{<\kappa} < \mu$) and every basis of X consisting of regular open sets has power $\geq \mu$ then $\hat{\mathfrak{s}}(X) \geq \mu$.

2) If μ is regular, X has a subspace Y whose topology is a refinement of λ_2.

Note: Theorem 2.14 was prooved by F. Argyros and A. Tsarpaleas independently of [Sh].

Proof: The proof are like [Sh] 4,2 , 4.4; we concentrate on 2.14(2), so μ is regular (anyhow we shall use only this part). Here $\bar{u}$ denote the closure of u. Really it is a repetition of [Sh 1] with one change; use of compactness for a family of sets $\overline{u_\alpha^2 - u_\beta^1}$.

2) Let $\underset{\sim}{B}$ be such a base. W.l.o.g. $(\forall u \in \underset{\sim}{B})[\bar{u}$ is compact] (otherwise replace $\underset{\sim}{B}$ by $\{u \in \underset{\sim}{B} : \bar{u}$ is compact$\}$). Let $\chi = (2^{2^{|X|}})^+$,$H(\chi)$ the family of sets of hereditary power $< \chi$. We define by induction on $i < \mu$, $N_i \prec (H(\chi), \in)$, such that $\underset{\sim}{B} \in N_i$, $\|N_i\| < \mu$, $\langle N_j : j \leq i \rangle \in N_{i+1}, N_j \prec N_i$ for $j < i$ and every sequence of $< \kappa$ member of N_i belong to N_i when i is a successor ordinal. (hence when $cf\ i \geq \kappa$). For each $i < \mu$, let $\underset{\sim i}{B} = \{u \in N_i : u$ regular open, $\bar{u}$ compact$\}$.

As $|\underset{\sim i}{B}| < \mu$ by a hypothesis it is not a basis of X, hence there are in N_{i+1} $p_i \in X, u_i^0 \in B$, $p_i \in u_i^0$, such that for no $v \in \underset{\sim i}{B}$, $p_i \in v \subseteq u_i^0$. We can find for $\ell < 3$, $u_i^\ell \in \underset{\sim i+1}{B}$, such that $p_i \in u_i^\ell$, $\overline{u_i^{\ell+1}} \subseteq u_i^\ell$. Restrict ourselves to case $cf\ i \geq \kappa$.

Let $J_i^2[I_i^\ell]$ be a maximal family of pairwise disjoint open sets $u \in \underset{\sim i}{B}, u \subseteq u_i^\ell$ $[u \cap u_i^\ell = \phi]$. So J_i^ℓ, I_i^ℓ, are subsets of N_i of power $< \kappa$ (as κ is the cellularity of X) hence $J_i^\ell, I_i^\ell \in N_i$. Let $A_i^\ell = X - \overline{\bigcup I_\ell^i}$, so A_i^ℓ is open, belongs to N_i (non empty) $u_i^\ell \subseteq A_i^\ell$ (as $X - u_i^\ell$ is closed, $\bigcup I_i^\ell \subseteq X - u_i^\ell$) and there is no open (non empty) $v \subseteq A_i^\ell - u_i^\ell$, $v \in N_i$. Also $A_i^\ell \in N_i$. Let $B_i^\ell = \bigcup J_i^\ell$, so $B_i^\ell \subseteq u_i^\ell, B_i^\ell$ is open belongs to N_i and there is no open $v \subseteq u_i^\ell - B_i^\ell$, $v \in N_i$. By Fodour's Lemma there are A^ℓ, B^ℓ such that $S = \{i : i < \mu, cf\ i < \kappa.$ $A_i^\ell = A^\ell, B_i^\ell = B^\ell$ for $\ell = 0,1,2\}$ is stationary. It is enough to prove

(*) for disjoint finite $w(1), w(2) \subseteq S$,

$$\bigcap_{\alpha \in w(1)} u_\alpha^2 \not\subseteq \bigcup_{\beta \in w(2)} u_\beta^1$$

As then for any non empty $w \subseteq S$, $\{\overline{u_\alpha^2 - u_\beta^1} : \alpha \in w, \beta \in S - w\}$ is a family of closed sets, the intersection of any finitely many is non empty and $\overline{u_\alpha^2}$ is compact for $\alpha \in w$, so there is q_w in the intersection. So $\{((q_{\{\alpha\}}, u_\alpha^2) : \alpha \in s\}$ exemplify $\hat{s}(X) > \mu$, and if $S = \{\xi_i : i < \mu\}$, let $H: {}^{\lambda}2 \to X$ be define by $H(\eta) = q_{\{\xi_1 + i:\, \eta(i) = 0\} \cup \{\xi_0\}}$, then $Y = \{H(\eta) : \eta \in {}^{\lambda}2\}$ is as required.

Let $RO(X)$ be the Boolean Algebra of regular open subsets of X. So in $RO(X)$ we identify $u \in \tau(X)$ with $int(\overline{u})$ (and so the operations are changed accordingly). So $RO(X)$ is complete, in $RO(X)$ $\bigcup\limits_{i<\alpha} A_i = int(\overline{\bigcup\limits_{i<\alpha} A_i})$ i.c. the interior of $\bigcup\limits_{i<\alpha} A_i$; $\bigcap\limits_{i<\alpha} A_i = int(\bigcap\limits_{i<\alpha} A_i)$. So $RO(X)$ satisfies the κ-chain condition and $RO(X) \cap N_i$ is a complete subalgebra.

So in $RO(X)$, A_i^ℓ is minimal such that $A_i^\ell \in N_i$, $u_i^\ell \subset A_i^\ell$ and B_i^ℓ is maximal such that $B_i^\ell \subset u_i^\ell$, $B_i^\ell \in N_i$.

Proof of (*): We shall work in $RO(X)$ and prove by induction on $n = |w(1)| + |w(2)|$;.

$(*)^+$ $RO(X) \models \bigcap\limits_{\alpha\in w(1)} u_\alpha^2 \not\subset \bigcup\limits_{\alpha\in w(2)} u_\alpha^1 \cup B^1$.

When n is zero the statement is obvious. Let $\alpha = Max((w(1) \cup w(2))$ and $Max(w(1) \cup w(2) - \{\alpha\}) \le \beta < \alpha$.

By the induction hypothesis $v = \bigcap\limits_{\substack{\gamma\in w(1)\\ \gamma\neq\alpha}} u_\gamma^2 - \bigcup\limits_{\substack{\gamma\in w(2)\\ \gamma\neq\alpha}} \overline{u_\gamma^1} \cup B^1$ is $\neq 0$ (in $RO(X)$). Clearly $v \in B_{\sim\alpha}$, and if (*) fails then $v \subset B_\alpha^1 = B^1$ (if $\alpha \in w(2)$) or $\phi = v \cap A_\alpha^2 = v \cap A^2$ (if $\alpha \in w(1)$). In both cases a contradiction follows.

2.18 Conclusion: For locally compact X, $w(X)^{<\hat{s}(X)} \le 2^{<\hat{s}(X)}$.

Proof : Suppose $w(X)^{<\hat{s}(X)} > 2^{<\hat{s}(X)}$, let $\mu = Min\{\mu^{<\kappa} \ge w(X)\}$, where κ is the cellularity of X. Clearly $\kappa \le \hat{s}(X), \mu \le w(X)$, and $(\forall\chi < \mu)(\chi^{<\kappa} < \mu)$ (as $(\chi^{<\kappa})^{<\kappa} = \chi^{<\kappa}$, κ being regular). So by 2.17 $\mu < \hat{s}(X)$ but $|w(X)|^{<\hat{s}(X)} \le (\mu^{<\kappa})^{<\hat{s}(X)} \le \mu^{<\hat{s}(X)} \le (2^{<\mu})^{<\hat{s}(X)} \le 2^{<\hat{s}(X)}$ contradiction. [if we want to use only the part of 2.17 actually prove, note that

a) $\mu = \hat{s}(X)$ is singular (by the previous argument).

b) if μ is not strong limit, let $\vartheta < \mu \le 2^{\vartheta}$, so

$\models w(X)^{<\mathfrak{s}(X)} \leq (\mu^{<\kappa})^{<\mathfrak{s}(X)} = \mu^{<\mathfrak{s}(X)} \leq (2^{\vartheta})^{<\mathfrak{s}(X)} = 2^{<\mathfrak{s}(X)}$ contradiction;

c) if μ is strong limit singular $\mathfrak{s}(X) = \mu$ is impossible (see [J2] or 3.4.]).

§3 Nice cardinal functions on a topological space.

3.1 Definition : 1) φ is nice for X if φ is a function from subsets of the topological space X to cardinals satisfying

(i) $\varphi(A) \leq \varphi(A \cup B) \leq \varphi(A) + \varphi(B) + \aleph_0$ (i.e. monotonicity and subadditivity)

2) We call φ (χ,μ)-complete provided that if $A_i \subseteq X$, $\varphi(A_i) < \chi$ for $i < \mu$ then $\varphi(\bigcup_{i<\mu} A_i) < \chi$.

Let $C(\varphi,\mu) = \{\chi : \varphi$ is (χ,μ)-complete$\}$.

3) We call φ $(<\lambda,\mu)$-complete, if for arbitrarily large $\chi < \lambda, \varphi$ is (χ,μ)-complete.

4) Let Ch_φ be the function from X to cardinals

$$Ch_\varphi(y) = Min\{\varphi(u): y \in u \in \tau(X)\}$$

3.1A Remark: 1) We can replace $i < \mu$ by $i < \alpha < \mu$ and made suitable changes later.

2) In our applications we can restrict the domain of φ to the Boolean Algebra generated by $\tau(X)$ and even more, e.g. in 3.2 to simple combinations of the $u_{i,\xi,\zeta}$.

3) We can change the definition of $(< \lambda,\mu)$-complete to

(*) if $A_i \subseteq X(i < \mu)$, $\underset{i<\mu}{Sup}\, \varphi(A_i) < \lambda$ then $\varphi(\bigcup_{i<\mu} A_i) < \lambda$

without changing our subsequence use. [we then will use: if $\varphi(A_\alpha) < \chi_i$ for $\alpha < \mu$ then $\varphi(\bigcup_{\alpha<\mu} A_\alpha) < \chi_{i+1}$].

3.2 Lemma : Suppose λ is singular of cofinality ϑ, $\lambda = \sum_{i<\vartheta} \chi_i$, $\chi_i < \lambda$,

$\vartheta < \lambda$ and $\mu = \beth_5(\vartheta)^+$ or even $\mu = \beth_2(\beth_2(\vartheta)^+)^+$. If

(i) φ is nice for X.

(iii) $X_{\chi_i} = \{y \in X: Ch_\varphi(y) \geq \chi_i\}$ has cardinality $\geq \mu$ for $i < \vartheta$.

(iii) φ is $(<\lambda,\mu)$-complete.

Then there are open $u_i \subset X(i < \vartheta)$ such that

$$\varphi(u_i - \bigcup_{j \neq i} u_j) \geq \chi_i$$

Remark: If $|\{y \in X: Ch_\varphi(y) \geq \chi_i\}| < \mu$ it essentially follows from (χ_i,μ)-completeness that $\varphi(X_{\chi_i}) \geq \lambda$ where $X_\chi = \bigcup\{v \in \tau(X): \varphi(v) < \chi\}$. Otherwise $\varphi(X - X_{\chi_i}) \geq \lambda$ by additivity, but $\varphi(X - X_{\chi_i}) \leq \Pi\{\varphi(\{y\}): y \in X - X_{\chi_i}\}$ so by (χ_i,μ)-completeness for some $y \in X$, $\varphi(\{y\}) \geq \chi_i$ which is impossible for the instances which interest us.

Proof: W.l.o.g. $\chi_i \in C$, $C \stackrel{def}{=} C(\varphi,\mu) \cap \lambda$. Choose dinstinct $y_{i,\xi} \in X - X_{\chi_i}$ for $i < \vartheta$, $\xi < \mu$.

Let $u^a_{i,\xi,\zeta}(i < \vartheta, \xi \neq \zeta < \mu)$ be open sets such that $y_{i,\xi} \in u_{i,\xi,\zeta}$ and $u_{i,\xi,\zeta} \cap u_{i,\zeta,\xi} = \phi$. Now

(*) for every $i < \vartheta, \xi(0) < \xi(1) < \xi(2) < \mu$, there is $x = x_{i,\xi(0),\xi(1),\xi(2)}$ such that :

(a) $x \in u_{i,\xi(1),\xi(0)} \cap u_{i,\xi(1),\xi(2)}$.

(b) if $P \subseteq \Gamma \stackrel{def}{=} \{u_{j,\xi,\zeta},\ X - u_{j,\xi,\zeta} : j < \vartheta,\ \xi \neq \zeta < \mu\}$,

$$|P| \leq \vartheta, \textit{ and } x \in \bigcap_{A \in P} A \quad \textit{then} \quad \varphi(\bigcap_{A \in P} A) \geq \chi_i$$

If (*) fail, (for $i,\xi(0),\xi(1),\xi(2)$) then for every $x \in u_{i,\xi(1),\xi(0)} \cap u_{i,\xi(1),\xi(2)}$ some P contradicts (b). So there are $P_i \subseteq \Gamma\ (i < \alpha)$, $|P_i| \leq \vartheta$, $\varphi(\bigcap_{A \in P_i} A) < \chi_i$, and $\bigcup_{i<\alpha} \bigcap_{A \in P_i} A \supseteq u_{i,\xi(1),\xi(0)} \cap u_{i,\xi(1),\xi(2)}$. As $\alpha \leq |\Gamma|^\vartheta \leq \mu^\vartheta = \mu$, by the (χ_i,μ)-completeness (as $\varphi(\bigcap_{A \in P_i} A) < \chi_i$):

$$\varphi(u_{i,\xi(1),\xi(0)} \cap u_{i,\xi(1),\xi(2)}) \le \prod_{i<\alpha} \varphi(\bigcap_{A \in \beta} A) < \chi_i$$

But $y_{i,\xi(1)} \in u_{i,\xi(1),\xi(0)} \cap u_{i,\xi(1),\xi(2)}$, $y_{i,\xi(1)} \notin X_{\chi_i}$, contradiction. So (*) holds and let $x_{i,\xi(0),\xi(1),\xi(2)}$ exemplify it. Now define a five place function F on $\{y_{i,\xi}: i < \vartheta, \xi < \mu\}$, if $i \ne j < \vartheta$, $\xi(0) < \xi(1) < \xi(2) < \mu$, $\zeta(0) < \zeta(1) < \mu$:

$$F(y_{i,\xi(0)}, y_{i,\xi(1)}, y_{i,\xi(2)}, y_{j,\zeta(0)}, y_{j,\zeta(1)})$$

is 0 if $x_{i,\xi(0),\xi(1),\xi(2)} \in u_{j,\zeta(0),\zeta(1)}$ and is 1 otherwise.

By Erdos Rado if $\mu = \beth_5(\vartheta)^+$ and [Sh 2] if $\mu = \beth_2(\beth_2(\vartheta)^+)^+$ (see remark 3.18 below) there are $\xi(i,\ell)$ ($i < \vartheta$, $\ell < 3$) such that for $i \ne j < \vartheta$:

$$F(y_{i,\xi(i,0)}, y_{i,\xi(i,1)}, y_{i,\xi(i,2)}, y_{j,\xi(j,0)}, y_{j,\xi(j,1)}) =$$
$$F(y_{i,\xi(i,0)}, y_{i,\xi(i,1)}, y_{i,\xi(i,2)}, y_{j,\xi(j,1)}, y_{j,\xi(j,2)})$$

(and $\xi(i,0) < \xi(i,1) < \xi(i,2)$)

We can conclude that

$$x_{i,\xi(i,0),\xi(i,1),\xi(i,2)} \notin u_{j,\xi(j,1),\xi(j,0)} \cap u_{j,\xi(j,1),\xi(j,2)}$$

(because $u_{i,\xi,\zeta} \cap u_{i,\zeta,\xi} = \phi$ for $\xi \ne \zeta$).

Let $u_i = u_{i,\xi(i,1),\xi(i,0)} \cap u_{i,\xi(i,1),\xi(i,2)}$. So $x_{i,\xi(i,0),\xi(i,1),\xi(i,2)} \in u_i - \bigcup_{j \ne i} u_j$, and by the choice of $x_{i,\xi(i,0),\xi(i,1),\xi(i,2)}$, u_i clearly $\varphi(u_i - \bigcup_{j \ne i} u_j) \ge \chi_i$, as required.

3.2A Remark: 1) The demand on μ is (see [Sh 2] Definition 1) to be able to use that $\langle (\mu)_\vartheta \rangle$ have $\langle (3)_\vartheta \rangle$-cannonization for $\{<2;3>^5_2, <3;2>^5_2\}$, but really $\{<2;3>^5_2, <3;2>^2_2\}$.

Really we can define F for any five tuples from $\{y_{i,\xi}: i < \vartheta, \xi < \mu\}$, and it is enough to find $\xi(i,\ell) < \mu$, $\alpha(i,\ell) < \vartheta$ (for $i < \vartheta, \ell < 3$) such that $\vartheta = \operatorname{Sup}_{i<\vartheta} (\operatorname{Min}_{\ell<3} \alpha(i,\ell))$, $[k \ne m \implies \xi(i,k) \ne \xi(i,m)]$ and for $i < j < \vartheta$,

$$F(y_{\alpha(i,0),\xi(i,0)}, y_{\alpha(i,1),\xi(i,1)}, y_{\alpha(i,2),\xi(i,2)}, y_{\alpha(j,0),\xi(j,0)}, y_{\alpha(j,1),\xi(j,1)})$$
$$= F(y_{\alpha(i,0),\xi(i,0)}, y_{\alpha(i,1),\xi(i,1)}, y_{\alpha(i,2),\xi(i,2)}, y_{\alpha(j,1),\xi(j,1)}, y_{\alpha(j,2),\xi(j,2)}.)$$

2) If $\vartheta > \aleph_0$ is weakly compact, $\mu = 2^\vartheta$ is o.k.; in fact we can use just $\{y_{i,0}: i < \vartheta\}$ by 3.2A(1).

3.2B Remark: How do we apply [Sh 2] in the proof of 3.2? By the compositioin claim [Sh 2, Claim 5, p. 349] it is enough to prove that:

(a) $\langle (\beth_2(\beth_2(\vartheta)^+)^+)_\vartheta \rangle$ has a $\langle (\beth_2(\vartheta)^{++}))_\vartheta \rangle$-canonization for $\{(2;3)^2_2, (3;2)^2_2\}$.

(b) $\langle (\beth_2(\vartheta)^{++})_\vartheta \rangle$ has a $\langle ((2^\vartheta)^+)_\vartheta \rangle$-canonization for $\{(2;1)^2_2\}$ really even for $\{(2;1)^2_{\beth_2(\vartheta)}\}$.

(c) $\langle ((2^\vartheta)^+)_\vartheta \rangle$ has a $\langle (3)_\vartheta \rangle$-canonization for $\{(1)^1_{2^\vartheta}\}$.

Now (c) is trivial, and (a) we get by e.g. applying [Sh 2, 6(B), p. 249] twice; Now to get (b) (and even for $\{(2;1)^2_{\beth_2(\vartheta)}\}$) we apply [Sh 2, 6(F)] with $S = \vartheta$, $\lambda_\xi = \beth_2(\vartheta)^{++}$, $\kappa_\xi = (2^\vartheta)^+$, and check the condition.

3.3 Theorem : 1) If $\mu = \beth_5(cf\ \lambda)^+ < \lambda$, or $\mu = \beth_2(\beth_2(cf\ \lambda)^+)^+ < \lambda$, X is a Hausdorff space, with spread λ, then the supremum is obtained, i.e., $\mathfrak{s}(X) \neq \lambda$.

2) The same apply to $h(Y), z(X)$.

Proof : Suppose X is a Hausdorff space, $\mathfrak{s}(X) \geq \lambda$. Let $\lambda = \sum_{i<cf\ \lambda} \chi_i$, $\chi_i < \lambda$, $\vartheta \stackrel{def}{=} cf\ \lambda$, let $|A_i| = \chi_i$, A_i discrete w.l.o.g. $X = \bigcup_{i<\vartheta} A_i$ and let $\varphi(A) = |A|$, and let C be the family of regular cardinals $< \lambda$ but $> \mu$. Now (i), (iii) are immediate. If (ii) fail for χ, by Hajnal free subset theorem the spread is λ. Otherwise we can find by lemma 3.2 open $u_i (i < cf\ \lambda)$, $|u_i - \bigcup_{j\neq i} u_j| \geq \chi_i$ w.l.o.g. each χ_i is regular $> cf\ \lambda$, so for each i for some $\alpha_i < cf\ \lambda$, $(u_i - \bigcup_{j\neq i} u_j) \cap A_{\alpha_i}$ has power χ_i. The rest is easy too.

3.4 Lemma: Suppose κ is a strong limit cardinal, X an infinite Hausdorff space, $o(X) \geq \kappa$. If $o(X)^{<\kappa} > o(X)$ then for some $Y \subseteq X$ and χ, $|X| \leq \chi = \chi^{<\kappa} < o(X)$, $|X - Y| < \kappa$, Y open, $o(Y) = o(X)$, $Y = \bigcup\{v \in \tau : o(v) < \chi\}$, so Y has a strong base of power χ.

Proof: For $\kappa = \aleph_0$, this is trivial; if κ is strongly inaccessible then κ is the limit of strong limit singular cardinals, and it suffice to prove it for each of them [let for $\sigma < \chi$

$\chi_\sigma = Min\{\chi: X - \bigcup\{u \in \tau: o(u) < \chi\}$ has cardinality $< \sigma\}$.

$Z_\sigma = \{y \in X : Ch_0(y) \geq \chi_\sigma\}$ $(= X - \bigcup\{u \in \tau : o(u) < \chi\})$

so when σ increases χ_0 decrease, (and χ_σ is well defined: $\chi_\sigma \leq o(X)$); so for some $\sigma(0) < \kappa$, $\chi_\sigma = \chi_{\sigma(0)}$ hence $Z_\sigma = Z_{\sigma(0)}$ whenever $\sigma(0) \leq \sigma < \kappa$. W.l.o.g. $\sigma(0)$ is strong limit singular; checking the definition of χ_σ.$\chi_\kappa = \chi_{\sigma(0)}$. (as $Ch_0(z) \geq \kappa$ for $z \in Z_{\sigma(0)}$) For every strong limit singular σ, $\sigma(0) < \sigma < \kappa$, as 3.4 is assumed to be proved for it, there are χ, Y as required; clearly (by the "Min" in the definition of χ_σ) $\chi_\kappa = \chi_\sigma \leq \chi = \chi^{<\sigma}$, so $\chi_\kappa^{<\sigma} < o(X)$. As $o(X) \geq \kappa > \sigma$, κ strong limit regular, clearly $o(X) \geq \kappa \geq 2^{<\kappa}$, hence $o(X) > 2^{<\kappa}$, so either $\chi_\kappa^{<\kappa} \leq (2^{<\kappa})^{<\kappa} = 2^{<\kappa} < o(X)$ or by 2.11. $\chi_\kappa^{<\kappa} = \chi_\kappa^\sigma$ for some $\sigma < \kappa$, hence $\chi_\kappa^{<\kappa} < o(X)$. Now $\chi = \chi_\kappa^{<\kappa}$ is as required (if $|X| > \chi$ use Hajnal free subset theorem.)]

So w.l.o.g. κ is a strong limit singular cardinal. Let X be a counterexample, i.e. $o(X)^{<\kappa} > o(X)$.

Let $\lambda = Min\{\lambda : \lambda^\kappa \geq o(X)\}$, so $\lambda^\kappa = o(X)^\kappa > o(X)$, and $\lambda \leq o(X)$. Also $[\sigma < \kappa, \chi < \lambda \Longrightarrow \chi^\sigma < \lambda]$ and $cf\ \lambda \leq \kappa$. Let $\vartheta = cf\ \lambda$, so $\vartheta \leq \kappa$ but ϑ is regular so $\vartheta < \kappa$, and also $\mu \overset{def}{=} \beth_5(\vartheta)^+$ is $< \kappa$, hence $(\forall \sigma < \kappa)\sigma^\mu < \lambda$.

We define the function φ:

$\varphi(A) = |\{u \cap A : u$ is an open subset of $X\}|$.

The family C of cardinals will be $\{(\chi^\mu)^+ : \chi < \lambda\}$.

Now we want to apply the lemma 3.2. Its conclusion clearly suffice by 2.3A (ii). Now "φ is nice for X" and "φ is $(<\lambda, \mu)$-complete" are immediate. So (ii) necessarily fail for some $\chi < \lambda$. So $Y = \bigcup\{v : o(v) < \chi\}$ satisfies $|X - Y| < \mu$, hence $o(Y) = o(X)$ [as $o(X-Y) \leq 2^\mu < \kappa \leq o(X)$]. Also $|Y| < \lambda$ [otherwise by Hajnal free subset theorem, $\mathfrak{s}(X) \geq \mathfrak{s}(Y) > \lambda$, hence $o(X) \geq 2^\lambda$, but $2^\lambda \geq o(X)$ so $o(X) = 2^\lambda$, hence $o(X)^\kappa = o(X)$ contr]. So Y (as a subspace) has a strong base $\underset{\sim}{B}$ of power $\leq \chi + |X| < \lambda$.

3.5 Conclusion: If X is Hausdorff space, κ strong limit cardinal $o(X) \geq \kappa$,

$o(X)^{<\kappa} > o(X)$, *then* for every base $\underset{\sim}{B}$ of X $|\underset{\sim}{B}|^{<\hat{s}(X)} \geq o(X)$.

Proof : See 3.3, and apply 2.6 to the space Y.

3.6 Conclusion: 1) If B is a Boolean Algebra , κ strong limit and $|B| \geq \kappa$ then $id(B)^{<\kappa} = id(B)$.

2) If X is locally compact Hausdorff space, κ strong limit, then $o(X)^{<\kappa} = o(X)$.

Proof : 1) By 3.5 applied to the space of ultrafilters of B, $|B|^{<\hat{s}(B)} \geq o(X)$. By 2.12 $|B|^{<\hat{s}(B)} = 2^{<\hat{s}(B)}$, and clearly $2^{<\hat{s}(B)} \leq o(X)$, so $|B|^{<\hat{s}(B)} = 2^{<\hat{s}(B)} = o(X)$. Now $cf\ \hat{s}(X) \geq \kappa$ by 3.4 (as $\hat{s}(X) \geq \beth_5(\mu)^+$ whenever $\mu < \kappa$), hence $(2^{<\hat{s}(X)})^{<\kappa} = 2^{<\hat{s}(X)}$. As $id(B) = o(X)$ we finish.

2) By 3.5 $|\underset{\sim}{B}|^{<\hat{s}(X)} \geq o(X)$ for every base $\underset{\sim}{B}$, but by 2.18 $w(X)^{<\hat{s}(X)} \leq 2^{<\hat{s}(X)}$. As $2^{<\hat{s}(X)} \leq o(X)$ we get $2^{<\hat{s}(X)} = o(X)$, as $\hat{s}(X) \geq \kappa$ (remember κ strong limit, $o(X) \geq \kappa$) by 3. 4 $cf\ \hat{s}(X) \geq \kappa$ hence $(2^{<\hat{s}(X)})^{<\kappa} = 2^{<\hat{s}(X)}$.

Remark: If you want to apply only the part of 2.18, 2.17 actually proved, separate the case λ is strong limit in 3.4.

§4 Further consequences.

4.1 Claim: Let B be a Boolean Algebra , χ a cardinal, and we define by induction on i, ideals $I_i = I_i^\chi(B)$ increasing continuous: $I_0 = \{0\}$, $I_{i+1} = \{x \in B : id((B/I_i) \restriction (x/I_i)) < \chi_i\}$ where χ_i is choose as a minimal cardinal $< \chi$ such that $I_{i+1} \neq I_i$.

1) For some $\gamma = \gamma(*) = i_\chi(B) < |B|^+, I_{\gamma(*)}$ is defined but not $\chi_{\gamma(*)}$ (nor $I_{\gamma(*)+1}$).

2) $B = I_{\gamma(*)}$ or for every $x \in B - I_{\gamma(*)}$, $id((B/I_{\gamma(*)}) \restriction (b/I_{\gamma(*)})) \geq \chi$.

3) The number of ideals $J \subset I_{\gamma(*)}$ of B has the form $\sum_{\alpha<\beta} \mu_\alpha^{\kappa(\alpha)}$ where $\beta \leq |B|^{<\hat{s}(B)}$, $\mu_\alpha < \chi$, $\kappa(\alpha) < \hat{s}(B)$.

This follows from:

4.2 Claim: For a Hausdorff space X with a base $\underset{\sim}{B}$ and cardinal χ, define by induction on i $u_i = u_i^\chi(X)$:

$u_0 = \phi$

$u_{i+1} = u_i \bigcup \{v : v \in \underset{\sim}{B} \ , o(v - u_i) < \chi_i\}$ where $\chi_i < \lambda$ is minimal such that $u_{i+1} \neq u_i$ $u_\delta = \bigcup_{i<\delta} u_i$ (so u_i is increasing continuous.)

1) For some $\gamma(*) = \gamma^\chi(X) < |X|^+$, (and $\gamma(*) < |w(X)|^+$) $u_{\gamma(*)}$ is defined but not $u_{\gamma(*)+1}$ and for every $y \in X - u_{\gamma(*)}, (\forall v)(y \in v \in \tau \rightarrow o(v - u_{\gamma(*)}) \geq \chi.)$

2) $o(u_{\gamma(*)})$ if $> |\underset{\sim}{B}|^{<\mathfrak{s}(X)}$, has the form $\sum_{\alpha<\beta} \mu_\alpha^{\kappa(\alpha)}$ where $\beta \leq |\underset{\sim}{B}|^{<\mathfrak{s}(B)}$, $[\mu_\alpha < \chi$, or $\mu_\alpha = \chi, \kappa(\alpha) \geq cf\ \chi]$ and $\kappa(\alpha) < \mathfrak{s}(X)$.

Proof : Like 2.6.

For every $u \subset u_{\gamma(*)}$ choose by induction on i, v_i, such that:

(i) $v_j \cap v_i \subset u$ for $j < i$.

(ii) $v_i \not\subset u$, $v_i \in \underset{\sim}{B}$.

(iii) $v_i \subset u_{\alpha(i)}$ for some $\alpha(i) \leq \gamma(*)$ but for no $\beta < \alpha(i)$ and $v' \subset v_i$, is $v' \not\subset u$, $v' \subset u_\beta$ and $v' \in \tau$.

So let β be first such that v_β is not defined. By (iii) for each $i < \beta$ $\alpha(i)$ is successor ordinal and $u_{\alpha(i)-1} \cap v_i \subset u$. As in 2.6 $\bar{v} = \langle v_j : j < \beta \rangle$, $u \cap v_j$ determine u, the number of u corresponding to $\bar{v}$ is $\prod_{j<\beta} o(v_i - u_{\alpha(i)-1} - \bigcup_{j \neq i} v_j)$ each multiplicant is $\leq o(v_i) \leq \chi_i < \chi$, $\beta < \mathfrak{s}(X)$ and the number of $\bar{v}$ is $\leq |B|^{<\mathfrak{s}(X)}$.

4.3 Remark: At least for compact spaces, this gives heavy restrictions on the relevant cardinals.

Let $\aleph_0 \leq \kappa_0 < \cdots < \kappa_n$ list the cardinals κ such that $2^\kappa < o(X)$, and for some $\lambda = \lambda[\kappa]$, $\kappa = cf\ \lambda$, and $\lambda^\kappa > o(X) > \lambda$ but $(\forall \chi < \lambda)\ [\chi^\kappa < \lambda]$ so $o(X)^{<\kappa_0} = o(X) < o(X)^{\kappa_0}$ (if there is no such κ we have no problem). As $\lambda[\kappa_a] = \lambda[\kappa_b]$ implies $\kappa_a = \kappa_b$, and $[\kappa_a < \kappa_b \Longrightarrow \lambda([\kappa_a]) > \lambda([\kappa_b])]$, clearly n is finite and trivially each κ_ℓ is regular and let for $\ell = 1,n$, $\lambda_\ell = Min\{\lambda\colon \lambda^{\kappa_\ell} \geq o(X)\}$; but $\lambda[\kappa_\ell] \geq \lambda_\ell$ (as $\lambda[\kappa_\ell]^\kappa \geq o(X)$)) and $\lambda[\kappa_\ell] \leq \lambda_\ell$ (as $(\forall \chi < \kappa[\kappa_\ell])\ [\chi^\kappa < \lambda[\kappa_\ell]]$), so $\lambda[\kappa_\ell] = \lambda_\ell$. Hence $cf\ \lambda_\ell = \kappa_\ell$, $\lambda_0 > \lambda_1 > \cdots > \lambda_n$, $(\forall \chi < \lambda_\ell)[\chi^{\kappa_\ell} < \lambda_\ell]$. Moreover (for $\ell < n$) $(\forall \chi < \lambda_\ell)(\chi^{<\kappa_{\ell+1}} < \lambda_\ell)$ [first suppose $\chi < \lambda$, $\kappa_\ell \leq \vartheta < \kappa_{\ell+1}$, if $\chi^\vartheta \geq \lambda_\ell$ then $\chi^\vartheta \geq \lambda^\vartheta \geq \lambda^{\kappa_\ell} \geq o(X)$, w.l.o.g. χ is minimal with this property, so $\chi^\vartheta \geq o(X) > 2^{\kappa_{\ell+1}} \geq 2^\vartheta$ hence $\chi > 2^\vartheta$. Clearly $(\forall \mu < \chi)(\mu^\vartheta < o(X))$ hence $(\forall \mu < \chi)(\mu^\vartheta < \chi)$, and $cf(\chi) \leq \vartheta$ (otherwise $\chi^\vartheta = \sum\limits_{\alpha<\chi} |\alpha|^\vartheta \leq \chi < \lambda_\ell \leq o(X)$ contr.). So $cf\ \chi \leq \vartheta < \kappa_{\ell+1}$ and by χ's minimality $(\forall \mu < \chi)(\mu^{cf\ \chi} \leq \mu^\vartheta < \chi)$. Lastly $cf\ \chi > \kappa_\ell$ [otherwise $\chi^\vartheta = \chi^{cf\ \chi} \leq \chi^{\kappa_\ell} < \lambda_\ell$ contradicting the assumption of ϑ]. So $\vartheta \in \{\kappa_0, \ldots, \kappa_n\}$, contr. Secondly suppose $\chi^{<\kappa_{\ell+1}} \geq \lambda_\ell$, for some $\chi < \lambda_\ell$, as $\vartheta < \kappa_{\ell+1} \Longrightarrow 2^\vartheta < \lambda_\ell$, by 2.11 for some $\vartheta < \kappa_{\ell+1}$, $\chi^\vartheta = \chi^{<\kappa_{\ell+1}}$ and we get the first case].

Let $\lambda_{n+1} = Min\{\chi\colon 2^\chi \geq o(X)\}$ and $\kappa_{n+1} = cf\ \lambda_{n+1}$; so $\lambda_{n+1} \leq \lambda_n$, hence, as above) $(\forall \chi < \lambda_n)(\forall \vartheta < \lambda_{n+1})[\chi^\vartheta < \lambda_n]$. By the proof of 3.4 $\beth_5(\kappa_\ell)^+ \geq \kappa_{\ell+1}$ (for $\ell < n$), otherwise using $\lambda_n, \kappa_\ell, \mu = \beth_5(\kappa_\ell)^+$ we get contradiction. If λ_{n+1} is singular, $\langle 2^\chi\colon \chi < \lambda_{n+1} \rangle$ is not eventually constant [as then $(\exists \chi < \lambda_{n+1}) 2^\chi = 2^{\chi_{n+1}}$], $2^{<\lambda_{n+1}} \leq o(X)$, $(2^{<\lambda_{n+1}})^{\kappa_{n+1}} = 2^{\lambda_n} > o(X)$, so $\lambda[\kappa_{n+1}] = 2^{<\lambda_{n+1}}$, so $\lambda_n = \lambda_{n+1}$ hence $\beth_{6(n+1)}(\kappa_0) \geq o(X)$, $o(X)^{<\kappa_0} = o(X)$. If λ_{n+1} is regular, then $(\forall \vartheta < \lambda_{n+1})\ (\forall \chi < \lambda_n)[\chi^\vartheta < \lambda_n]$ hence $\beth_5(\kappa_n)^+ \geq \lambda_{n+1}$, so we get the same conclusion.

4.4 Lemma: Suppose X is a Hausdorff space, λ a singular cardinal, $\vartheta = cf\ \lambda$, $\lambda = \sum\limits_{i<\vartheta} \chi_i, \chi_i < \lambda$, $\mu < \lambda$ and (i), (ii), (iii) of 3.2 holds (for φ).

1) If $\mu = \beth_2(\vartheta)^+$ (or even $\sum\limits_{\sigma<\vartheta} \beth_2(\sigma)^+$) *then* there are open sets $u_i (i < \vartheta)$ such that $\varphi(u_i - \bigcup\limits_{j>i} u_j) \geq \chi_i$.

2) If $X = \bigcup\{u\colon o(u) < \lambda\}$, μ as in 1) then there are open sets u_i such

that $\varphi(u_i - \bigcup_{j \neq i} u_j) \geq \chi_i$.

3) If $\mu \geq \beth_3(2^{<\vartheta})^+$, φ is $(<\chi_0, \mu)$-complete, then there are $u_i (i < \vartheta)$ such that $\varphi(u_i - \bigcup_{j \neq i} u_j) \geq \chi_0$ (so λ, $\chi_i (0 < i < \vartheta)$, are irrelevant).

Remarks: 1) Part 1) of the lemma is suitable to deal with Boolean algebras, part 2) with existence of $\{x_\alpha : \alpha < \lambda\}$ such that for every $\alpha < \lambda$ for some u, $x_\alpha \in u \cap \{x_\beta : \beta < \lambda\} \subseteq \{x_\beta : \beta \leq \alpha\}$.

Proof: 1) We repeat the proof of 3.2, for $\mu = \beth_2(\vartheta)^+$, but cannot use the partition relation used there, but we can use a weaker one. We choose by induction on $j < \vartheta$, $\xi(j,0) < \xi(j,2) < \xi(j,2) < \mu$ such that for $i < j$:

$$F(y_{i,\xi(i,0)}, y_{i,\xi(i,1)}, y_{i,\xi(i,2)}, y_{j,\xi(j,0)}, y_{j,\xi(j,1)}) =$$
$$F(y_{i,\xi(i,0)}, y_{i,\xi(i,1)}, y_{i,\xi(i,2)}, y_{j,\xi(j,1)}, y_{j,\xi(j,2)})$$

This is clearly possible by the assumption on μ.

We can conclude that, letting $u_i = u_{i,\xi(i,1),\xi(i,0)} \cap u_{i,\xi(i,1),\xi(i,2)}$ then $x_{i,\xi(i,0),\xi(i,1),\xi(i,2)} \in u_i - \bigcup_{j > i} u_j$, so we can get the desired conclusion.

2) In the proof of 1) we can take care that for every $i < \vartheta$, $\xi \neq \zeta < \mu$, $u_{i,\xi,\zeta}$ satisfies $o(u_{i,\xi,\zeta}) < \lambda$; hence we shall get $o(u_i) < \lambda$. So by thinning the sequence $\langle u_i : i < \vartheta \rangle$, as $o(u_i) \geq \chi_i$, $\lambda = \sup_{i < \vartheta} \chi_i$ we can assume:

$$[i < j \Rightarrow o(u_i) < u_j].$$

As φ is $(<\chi_i, \mu)$-complete, $\vartheta \leq \mu$, necessarily $o(\bigcup_{i<j} u_i) < \chi_j$. Hence

$$o(u_i - \bigcup_{j \neq i} u_j) = o((u_i - \bigcup_{j<i} u_j) - \bigcup_{j<i} u_j) \geq \chi_j$$

as required.

3) Really the proof is as in 3.2, but we use (for $\sigma = 2$, κ finite large enough, note $\mu = \beth_2(\sigma^{<\vartheta})^+$; is O.K. in 4.5):

4.5 Observation: If F is a 5-place function from μ to σ, $\sigma \geq 2, \vartheta \geq \aleph_0$ $\mu \rightarrow (\kappa)^3_\psi$, $\psi = 2^{(\sigma^{<\vartheta})+\kappa}$ $\kappa \rightarrow (3)^2_\sigma$ [e.g. $\mu > \beth_1(\psi^{2^{<\sigma}}) = \beth_3(\sigma^{<\vartheta}+\kappa)$ $\kappa = (2^{<\sigma})^+ + \aleph_0$],

$\kappa \to (3)^2_\vartheta$ [e.g. $\mu > \beth_1(2^{\psi+2^{<\vartheta}}) = \beth_3(\sigma^{<(\vartheta+\kappa^+)})$, $\kappa = (2^{<\vartheta})^+$, or κ is finite large enough] then, there are distinct $\xi(i,\ell)(i < \vartheta, \ell > 3)$ such that, for $i \neq j$:

$$F(\xi(i,0),\xi(i,1),\xi(i,2),\xi(j,0),\xi(j,1)) =$$
$$F(\xi(i,0),\xi(i,1),\xi(i,2),\xi(j,1),\xi(j,2))$$

Remark: We can get of course more general theorem.

Proof : We choose by induction on $i<\vartheta$, $Y_i \subseteq \mu$, $|Y_i| \leq \sigma^{|i|+\kappa} + \aleph_0$, Y_i increasing and all "types" of cardinality $< |i|^+ + \kappa^+$ realized in μ are realized in Y_{i+1}. Let $Y = \bigcup_{i<\vartheta} Y_i$. Now we can find distinct $\xi^*(\ell) \in \mu - Y$ for $\ell < \kappa$ such that for every $\xi_0,\xi_1,\xi_2 \in \bigcup_{i<\vartheta} Y_i$ there are $c_1(\xi_0,\xi_1,\xi_2), c_2(\xi_0,\xi_1)$ such that

$(*)_a$ for every $\ell < m < \kappa$ $F(\xi_0,\xi_1,\xi_2,\xi^*(\ell),\xi^*(m)) = c_1(\xi_0,\xi_1,\xi_2)$

$(*)_b$ and for every $\ell < m < n < \kappa$ $F(\xi^*(\ell),\xi^*(m),\xi^*(k),\xi_0,\xi_1) = c_2(\xi_0,\xi_1)$.

Why we can do this? We want to apply the partition relation $\mu \to (\kappa)^3_\psi$, for this we have to check what is the number of " colours", clearly it is $\leq 2^{(\kappa^2|Y|^3+\kappa^3|Y|^2)} \leq 2^{\aleph_0+\kappa+(\sigma^{<(\vartheta+\kappa^+)})} = \psi$. Now we choose by induction on $i < \vartheta$, $\xi(i,\ell), \ell < \kappa$ such that :

(i) $\xi(i,0),\xi(i,2),\xi(i,2)$ are distinct.

(ii) $\xi(i,\ell) \in Y_{i+1} - Y_i$.

(iii)
$F(\xi(j,0),\xi(j,1),\xi(j,2),\xi(i,\ell),\xi(i,m)) = F(\xi(j,0),\xi(j,1),\xi(j,2),\xi^*(\ell),\xi^*(m))$, when $j < i$, and $\ell, m < \kappa$.

(iv) $F(\xi(i,\ell_1),\xi(i,\ell_2),\xi(i,\ell_3),\xi(j,\ell_4),\xi(j,\ell_5)) =$

$$F(\xi^*(\ell_1),\xi^*(\ell_2),\xi^*(\ell_3),\xi(j,\ell_4),\xi(j,\ell_5))$$

when $j < i$, $\ell_1 < \cdots < k$.

There is no problem in doing this:

For each $i < \vartheta$, as $\kappa \to (3)^2_\sigma$ there are $\ell_0(i) < \ell_1(i) < \ell_2(i) < \kappa$ such that:

$$F(\xi^*(0),\xi^*(1),\xi^*(2),\xi(i,\ell_0(i)),\xi(i,\ell_1(i))) =$$
$$F(\xi^*(0),\xi^*(1),\xi^*(2),\xi(i,\ell_1(i)),\xi(i,\ell_2(i)))$$

Now $\xi'(i,m) = \xi(i,\ell_m(i))$ $(i < \vartheta, m < 3)$ are as reequired.

4.4A Remark: Assume (i), (ii), (iii) of 3.2. We try to decrease μ. Let $Z_i = \{y \in X\colon Ch_\varphi(y) \geq \chi_i\}$, so $|Z_i| \geq \mu$, and let $X_{<\lambda} = \bigcup\{u : \varphi(u) < \lambda\}$. If $|X - X_{<\chi}| < \mu$ then necessarily $|Z_i \cap X_{<\chi}| \geq \mu$, so we can continue as in 4.4(2). So we assume $|X - X_{<\chi}| \geq \mu$ and let $y_\xi \in X - X_{<\lambda}$ $(\xi < \mu)$ be distinct. Choose for $\xi < \zeta$, open disjoint sets $u_{\xi,\zeta}, u_{\zeta,\xi}$ such that $y_\xi \in u_{\xi,\zeta}$, $y_\zeta \in u_{\zeta,\xi}$. As in 3.2's proof we can choose for distinct $\xi(0),\xi(1),\xi(2) < \mu$, $x_{i,\xi(0),\xi(1),\xi(2)} \in u_{\xi(1),\xi(0)} \cap u_{\xi(1),\xi(2)}$ such that: for every $P \subseteq \{u_{\xi,\zeta}, x - u_{\xi,\zeta}\colon \xi,\zeta < \mu\}$, $|P| \leq \vartheta$,

$$[x_{i,\xi(0),\xi(1),\xi(2)} \in \bigcap_{a \in P} a \Longrightarrow \varphi(\bigcap_{a \in P} a) \geq \chi_i]$$

We need the parallel of 4.5 for ϑ functions simultaneously or, what is equivalent, the range of F has cardinality 2^ϑ, so $\sigma = 2^\vartheta$, and we get $\mu \geq \beth_5(\vartheta)^+$ but this is not interesting.

§5 When the spread is obtained and how helpful is regularity of the space

5.1 Lemma : 1) Suppose X is a regular (i.e. T_3) topological space, $\underset{\sim}{B}$ a base of X, $\lambda = \sum_{i<\vartheta} \chi_i$, $\vartheta < \chi_i < \lambda$, $\mu = (2^\vartheta)^+$ and

(i) φ is nice for X,

(ii) for every (closed) $Y \subseteq X$ with $\varphi(Y) \geq \lambda$ and $i < \vartheta$, there are $y_\alpha \in Y$ $(\alpha < \mu)$, $Ch_{\varphi \restriction Y}(y_\alpha) \geq \chi_i$ and $\{y_\alpha : \alpha < \mu\}$ is a discrete set,

(iii) φ is $(<\lambda,\mu)$-complete.

Then for some $u_i \in \underset{\sim}{B}$ $(i < \vartheta)$, $\varphi(u_i - \bigcup_{j \neq i} u_j) \geq \chi_i$.

2) Instead $\mu = (2^\vartheta)^+$ it suffices that $\mu = \mu^\vartheta > 2^\vartheta$ (and (i), (ii), (iii)).

3) We can replace (ii) above by

(ii)' for each $i < \vartheta$ there are $u_\alpha \in \underset{\sim}{B}(\alpha<\mu)$ such that:

$(\forall g:\mu \to 2^\vartheta)(\exists\alpha \neq \beta)[g(\alpha) = g(\beta) \wedge \varphi[(u_\alpha - \overline{u}_\beta) \cap Y) \geq \chi_i]$.

or

(ii)" there are $u_\alpha, y_\alpha \in u_\alpha \in \underset{\sim}{B}$, such that: $Ch_\varphi(y_\alpha) \geq \chi_i$ and

$(\forall g;\mu \to 2^\vartheta)(\exists\alpha \neq \beta)[g(\alpha) = g(\beta) \wedge y_\alpha \notin \overline{u_\beta}]$.

Proof : 1) W.l.o.g. φ is (χ_i,μ)-complete for $i < \vartheta$. We first try to choose a family K of open subsets of X, (or even $\subseteq \underset{\sim}{B}$), and a $Y \subseteq X$ such that:

(A) $|K| = |Y| = (2^\vartheta)^+$.

(B) *if* u is the union of $< \vartheta$ members of K, $\varphi(X-u) \geq \lambda$ and $i < \vartheta$ *then* there is a sequence $\langle y_\alpha, v_\alpha^0, v_\alpha^1 : \alpha < (2^\vartheta)^+\rangle$ such that: $y_\alpha \in Y-u$, $[y_\alpha \in v_\beta^1 \Leftrightarrow \alpha = \beta]$, $v_\alpha^0, v_\alpha^1 \in K$, $y_\alpha \in v_\alpha^0 \subseteq \overline{v}_\alpha^0 \subseteq v_\alpha^1$, and $(\forall v \in \tau(X))$ $[y_\alpha \in v \to \varphi(v-u) \geq \chi_i]$.

It is easy to find such K, Y (by (ii)). Let for $i < \vartheta$,

$Z_i(K) \overset{def}{=} \{z \in X$: if $u_\alpha \in K(\alpha < \vartheta)$, and $u_\alpha^* \in \{u_\alpha, X-u_\alpha\}$ and $z \in u_\alpha^*$ for each $\alpha < \vartheta$ then $\varphi(\bigcap_{\alpha<\vartheta} u_\alpha^{t(*)}) \geq \chi_i\}$.

By the proof of 3.2 for each $i < \vartheta$ there is $z_i \in Z_i(K)$. Now we choose by induction on i, x_i, u_i such that:

(a) $u_i \in K$, $x_i \in Z_i(K)$,

(b) $x_i \in u_i, (\forall\varepsilon < i)(x_\varepsilon \notin u_i \wedge x_i \notin u_\varepsilon)$,

(c) $z_\varepsilon \notin u_i$ when $i < \varepsilon < \vartheta$.

Suppose x_j, u_j are defined for $j < i$. We want to apply (B) to $\bigcup_{j<i} u_j$, now for each ε, if $i \leq \varepsilon < \vartheta$ then $\varphi(X - \bigcup_{j<i} u_j) \geq \chi_\varepsilon$ as $\{u_j : j < i\} \subseteq K$, $z_\varepsilon \notin \bigcup_{j<i} u_j$ and $z_\varepsilon \in Z_\varepsilon(K)$. Hence $\varphi(X - \bigcup_{j<i} u_j) \geq \lambda$. So by (B) above there is $\langle y, v_\alpha^0, v_\alpha^1 : \alpha < (2^\vartheta)^+\rangle$ as mentioned there. By cardinality consideration, for some $\alpha \neq \beta$,

$$v_\alpha^0 \cap (\{z_j : j < \vartheta\} \cup \{y_j : j < i\}) = v_\beta^0 \cap (\{z_j : j < \vartheta\} \cup \{y_j : j < i\})$$

So $u_i \stackrel{def}{=} v_\alpha^0 - \overline{v_\beta^0}$ is open, is disjoint to $\{z_j : j < \vartheta\} \cup \{y_j : j < i\}$, and y_α belongs to it (as $y_\alpha \notin v_\beta^1, \overline{v_\beta^0} \subset v_\beta^1$). As (by (B)) $(\forall v \in \tau(X))[y_\alpha \in v \to \varphi(v - \bigcup_{j<i} u_j) \geq \chi_i]$, clearly $\varphi(u_i - \bigcup_{j<i} u_j) \geq \chi_i$, hence (as in 3.2) there is $x_i \in Z_i(K) \cap (u_i - \bigcup_{j<i} u_j)$. So we succeed in the induction. In the end as $u_i \in K$, $x_i \in Z_i(K) \cap (u_i - \bigcup_{j \neq i} u_j)$ clearly $\varphi(u_i - \bigcup_{j \neq i} u_j) \geq \chi_i$, so we finish.

2),3) Similar.

5.2 Lemma : Suppose X is a Hausdorff space, $\lambda = \sum_{i<\vartheta} \chi_i$, $\chi_i < \lambda$ and $\mu = \beth_2(\vartheta)^+$, $\underset{\sim}{B}$ a base for X, and

(i) φ is nice for X.

(ii) for every (closed) $Y \subset X, \varphi(Y) \geq \lambda$, and $i < \vartheta$ there are at least μ points $y \in Y$ with $Ch_{\varphi \restriction Y}(y) \geq \chi_i$.

(iii) φ is $(<\lambda, \mu)$-complete.

Then for some $u_i \in \underset{\sim}{B}, (i < \vartheta)$ $\varphi(u_i - \bigcup_{j \neq i} u_j) \geq \chi_i$.

Proof : Like the previous one, replacing (B) by (B)', (C)' (D)':

(B)' if u is the union of $< \vartheta$ members of $K, \varphi(X - u) \geq \lambda$ and $i < \vartheta$ then there are $\beth_2(\vartheta)^+$ points $y \in Y - u$ such that $(\forall v \in \tau(X))(y \in v \to \varphi(v - u) \geq \chi_i]$.

(C)' if $y_1 \neq y_2 \in Y$ then for some $u, v \in K$, $y_1 \in u, y_2 \in v, u \cap v = \phi$.

(D)' K is closed under finite intersections.

Then having defined $u_j, x_j (j < i)$ and shown $\varphi(X - \bigcup_{j<i} u_j) \geq \lambda$, we can find distinct $y_\alpha \in Y - \bigcup_{j<i} u_j (\alpha < \beth_2(\vartheta)^+)$ such that $Ch_{X - \bigcup_{j<i} u_j}(y_\alpha) \geq \chi_i$. We let $A = \{z_j : j < \vartheta\} \cup \{x_j : j < i\}$, $I_\alpha = \{v \cap A : y_\alpha \in v \in K\}$, so for some $\alpha \neq \beta < \beth_2(\vartheta)^+$,

$I_\alpha = I_\beta$, and using (C)' there is $u_i \in K$, such that $y_\alpha \in u_i$, obviously $A \cap u_i = \phi$. As $y_\alpha \in u_i$ $\varphi(u_i - \bigcup_{j<i} u_j) \geq \chi_i$, hence there is $x_i \in Z_i(K) \cap (u_i - \bigcup_{j<i} u_j)$.

We may remember:

5.3 Fact : 1) Suppose $\kappa = \kappa^{<\kappa}$, $\chi = \sum_{i<\vartheta} \chi_i$, χ_i increasing continuous $\kappa < \vartheta < \chi_i$.

Then for some forcing notion P:

a) P is κ-complete satisfying the κ^+-chain condition.

b) In V^P there is a topological space X with a basis of clopen sets such that $\hat{h}(X) = \hat{z}(X) = \hat{s}(X) = \chi$, $o(X) = \sum_{i<\vartheta} 2^{\chi_i}$ and $|X| = \chi$.

2) In fact we can get that X is the dual of a Boolean algebra and there is no set of pairwise incomparable members of the Boolean algebra, of cardinality χ.

Proof: Let $p \in P$ be a set of $< \kappa$ atomic conditions with no two contradictory ones, where an atomic condition is $\alpha \in u_\beta$ or $\alpha \notin u_\beta$, where $\alpha, \beta < \chi$, and $\alpha \in [\chi_i, \chi_{i+1}) \Longrightarrow \beta < \chi_i \vee \beta = \alpha \vee \beta \geq \chi_{i+1}$.

Two conditions are contradictory if they have the form $\alpha \in u_\beta, \alpha \notin u_\beta$. The order is inclusion.

Now (a) is obvious.

In V^P we define:

$$u_\beta^P = \{\alpha < \lambda : \alpha \in u_\beta \text{ belong to some } p \in \underset{\sim P}{G}\}$$

On χ we define a topological space X: by having $\{u_\beta^P : \beta < \chi\}$ be a basis of clopen sets.

The rest is easy too.

2) Similar (just as in [Sh 9] 4.4) . i.e. let $P = \{(B, W)$: B a Boolean algebra of cardinality $< \kappa$ generated by $\{x_i : i \in W\}$, W a subset of χ of cardinality $< \kappa$, and if $\alpha_0, \ldots, \alpha_n$ are distinct members of $W \cap [\chi_i, \chi_i^+)$ then $B \models x_{\alpha_0} \not\subseteq \bigcup_{\ell=1}^{n} x_{\alpha_\ell}\}$.

5.4 Conclusion: 1) If X is Hausdorff $\mathfrak{s}(X)$ is singular of cofinality ϑ then $cf\ (\ \mathfrak{s}(X)) < 2^{2^\vartheta}$. [repeat the proof of 3.3 but instead of 3.2 use 5.1 remembering $cf\,(2^\kappa) > \kappa$].

2) If X is regular (i.e. T_3) $\mathfrak{s}(X)$ singular of cofinality ϑ then $cf\,(\mathfrak{s}(X)) < 2^\vartheta$. [repeat the proof of 2.3 but instead of 3.2 use 5.2 remembering $cf\,(2^\vartheta) > \vartheta$].

3) Both results are best possible in the sense of complementary consistency results. (see [JSh] and 5.3).

4) We can replace above s by z or h.

5.5 Lemma : Suppose λ is singular of cofinality ϑ, $\lambda = \sum_{i<\vartheta} \chi_i$, $\chi_i < \lambda$, and $\mu \geq 0$. Assume further (for a topological space X and function φ):

(i) φ is nice for X.

(ii) $\{y \in X$: $Ch_\varphi(y) \geq \chi_i\}$ has power $\geq \mu_1$ for $i < \vartheta$.

(iii) φ is $(< \lambda, \mu_0)$-complete.

1) If X is Hausdorff, $\mu_0 = \mu_1 = \sum_{\kappa<\vartheta} \beth_2(\kappa)^+$, then for some $u_i \in \tau(X)$ (for $i < \vartheta$) for each i, $\varphi(u_i - \bigcup_{j<i} u_j) \geq \chi_i$.

2) If X is regular, $\mu_0 = \mu_1 = \sum_{\kappa<\vartheta} (2^\kappa)^+$ *then* for some $u_i \in \tau(X)$ (for $i < \vartheta$) for each i $\varphi(u_i - \bigcup_{j<i} u_j) \geq \chi_i$.

Remark: The proofs are similar to those of 5.1, 5.2.

Proof : 1) W.l.o.g. φ is (χ_i, μ_0)-complete for each i. We define K, Y:

(A) K is a family of open subsets of X of power $\leq \mu_0$.

(B) Y is a subset of X of power $\leq \mu_1$.

(C) there are μ_0 distinct $y \in Y$ such that $Ch_\varphi(y) \geq \chi_i$.

(D) for any distinct $y_1, y_2 \in Y$ for some disjoint $u_1, u_2 \in K$, $y_1 \in u_1$ and $y_2 \in u_2$.

(E) K is closed under finite unions of intersections

There is no problem to carry this definition. Let $Z_i(K) = \{z \in X$: if for $j < \vartheta$ $a_j \subseteq X$, $a_j \in K \vee X - a_j \in K$, and $z \in a_j$ *then* $\varphi(\bigcap_{j<\vartheta} a_j) \geq \chi_i\}$. Now we choose by induction on $i < \vartheta$, x_i and u_i such that :

(a) $u_i \in K$, $x_i \in Z_i(K)$.

(b) $x_i \in u_i, (\forall j < i)\ (x_i \notin u_j)$.

Suppose we have defined x_j, u_j for $j < i$.

By (C) above there are distinct $y^i_\alpha \in Y$ for $\alpha < \mu_0$, with $Ch_\varphi(y^i_\alpha) = \chi_i$. By (E) above there are, for $\alpha \neq \beta$ $u_{\alpha,\beta} \in K_{\zeta+1}$, such that $y^i_\alpha \in u^i_{\alpha,\beta}$, and $u^i_{\alpha,\beta} \cap u^i_{\beta,\alpha} = \phi$. Now as $\mu_0 \rightarrow (3)^2_{2^{\aleph_0}}$ for some $\alpha < \beta < \gamma < \mu_0$:

$$u^i_{\alpha,\beta} \cap \{x_j : j < i\} = u^i_{\beta,\gamma} \cap (\{x_j : j < i\})$$

As $u^i_{\beta,\alpha} \cap u^i_{\alpha,\beta} = \phi$, clearly $u_i = u^i_{\beta,\alpha} \cap u^i_{\beta,\gamma}$ is disjoint to $\{x_j : j < i\}$. Also $y^i_\beta \in u^i_{\beta,\alpha} \cap u^i_{\beta,\gamma}$, so $\varphi(u_i) \geq \chi_i$, hence as in the proof of 3.2 there is $x_i \in u_i \cap Z_i(K)$. In the end x_i witnesses $\varphi(u_i - \bigcup_{j>i} u_j) \geq \chi_i$ as $x_i \in u_i$, $(\forall j > i)(x_i \notin u_j)$.

2) Similarly (remembering the proof of 5.2).

References

[HJ1]

A. Hajnal and I. Juhasz, Some remarks on a property of topological cardinal functions, *Acta Math. Acad. Sci. Hungar.* 20 (1969), 25-37.

[HJ2]

A. Hajnal and I. Juhasz, On the number of open sets, *Ann. Univ. Sci. Budapest*, 16 (1973), 99-102.

[J1] I. Juhasz, Cardinal functions in topology, *Math. Center Tracts*. Amsterdam, 1971.

[J2] I. Juhasz, Cardinal functions in topology - ten year later, *Math. Center Tracts*. Amsterdan, 1980.

[JSh]

I. Juhasz and S. Shelah, How large can a hereditary separable or hereditarily Lindelof space by? *Israel J. of Math*, submitted.

[KR] K. Kunen and J. Roitman, Attaining the spread of cardinals of cofinality ω, *Pacific J. Math.* 70 (1977), 199-205.

[R] J. Roitman, Attaining the spread at cardinals which are not strong limit, *Pacific J. Math.* 57 (1975), 545-551.

[Sh1]

S. Shelah, Remarks on Boolean algebra, *Algebra Universalis*, 11 (1980), 77-89.

[Sh2]

---------, Canonization theorems and applications, *J. of Symb. Logic*, 46 (1981), 345-353.

[Sh3]

------, On cardinal invariants in topology, *General Topology and its applications*, 7 (1977), 251-259.

[Sh4]

-------, On some problem in general topology, a preprint, Jan. 1976.

[Sh5]

------, If $\diamondsuit_{\aleph_1}$ + "there is an $\aleph_1$-Kurepa tree with κ-branches" then some B.A. of power $\aleph_1$ has λ filters and $\lambda^{\aleph_0}$-ultrafilters. Mimeograph Notes from Madison, Fall 77.

[Sh6]

-------, On P-points, $\beta(\omega)$ and other results in general topology, *Notices of A.M.S. (1984)* 25 (1978), A-365.

[Sh7]

-------, Number of open sets and Boolean algebras with few endomorphisms. *Abstracts of A.M.S. 5(1984)*

[Sh8]

———, Boolean algebras, General topology and independence results, *Abstracts of A.M.S.* 5(1984).

[Sh9]

————, Constructions of many complicated uncountable structures and Boolean Algebra, *Israel J. Math.* 45 (1983), 100-146.

The Existence of Coding Sets.

Lately Zwicker (see [Z]), generalizing theorems on regular κ (and the filter of $\mathcal{D}_\kappa$ generated by the closed unbounded subsets) to $\mathcal{P}_{<\kappa}(\lambda)$ (and $\mathcal{D}_\kappa(\lambda)$) find that many times we can generalize if we restrict ourselves to a coding subset of $\mathcal{P}_{<\kappa}(\lambda)$. He shows existence for κ super-compact, (see Solovay [So] and Menas [M]) and Stanley and Vellman (independently) show existence for $\lambda = \kappa^+$ assuming a suitable morass exists.

During the meeting in Colorado, I heard about this (and the two other variants) and prove some existence theorems mainly that: for $\kappa > \aleph_2$ regular there is a coding set for $\mathcal{P}_{<\kappa}(\lambda)$. We present here somewhat improved versions written in Aug 83 announed in [Sh 2]. Here is a summary.

More information will appear in "More on Stationary Coding".

Definition : (1) We call S a weak (κ,λ)-stationary coding set (or weak (κ,λ)-SC or (κ,λ)-WSC) if S is a stationary subset of $\mathcal{P}_{<\kappa}(\lambda) = \{a \subset \lambda : |A| < \kappa\}$, and for no $a \neq b$ in S, $a \cap \kappa = b \cap \kappa$, $a \subset b$.

(2) We call S a (κ,λ)-stationary coding set (or (κ,λ)-SC) if S is a stationary subset of $\mathcal{P}_{<\kappa}(\lambda)$, and for some one-to-one function $h : S \to \lambda$, for every $a,b \in S$, $[a \neq b \wedge a \subset b \Longrightarrow h(a) \in b]$.

(3) We call S a strong (κ,λ)-stationary coding set (or strong (κ,λ)-SC or (κ,λ)-CD) if for $h(a) = \sup(a)$, (2) holds.

The simplest cases of our results are:

Theorem A: If $\kappa \geq \aleph_{n+1}$, then there is a (κ,κ^{+n})-WSC (on e.g. $\kappa^{+(\omega+1)}$ we could have weaker results.) (see (19).)

Theorem B: 1) If κ is an ineffable cardinal (or just $\diamondsuit_{\{\alpha\mu<\kappa:\mu \text{ inaccessible}\}}$), $\kappa < \lambda = \lambda^{<\kappa}$, and $\lambda \overset{\omega}{\not\to} (\omega)_{\kappa}^{<\omega}$ (Silver's relation) (see 11). then there is a (κ,λ)-SC

2) If $\lambda > \kappa > \aleph_1$, $\diamondsuit_{\{\delta<\lambda:\, cf\ \delta<\kappa\}}$ *then* there is a (κ,λ)-CD (see 7).

3) If $\lambda > \aleph_1$, $\diamondsuit_S$, $S \subseteq \lambda$, $(\forall\delta \in S)\ cf\ \delta = \aleph_0$, S does not reflect in any α of cofinality $\aleph_1$ then there is an $(\aleph_1,\lambda)$-CD (see 7).

Theorem C: If $\vartheta^{\mu} = \vartheta$, $\lambda < \vartheta^{+(\mu^+)}$, $\kappa = \vartheta^+$ and in $\kappa^{\kappa}/\mathcal{D}_{\kappa}$ there is an increasing sequence of length $\kappa^+ + 1$, then there is a (κ,λ)-WSC (see (13).

For Theorem 1 we use:

Theorem D: If D is a normal fine filter on $\mathcal{P}_{<\kappa}(\lambda)$, $\kappa = \mu^+,\lambda$ regular and $\{a \in \mathcal{P}_{<\kappa}(\lambda) : cf\ (\sup a) \neq cf\ \mu\} \in D$ *then* D is not λ^+-saturated.

By Lemma 20, and later result of Foreman, Magidor and Shelah [FMS], it is consistent then that there is no $(\aleph_1,\aleph_2)$-WSC.

1 Notation: κ will be a regular uncountable cardinal, λ a cardinal $\geq \kappa$, μ,χ infinite cardinals, $\mathcal{D}$ a fine normal filter on some $\mathcal{P}_{<\kappa}(A)$.

2 Definition : 1) $\mathcal{P}_{<\mu}(A) = \{a : a \text{ a subset of } A \text{ of power } < \mu\}$.

2) A filter $\mathcal{D}$ on $\mathcal{P}_{<\kappa}(A)$ is *fine* if for $x \in A$ $\{a : a \in \mathcal{P}_{<\kappa}(A),\ x \in A\}$ belong to $\mathcal{D}$ We say $\mathcal{D}$ is finer if for $b \subset \mathcal{P}_{<\kappa}(A)$, $\{a : \alpha \subseteq a,\ a \in \mathcal{P}_{<\kappa}(\lambda)\}$ belongs to $\mathcal{D}$

3) A filter $\mathcal{D}$ on $\mathcal{P}_{<\kappa}(A)$ is normal if for any $C_x \in \mathcal{D}$ (for $x \in A$) the set

$$\{a : a \in \mathcal{P}_{<\kappa}(A),\ a \in \bigcap_{x \in a} C_x\}$$

belongs to $\mathcal{D}$

4) $\mathcal{D}_{<\kappa}(A)$ is the minimal finer normal filter on $\mathcal{P}_{<\kappa}(A)$ (if $|A| \leq \kappa$ it is trivial) (normal include κ-complete).

3 Fact: 1) The set of ordinals $< \kappa$ belong to $\mathcal{D}_{\kappa}(\kappa)$ so $\mathcal{D}_{\kappa}(\kappa)$ can be identified with the filter of $\mathcal{D}_{\kappa}$ of closed unbounded subsets of κ.

2) If $\mathcal{D}$ is a (fine normal) filter on $\mathcal{P}_{<\kappa}(A)$, F a function from $\mathcal{P}_{<\aleph_0}(A)$

to $\mathcal{D}$ then the set $\{a : a \in \mathcal{P}_{<\kappa}(A)$ and for every $w \in \mathcal{P}_{<\kappa}(a)$, $a \supseteq F(w)\}$ belongs to $\mathcal{D}$

3) $\mathcal{D}_\kappa(A)$ is the filter of closed unbounded subsets of $\mathcal{P}_{<\kappa}(A)$, hence $\phi \notin \mathcal{D}_\kappa(A)$.

4 Definition : 1) $S \subseteq \mathcal{P}_{<\kappa}(A)$ is called $\mathcal{D}$- stationary if $\mathcal{P}_{<\kappa}(A)-S \notin \mathcal{D}$ If $\mathcal{D} = \mathcal{D}_\kappa(A)$ (and the identity of A,κ is clear) we omit $\mathcal{D}$.

2) S_1, S_2 are $\mathcal{D}$-almost disjoint if $(\mathcal{P}_{<\kappa}(A) - S_1 \cap S_2) \in \mathcal{D}$

3) We say $\mathcal{D}$ is μ-saturated if there are no μ $\mathcal{D}$-stationary pairwise $\mathcal{D}$-almost disjoint subsets of $\bigcup \mathcal{D}$

5 Main Definition : Let $\mathcal{D}$ be a filter on $\mathcal{P}_{<\kappa}(\lambda)$ and $S \subseteq \mathcal{P}_{<\kappa}(\lambda)$ is $\mathcal{D}$-stationary.

1) S is a weak stationary coding set (WSC) if $a,b \in S$, $a \neq b$, $a \cap \kappa = b \cap \kappa$ implies $a \not\subseteq b$.

2) S is a stationary coding set (SC) if there is a one-to-one function h from S into λ such that: $[a \in S \wedge b \in S \wedge a \neq b \wedge a \subset b \Longrightarrow h(a) \in b]$; We call h a witness.

3) S is a strong stationary coding set (CD) if the function $h(a) = \sup(a)$ is a witness (for its being a stationary coding set).

4) We shall say S is a $\mathcal{D}$-WSC if $S \neq \phi \bmod \mathcal{D}$ and similarly for SC, CD; we may also say: for $\mathcal{D}$ there is a WSC; when $\mathcal{D} = \mathcal{D}_\kappa(\lambda)$ we may write (κ,λ) instead of $\mathcal{D}$

6 Fact: 1) For $\mathcal{D}_\kappa(\kappa)$ there is a strong stationary coding set (κ itself!).

2) A strong stationary coding set is a stationary coding set, and a stationary coding set is a weak stationary coding set (for any fixed $\mathcal{D}$).

3) If λ is a singular, $\mathcal{D}$ a fine normal filter on $\mathcal{P}_{<\kappa}(\lambda)$ then there is no strong stationary coding set for $\mathcal{D}$

4) If $2^{<\kappa} \leq \lambda < \lambda^{<\kappa}$, $\mathcal{D}$ a fine normal filter on $\mathcal{P}_{<\kappa}(\lambda)$ then there is no SC for $\mathcal{D}$.

Proof : 3) Suppose S is a $\mathcal{D}$-CD, and let $C \subseteq \lambda$ be closed unbounded of cardinality $< \lambda$. Clearly $S_1 = \{a \in \mathcal{P}_{<\kappa}(\lambda)$: $a \cap c$ is unbounded in $a\} \in \mathcal{D}$, hence $S_2 = \{a \in S : \sup a \in c \cup \{\lambda\}\} \neq \phi \; mod \; \mathcal{D}$, but $|S_2| < \lambda$ so we get contradiction to "$\mathcal{D}$ is fine".

4) If S is such a set, if h witnesses its being SC then it shows $|S| \leq \lambda$, but every $a \in \mathcal{P}_{<\kappa}(\lambda)$ is a subset of some $b \in S$ hence $\lambda^{<\kappa} \leq |\mathcal{P}_{<\kappa}(\lambda)| = \sum_{b \in S} |\mathcal{P}(b))| \leq |S| \cdot 2^{<\kappa} \leq \lambda$ contradiction.

7 Claim: Suppose $\lambda > \kappa$ is regular, $\mathcal{D} = \mathcal{D}_\kappa(\lambda)$, $T \subseteq \lambda$ is stationary, $(\forall \delta \in T)\, [cf \;\delta < \kappa]$, and $\diamondsuit_T$ holds. Suppose also: $\kappa > \aleph_1$ or for some normal filter $\mathcal{D}$ over $\aleph_1$ for every increasing continuous $h : \omega_1 \to \lambda$, $\{i < \delta : h(i) \notin T\} \in \mathcal{D}$. *Then* there is a CD for $\mathcal{D}$.

7A Remark: Really (κ,λ)-CD for $\lambda > \kappa$ is a weak form of $\diamondsuit_{\{\delta<\lambda : cf \;\delta<\kappa\}}$. In fact if (κ,λ)-CD exists, $\lambda = \lambda^{<\lambda} > 2^{<\kappa}$ then $\diamondsuit_{\{\delta<\lambda : cf \;\delta < \kappa\}}$ holds.

Proof: We know that $\mathcal{D}_0 = \{A \subseteq \lambda : \diamondsuit_{\lambda - A}$ does not hold$\}$ is a normal (fine) filter on λ. As $\diamondsuit_T$ holds, $T \neq \phi \; mod \; \mathcal{D}_0$, hence for some $\vartheta < \kappa$, $T_0 = \{\delta \in T : cf \;\delta = \vartheta\} \neq \phi \; mod \; \mathcal{D}_0$. So $\diamondsuit_{T_0}$ holds hence there are $\langle M_\delta : \delta \in T_0 \rangle$, M_δ a model with universe δ and $\aleph_0$ functions such that or every model M with universe λ and $\aleph_0$ functions $\{\delta \in T_0 : M \upharpoonright \delta = M_\delta\}$ is stationary. Now we define by induction on $\delta \in T_0$, a set $A_\delta \subset \delta$ such that:

(a) A_δ is closed under the functions of $M_\delta\}$,

(b) $\sup A_\delta = \delta$.

(c) $|A_\delta| < \kappa$.

(d) if $\delta_1 \in \;\delta \cap T_0$, $A_{\delta_1} \subseteq A_\delta$ then $\delta_1 \in A_\delta$.

(This clearly suffices). We can even strengthen (d) to

(d') if $\delta_1 \in T_0 \cap \delta$ is in the closure of A_δ then $\delta_1 \in A_\delta$.

If $\kappa > \aleph_1$ let $\sigma < \kappa$ be regular cardinal such that $\sigma \neq \vartheta$, $(\sigma \geq \aleph_0)$, and then define by induction on $\xi < \sigma$, a set $A_\xi^\delta \subseteq \delta$, such that A_ξ^δ has cardinality $< \kappa$, is increasing with ξ, the closure of A_ξ^δ under the functions of M_δ is $\subseteq A_{\xi+1}^\delta$ and also every accumulation point of A_ξ^δ which is $< \delta$ is in $A_{\xi+1}^\delta$ and $\sup(A_0^\delta) = \delta$. Then $\bigcup \{A_\xi^\delta : \xi < \sigma\}$ is as required. The case $\kappa = \aleph_1$ is similar.

The following (8,9) is a variant of Silver [Si].

8 Definition: 1) $\mathit{Si}(\kappa,\lambda)$ means that for every algebra M with universe λ and countably many functions there are isomorphic subalgebras M_1, M_2 of power $< \kappa$, $M_1 \subseteq M_2$, $M_1 \neq M_2$ and $M_1 \cap \kappa = M_2 \cap \kappa =$ an ordinal.

2) For $\mathcal{D}$ a fine normal filter on $\mathcal{P}_{<\kappa}(\lambda)$, $\mathit{Si}(\mathcal{D})$ $[\mathit{SSi}(\mathcal{D})]$ means that for every $T \in \mathcal{D}$ $[T \neq \phi \bmod \mathcal{D}]$ and M above we can find M_1, M_2 as above $M_1 \in T$, $M_2 \in T$.

3) The negation of $\mathit{Si}(\kappa,\lambda)$, $\mathit{Si}(S)$, $\mathit{SSi}(\mathcal{D})$ are denoted by $N\,\mathit{Si}(\kappa,\lambda)$, $N\mathit{Si}(\mathcal{D})$, $N\mathit{SSi}(\kappa,\lambda)$ $N\mathit{SSi}(\mathcal{D})$ resp.

9 Fact: 1) If $\mathit{Si}(\kappa,\lambda)$ and $\lambda \leq \lambda^*$ then $\mathit{Si}(\kappa,\lambda^*)$.

2) If $\mathit{Si}(\mathcal{D})$ then $\mathit{Si}(\mathcal{D}_\kappa(\lambda))$

3) The first $\lambda \geq \kappa$ for which $\mathit{Si}(\kappa,\lambda)$ holds, is a strongly inacessible cardinal.

4) $\mathit{Si}(\mathcal{D}_\kappa(\lambda))$ is equivalent to $\mathit{Si}(\kappa,\lambda)$.

5) If $\mathit{Si}(\mathcal{D})$ then there is a minimal normal $\mathcal{D}_1$ extending $\mathcal{D}$ for which $\mathit{SSi}(\mathcal{D})$.

10 Claim: 1) Suppose $\diamondsuit_\kappa$ and $N\,\mathit{Si}(\kappa,\lambda)$, then there is a WSC for $\mathcal{D}_\kappa(\lambda)$.

2) If in 1), T is a set of strongly inaccessible cardinals (hence κ is Mahlo), $\diamondsuit_T$ and $\lambda = \lambda^{<\kappa}$ *then* there is a SC for $\mathcal{D}$.

3) Suppose $\mathcal{D}$ is a normal fine filter on $\mathcal{P}_{<\kappa}(\lambda)$, $\mathcal{D}_0 = \{A \subseteq \kappa : \{a \in \mathcal{P}_{<\kappa}(\lambda) : a \cap \kappa \in A\} \in \mathcal{D}\}$ (so $\mathcal{D}_0$ is necessarily a normal filter on κ). Suppose further $\diamondsuit(\mathcal{D}_0)$ which means: there is $\langle A_\delta : \delta < \kappa \rangle$ such that for every $A \subseteq \kappa$,

$\{\delta < \kappa : A \cap \delta = A_\delta\} \neq mod\ \mathcal{D}_0$. If $NSi(\mathcal{D})$ then there is a $\mathcal{D}$-WSC; and when in addition $\{\vartheta < \kappa : \vartheta$ strongly inaccessible$\} \neq mod\ \mathcal{D}_0$ then there is a $\mathcal{D}$-CD.

Proof : 1) Let $T \subseteq \kappa$ be stationary such that $\diamondsuit_T$ holds. There is an algebra M with countably many functions exemplifying $NSi(\mathcal{D}(\kappa,\lambda))$. As $\diamondsuit_T$, we can find models M_α ($\alpha \in T$) such that:

i) M_α is an algebra with countably many functions, and universe γ_α $\alpha \leq \gamma_\alpha < \kappa$.

ii) *if* $\langle N_i : i < \kappa \rangle$ is an increasing continuous sequence of algebras with countably many functions, $\|N_i\| < \kappa$, and $\kappa \subseteq \bigcup_{i<\kappa} N_i$ *then* for stationary many $i \in T$, N_i, M_i are isomorphic over i.

Let $S^* = \{a \in \mathcal{P}_{<\kappa}(\lambda) : M \restriction a$ is isomorphic to $N_{a \cap \kappa}$ where $a \cap \kappa \in T\}$. Now S^* is a WSC, in fact if $a \in S^*$, $b \in S^*$, $a \subset b$ (but $a \neq b$) then $a \cap \kappa < b \cap \kappa$.

2) Straightforward, let h be any one-to-one function from $\mathcal{P}_{<\kappa}(\lambda)$ into λ, then $S^{**} = \{a \in \mathcal{P}_{<\kappa}(\lambda): a \cap \kappa \in T,\ |a| \leq |a \cap \kappa|$ and $(\forall b)(b \subseteq a \wedge |b| < a \cap \kappa \rightarrow h(b) \in a)\} \neq \phi\ mod\ \mathcal{D}_\kappa(\lambda)$. Now also $S^* \cap S^{**} \neq \phi\ mod\ \mathcal{D}_\kappa(\lambda)$ and $S^* \cap S^{**}$ is a strong stationary coding set.

3) Left to the reader.

11 Conclusion: If κ is an ineffable cardinal (see e.g. [KM]), $N\ Si(\mathcal{D}_\kappa(\lambda))$ and $\lambda = \lambda^{<\kappa}$ then there is a SC for $\mathcal{D}_\kappa(\lambda)$.

Proof : It is known that for ineffable κ, $\diamondsuit_\kappa$, holds, moreover $\diamondsuit_T$, where $T = \{\mu < \kappa : \mu$ strongly inaccessible$\}$. By 10(2) we finish.

12 Observation : The following properties for a succesor cardinal κ and stationary $T \subseteq \kappa$ are equivalent:

i) in $\kappa^\kappa/(\mathcal{D}_\kappa + T)$ there is an increasing sequence sequence $\langle g_\alpha/(\mathcal{D}_\kappa+T) : \alpha < \kappa^+ \rangle$ and $g/(\mathcal{D}_\kappa+T)$ such that $g/(\mathcal{D}_\kappa+T) \not\leq g_\alpha/(\mathcal{D}_\kappa+T)$ for every $\alpha < \kappa^+$.

ii) there is $g : \kappa \rightarrow \kappa$ such that for any well-ordering $<^*$ of

κ, $\{\alpha \in T : (\alpha, <^* \restriction \alpha)$ has order-type $< g(\alpha)\}$ is stationary.

iii) there is $g : \kappa \to \kappa$ such that the set $\{a \in \mathcal{P}_{<\kappa}(\kappa^+) : a \cap \kappa \in T$, $(a, < \restriction a)$ has order-type exactly $g(a \cap \kappa)\}$ is stationary.

iv) *for any* cardinal $\mu, 1 \le \mu < \kappa$ such that $(\forall \delta \in T)[cf\ \delta < \mu \wedge |\delta|^{\mu} < \kappa]$, cardinal $\lambda > \kappa$ and subsets $P_i \subseteq \lambda (i < \mu)$, *there are* functions $g_i : \kappa \to \kappa$ $(i < \mu)$, such that the set

$\{a : a \in \mathcal{P}_{<\kappa}(\lambda)$, $a \cap \kappa$ an ordinal from T and for $i < \mu$ the order type of $a \cap P_i$ is $g_i(a \cap \kappa)\}$

is stationary.

Proof: Trivially (iv) $\Longrightarrow$ (iii). Next we show (iii) $\Longrightarrow$ (ii): if g exemplify (iii), $<^*$ a well ordering of κ, then for some α, $\kappa \le \alpha < \kappa^+$, $(\alpha, <)$ is isomorphic to $(\kappa, <^*)$, and let h be such an isomorphism. Let $\alpha + 1 = \bigcup_{i<\kappa} a_i$, a_i increasing continuous, $|a_i| < \kappa$, so for some closed unbounded $C \subseteq \kappa$, $(\forall \delta \in C)$ $[a_\delta \cap \kappa = \delta$, and h is isomorphism from a_δ onto $(\delta, <^* \restriction \delta)]$. If (ii) fail (for this g, for this $<^*$) we can assume $(\forall j \in C \cap T)[(\delta, <^* \restriction \delta)$ has order type $> g(\alpha)$, but then $\{a \in \mathcal{P}_{<\kappa}(\kappa^+) : \alpha + 1 \in a, a \cap (\alpha+1)$ is a_δ for some $\delta \in C\}$ belongs to $\mathcal{D}_\kappa(\kappa^+)$ contradicting the choice of g.

Now we show (ii) $\Longrightarrow$ (i), let for $\alpha \in (\kappa, \kappa^+)$, $<_\alpha$ be a well ordering of κ of order type α, and let $g_\alpha(i) \overset{def}{=}$ "order type of $(i, <_\alpha \restriction i)$", the checking is easy.

Now if (i) holds for g, $g_\alpha (\alpha < \kappa^+)$, also (ii) holds for g: let $<_\alpha$ be any well ordering of κ of order type α (for $\kappa \le \alpha < \kappa^+$), $g^*_\alpha : \kappa \to \kappa$ be defined by $g^*_\alpha(i) =$ "the order type of $(i, <_\alpha \restriction i)$", for $\alpha < \kappa$ let $g^*_\alpha(i) = i$; we can proved by induction on $\alpha < \kappa^+$ that $g^*_\alpha / \mathcal{D}_\kappa \le g_\alpha / \mathcal{D}_\kappa$ so (ii) is clear.

Lastly assume (ii) holds for g, $\mu < \kappa$ $(\forall \delta \in T)[cf\ \delta > \mu \wedge |\delta|^{\mu} < \kappa]$, $P_i \subseteq \lambda (i < \mu)$ and we shall prove (iv) (for those $P_i (i < \mu)$). Let for $\delta \in T$ (w.l.o.g. $|g(\delta)| \le |\delta|$), h_δ be a one-to-one function from δ onto $g(\delta)$. For any sequence $\bar{\beta} = \langle \beta_j : j < \mu \rangle$ of ordinals $< \kappa$, we define function $g_{\bar{\beta}, j} : \kappa \to \kappa$ by $g_{\bar{\beta}, j}(i) = h_i(\beta_j)$. If $\langle g_{\bar{\beta}, i} : i < \mu \rangle$ is not as required then there is a

$C_{\bar{\beta}} \in \mathcal{D}_\kappa(\lambda)$ such that $(\forall a \in C_{\bar{\beta}})(\exists j < \mu)$ $[a \cap P_j$ has order-type $\neq g_{\bar{\beta},j}(a \cap \kappa)]$. Let $\{\bar{\beta}^\xi : \xi < \kappa\}$ list all such sequences $\bar{\beta}$, then

$C = \{a \in \mathcal{P}_{<\kappa}(\lambda)$: if $\xi \in a \cap \kappa$ then $a \in C_{\bar{\beta}^\xi}$ and if $\beta_j < a \cap \kappa$ for $j < \mu$ then for some $\xi < a \cap \kappa, \bar{\beta}^\xi = \langle \beta_j : j < \mu \rangle\}$

is in $\mathcal{D}_\kappa(\lambda)$. But $S = \{a \in \mathcal{P}_{<\kappa}(\lambda) : a \cap \kappa \in T$, and the order-type of a is $< g(a \cap \kappa)\}$ is $\mathcal{D}$-stationary, so there is $a^* \in S \cap C$. Let $\beta_j^* =$ order-type of $P_j \cap a$, so necessarily $\beta_j^* < g(a \cap \kappa)$, hence for some $\beta_j < a \cap \kappa$, $h_{a \cap \kappa}(\beta_j) = \beta_j^*$; now we get contradiction to $a^* \in C_{\langle \beta_j : j < \mu \rangle}$.

Remark: At least (i) $\Longleftrightarrow$ (ii) is well known.

12A Remark: We can omit the assumption "κ successor" if we add in (i), (ii), (iii) $g(\alpha) < \omega + |\alpha|^+$, and in (iv) $g_i(\alpha) < \omega + |\alpha|^+$.

13 Claim: Suppose (i) of Fact 12 holds so κ is a successor; or at least (i) of 12A holds. Suppose further that $\lambda = \kappa^{+\alpha}$ (i.e. $\kappa = \aleph_\beta$, $\lambda = \aleph_{\beta+\alpha}$) and $|\alpha|^+ < \kappa$, $(\forall \gamma < \kappa)\ \gamma^{|\alpha|} < \kappa$. Then there is a WSC for $\mathcal{D}_{<\kappa}(\lambda)$.

Proof : Clearly we can find $g_i (i \le \alpha)$ as in 12 (iv) replacing μ by α, and P_i by μ_i.

Let $M = (\lambda, f, g)$ f a two place function such that for every $i < \lambda$, $i = \{f(i,j) : j < |i|\}$, and for $j < |i|$, $g(i, f(i,j)) = j$, and $S = \{a : a \in \mathcal{P}_{<\kappa}(\lambda)$, a closed under f and g, $a \cap \kappa$ an ordinal, and for every $i \le \alpha$, the order type of $a \cap \kappa^{+i}$ is $g_i(a \cap \kappa)\}$. By 12 (iv) S is $\mathcal{D}_\kappa(\lambda)$-stationary. Suppose $a \neq b$ are in S, $a \cap \kappa = b \cap \kappa$, and $a \subseteq b$. We know (as $a,b \in S$, $a \cap \kappa = b \cap \kappa$) that for each $i \le \alpha$, $a \cap \kappa^{+i}$, $b \cap \kappa^{+i}$ has the same order type. Now we prove by induction on i that $a \cap \kappa^{+i} = b \cap \kappa^{+i}$. For $i = 0$ this is given; for i limit by the induction hypothesis; for $i = j + 1$: $a \cap \kappa^{+i}$ is unbounded in $b \cap \kappa^{+i}$ as they have the same order type, now apply the functions f,g under which a,b are closed (and $a \cap \kappa^{+j} = b \cap \kappa^{+j}$). For $i = \alpha$ we get the desired conclusion.

14 Lemma : 1) If $\kappa = \mu^+$, $\mathcal{D}$ a fine normal filter on $\mathcal{P}_{<\kappa}(\lambda)$, λ regular $\{a : cf\ (\sup a) \neq cf\ (\mu)\} \in \mathcal{D}$ then $\mathcal{D}$ is not λ^+-saturated (see Definition 4(3).).

2) $\mathcal{P}_{<\kappa}(\lambda) - \{a : cf\ (\sup\ a) \neq cf\ \mu\} \neq \mathcal{D}$ is enough in 14(1).

Proof: 1) Let P be the set of $\mathcal{D}$-stationary sets ordered by inverse inclusion. Suppose $\mathcal{D}$ is λ^+-saturated, so P satisfies the λ^+-Chain condition. We shall prove that λ^+ is a cardinal of V^P, all V-cardinals $< \kappa$ are V^P- cardinals, $V^P \models$ " $|\lambda| = \mu$, $cf\ \lambda \neq cf\ \mu$", thus contradicting [Sh 1], XIII 4.9, p. 440. The following facts fulfilling the above, are folklore at least for $\lambda = \kappa$, and straightforward generalization generally.

Fact A: For every P-name $\underset{\sim}{\alpha}$ of an ordinal $< \alpha^*$, there is a function $g : \mathcal{P}_{<\kappa}(\lambda) \to \alpha^*$, such that $[a \neq \phi \wedge \alpha^* \leq \lambda \Longrightarrow g(a) \in a]$, and ($\underset{\sim}{G}_P$ - the P-name of the generic set) $\Vdash_P$ "$\underset{\sim}{\alpha}$ is the unique α such that $\{a : g(a) = \alpha\} \in \underset{\sim}{G}_P$".

Proof : Let $\langle S_i : i < \beta \rangle$ be a maximal antichain of P such that for each i for some α_i, $S_i \Vdash_P$ "$\underset{\sim}{\alpha} = \alpha_i$". Now $|\beta| \leq \lambda$ (by the λ^+-chain condition) so w.l.o.g. $\beta = \lambda$, (we allow $S_i = \phi$) so for $i \neq j$, $S^{i,j} \overset{def}{=} \mathcal{P}_{<\kappa}(\lambda) - S_i \cap S_j$ belong to $\mathcal{D}$. Let $C = \{a \in \mathcal{P}_{<\kappa}(\lambda)$: for every $i \in a [\alpha_i < \lambda \Longrightarrow \alpha_i \in a]$, and for every $i \neq j \in a$, $a \in S^{i,j}\}$.

By the normality of $\mathcal{D}$, $C \in \mathcal{D}$.

Define a partial function h on C: $h(a) = i$ if $a \in S_i$, $i \in a$. By C's definition h has at most one value.

If $S \overset{def}{=} C - \mathrm{Dom}\ h$ is $\mathcal{D}$-stationary then remember $S \cap S_i \subseteq \mathcal{P}_{<\kappa}(\lambda) - \{a : i \in a\}$, so S contradict the "$\langle S_i : i < \alpha \rangle$ is a maximal antichain". So h is defined on some $C^* \in \mathcal{D}$, and is as required (it does not matter how we complete it on $\mathcal{P}_{<\kappa}(\lambda) - \{\phi\}$, as long as $\alpha^* \leq \lambda$, $a \neq \phi \Longrightarrow h(a) \in a$.

Fact B: Forcing by P does not collapse any $\vartheta < \kappa$.

Proof : Let $S \in P$, $S \Vdash_P$" ϑ is collapsed". Choose minimal ϑ, so ϑ is regular in V and (maybe changing S) for some regular $\sigma < \vartheta$:

$\Vdash_P$"$\underset{\sim}{f}$ is a function from σ to ϑ"

$S \Vdash_P$"$\underset{\sim}{f}$ has an unbounded range".

W.l.o.g. for each $i < \vartheta$, we apply Fact A (with $\alpha^* = \vartheta \le \lambda$) on the P-name $\underset{\sim}{f}(i)$ and get g_i. For every $a \in \mathcal{P}_{<\kappa}(\lambda) - \{\phi\}$, as (in V) ϑ is regular $> \sigma$, clearly $g(a) = \sup\{g_i(a) : i < a\}$ is an ordinal $<\vartheta$. As $\mathcal{D}$ is ϑ^+-complete and fine, $\{a : \vartheta \subseteq a\} \in \mathcal{D}$ and as Rang $(g_i) \subseteq \vartheta$ and $\mathcal{D}$ is normal clearly for some $\gamma < \vartheta$, $S^* = \{a \in S : g(a) = \gamma\}$ is $\mathcal{D}$-stationary; trivially

$$S^* \Vdash_P \text{" Rang } \underset{\sim}{f} \subseteq \gamma\text{"}$$

contradiction.

Fact C: $S \Vdash_P$" $cf\ (\chi) = \vartheta$" if χ is a regular cardinal $\le \lambda$, $S \in P$, and $S \subseteq \{a \in \mathcal{P}_{<\kappa}(\lambda) : \vartheta = cf\ (\sup\ (a \cap \chi))\}$.

For every $a \in S$ let $\langle \beta(a,i) : i < \vartheta \rangle$ be an increasing sequence of ordinals from $a \cap \chi$ with limit $\sup(a \cap \chi)$. We know by $\mathcal{D}$'s normality that for every $i < \vartheta$, and $S' \subseteq S$, $S' \in P$, for some $S'' \subseteq S'$, $S'' \in P$, and $\beta(a,i)$ is constant for $a \in S''$.

Let $\underset{\sim}{\beta}_i$ be the unique ordinal β such that $\{a : \beta(a,i) = \beta\} \in \underset{\sim}{G}_P$ (this is a P-name). So $S \Vdash_P$"$\underset{\sim}{\beta}_i$ is an ordinal $< \kappa$ and $\underset{\sim}{\beta}_i < \underset{\sim}{\beta}_j$ for $i < j < \vartheta$."

Also we shall show that $S \Vdash_P$"$\{\underset{\sim}{\beta}_i : i < \vartheta\}$ is unbounded below χ (hence $S \Vdash$" $cf\,(\chi) = \vartheta$" and we shall finish). This holds because for every $S' \in P$,($S' \subseteq S$) and $\beta < \chi$ w.l.o.g. $(\forall a \in S)(\beta \in a)$, and so for every $a \in S'$ there is $i_a < \vartheta$ such that $\beta(a,i_a) > \beta$, and the function i_a has a constant value j on some $\mathcal{D}$-stationary $S'' \subseteq S'$ and $S'' \Vdash$ "$\underset{\sim}{\beta}_j > \beta$". So we finish.

Remark: This is essentially Ulam argument for "$\mathcal{D}_{\omega_1}$ is not $\aleph_1$-saturated".

Fact D: $\Vdash_P$" the power of λ is μ". It is enough to prove that every regular $\chi, \kappa \le \chi \le \lambda$ is collapsed. As the number of possible cf $(\sup(a \cap \chi))$ is

$\leq |\{\vartheta : \vartheta < \kappa\}| = |\{\vartheta : \vartheta \leq \mu < \kappa\}| < \kappa$ we can use Fact C.

Fact E: $\Vdash_P$" $cf\ \lambda \neq cf\ \mu$".

By a hypothesis $S_0 = \{a \in \mathcal{P}_{<\kappa}(\lambda) : cf\,(\sup a) \neq cf\ \mu\} \in \mathcal{D}$ is in $\mathcal{D}$. As $\{cf\ \sup a : a \in S_0\} \subseteq \{\vartheta : \vartheta \leq \mu\}$ has power $< \kappa$ for every $S \in P$, for some ϑ, $S = \{a \in S \cap S_0 : cf\,(\sup a) = \vartheta\} \neq \phi \bmod \mathcal{D}$. As $S \subseteq S_0$, necessarily $\vartheta \neq cf\ \mu$. By Fact C $S_1 \Vdash_P$ "$cf\ \lambda = \vartheta$" and by Fact C, $(cf\,\mu)^V = (cf\ \mu)^{V^P}$ so $S_1 \Vdash_P$"$cf\ \ \lambda \neq \mu$". Hence $S \nVdash^P$"$cf\ \lambda = cf\ \mu$".

As $S \in P$ was arbitarily $\Vdash_P$" $cf\ \mu \neq cf\ \lambda$". So we finish the proof of 1.14.

2) Trivial by 14(1).

15 Definition : For W a set of regular cardinals $< \kappa$, let $\mathcal{D}^W_{<\kappa}(A)$ be the minimal (fine normal) filter on $\mathcal{P}_{<\kappa}(A)$, such that for every well ordering $<^*$ of A the following set belongs to $\mathcal{D}$:

$S(<^*) = S^W(<^*, A) = \{a \in \mathcal{P}_{<\kappa}(A)$: if $\mu \in W$, $\bar{x} = \langle x_i : i < \mu \rangle$ is an $<^*$-increasing but bounded (in $(A, <^*)$ but not necessarily in a) and $x_i \in a$ then the limit of $\bar{x}$ belong to $a\}$.

Remark: 1) We can ask only that each $\mathcal{D}^{\{\mu\}}_{<\kappa}(A)$ is included; the difference is small.

2) $\mathcal{D}^W_{<\kappa}(A)$ may be trivial, i.e. $\phi \in \mathcal{D}^W_{<\kappa}(A)$.

16 Fact: For every $S \in \mathcal{D}^W_{<\kappa}(A)$ there is a function $G : \mathcal{P}_{<\aleph_0}(A) \to \mathcal{P}_{<\kappa}(A)$ and well orderings $<^*_x$ of A for $x \in A$ such that S includes:

$Cn(G, <^*) = \{a \in \mathcal{P}_{<\kappa}(A)$: for every $w \in \mathcal{P}_{\aleph_0}(a), G(w) \subseteq a$ and for every $x \in A$, $a \in S^W(<^*_x, A)\}$.

17 Fact: If $\lambda \geq \kappa, W$ is a set of regular cardinals $< \kappa$, $\vartheta < \kappa$ is regular, $\vartheta \notin W$ then $\phi \notin \mathcal{D}^W_{<\kappa}(A)$.

Proof : Let G, $<^*_x(x \in A)$ be as in 16.

It suffices to prove that $\{a : (\forall x \le a)[a \in Gn(G,<_x^*)\} \ne \phi$. We define by induction on $i < \vartheta$, a set $a_i \in \mathcal{P}_{<\kappa}(A)$.

Let $a_0 = \{0\}$.

For i limit $a_i = \bigcup_{j<i} a_j$.

For $i = j+1$, let $a_i = a_j \cup \bigcup\{G(w) : w \in \mathcal{P}_{<\kappa}(a_j)\} \cup \{y:$ for some $x \in a_j$, y is an accumulation point of a_j in $(A,<_x^*)\}$, the last part contributes $\le \aleph_0 + |a_i|$ elements as each $<_x^*$ is well ordering.

Now $\bigcup_{i<\vartheta} a_i$ is as required.

18 Lemma : Suppose $W \cup \{\vartheta\}$ is a set of regular cardinals $< \kappa$, $\vartheta \notin W$, the set S is a WSC for $\mathcal{D}_{<\kappa}^{W\cup\{\vartheta\}}(\lambda)$ (which is not trivial), and $\mathcal{D}_{<\kappa}^{W\cup\{\vartheta\}}(\lambda) + S$ is not λ^+-saturated. Then there is a WSC set for $\mathcal{D}_{<\kappa}^{W}(\lambda^+)$.

Proof : Let $\langle S_\alpha : \alpha < \lambda^+ \rangle$ exemplify $\mathcal{D}_{<\kappa}^{W\cup\{\vartheta\}} + S$ is not λ^+-saturated, and w.l.o.g. $S_\alpha \subseteq S$ for each α.

W.l.o.g. let when $\lambda \le \alpha < \lambda^+$, g_α be a one-to-one function from α onto λ, let for a set a and function h, $h''(a) = \{h(x) : x \in a\}$.

We now define for every $\alpha \in T \overset{def}{=} \{\beta : \lambda \le \beta < \lambda^+$ and $cf\ \beta = \vartheta\}$ a subset S^α of $\mathcal{P}_{<\kappa}(\alpha)$ such that:

(i) S^α is stationary for $\mathcal{D}_{<\kappa}^{W\cup\{\vartheta\}}(\alpha)$.

(ii) $\{g''_\alpha(a) : a \in S^\alpha\} \subseteq S_\alpha$

(iii) each $a \in S^\alpha$ is an unbounded subset of α and $(\forall a \in S^\alpha)$ $(a \cap \lambda = g''_\alpha(a)$ and $a \cap \kappa \in \kappa)$.

(iv) if $a \ne b \in \bigcup\{S^\beta : \beta \in T, \beta \le \alpha\}$ and $a \cap \kappa = b \cap \kappa$ then $a \not\subseteq b$.

If we succeed then $S^* = \bigcup_{\alpha \in T} S^\alpha$ is as required. As T is stationary (in λ^+), $\beta \in T \to cf(\beta) \notin W$, by (i) and 16 easily S^* is $\mathcal{D}_{<\kappa}^{W}(\lambda^+)$-stationary. The other requirement for being a WSC of $\mathcal{D}_{<\kappa}^{W}(\lambda^+)$ follows by iv).

So we concentrate on the induction step.

Let $S_0^\alpha = \{a \in \mathcal{P}_{<\kappa}(\alpha) : g_\alpha''(a) \in S_\alpha\}$ and $S_1^\alpha = \{a \in \mathcal{P}_{<\kappa}(\alpha) : g_\alpha''(a) \in S\}$, clearly $S_0^\alpha \subseteq S_1^\alpha$ and both are clearly $\mathcal{D}_{<\kappa}^{W\cup\{\vartheta\}}(\alpha)$-stationary.

Now the set

$C^\alpha = \{a \in \mathcal{P}_{<\kappa}(\alpha)$: $a \cap \kappa$ an ordinal $< \kappa$, every accumulation point $\delta < \alpha$ of a, $\delta < \alpha$, $cf\ \delta \in W \cup \{\vartheta\}$ belong to a, a is closed under g_β, g_β^{-1} for $\beta \in a \cup \{\alpha\}$, and a is unbounded in $\alpha\}$

belong to $\mathcal{D}_{<\kappa}^{W\cup\{\vartheta\}}(\alpha)$. As $S_\alpha \cap S_\beta$ is not $\mathcal{D}_{<\kappa}^{W\cup\{\vartheta\}}(\lambda)$-stationary by (ii) easily for $\beta < \alpha$, $\beta \in T$ $S^\beta \cap \{a \cap \beta : a \in S^\alpha\}$ is not $\mathcal{D}_{<\kappa}(\beta)$-stationary hence $S^\alpha \cap S^{\beta,\alpha}$ is not $\mathcal{D}_{<\kappa}(\beta)$-stationary where

$$S^{\alpha,\beta} = \{a \in \mathcal{P}_{<\kappa}(\alpha) : a \cap \beta \in S^\beta\}.$$

Hence

$$S^\alpha = \{a \in \mathcal{P}_{<\kappa}(\alpha) : a \in S_0^\alpha,\ a \in C^\alpha \text{ and } (\forall \beta \in a)\ a \notin S^{\alpha,\beta}\}$$

in $\mathcal{D}_{<\kappa}^W(\alpha)$-stationary. We shall show that (iv) holds, thus finishing. [(ii) holds as $S^\alpha \subseteq S_0^\alpha$, (iii) as $S^\alpha \subseteq C^\alpha$).

If $a, b \in \bigcup\{S^\beta : \beta < \alpha, \beta \in T\}$ this is by the induction hypothesis. If $a, b \in S^\alpha$, then (remembering $a, b \in C^\alpha$), $a \cap \lambda = g_\alpha''(a) \cap \lambda$, $b \cap \lambda = g_\alpha''(b) \cap \lambda$, hence we use the assumption "S_α in a WSC for $\mathcal{D}_{<\kappa}^{W\cup\{\vartheta\}}(\lambda)$" and $S^\alpha \subseteq S_0^\alpha$. Now $a \in S^\alpha$, $b \in S^\beta$, $\beta < \alpha, a \subseteq b$ is impossible as a is unbounded in α.

We are left with the care $a \in S^\beta$, $b \in S^\alpha$, $\beta < \alpha$, and assume $a \subseteq b$; as $\sup a = \beta$ and $cf\ \beta = \vartheta \in W \cup \{\vartheta\}$ clearly $\beta \in b$. But by the definition of S^α, $\beta \in b \Rightarrow b \notin S^{\alpha,\beta} \Rightarrow b \cap \beta \notin S^\beta$, hence $b \cap \beta \neq a$. But as $b \in C^\alpha$, $\beta \in b$ clearly $b \cap \beta \in S_1^\beta$; so $a \subseteq b \cap \beta$, $a \neq b \cap \beta$ are both in S_1^β, and then $g''(a)$, $g''(b \cap \beta)$ will contradict "S is a WSC".

19 Conclusion; If $\kappa > \aleph_n$, κ successor then there is a WSC for $\mathcal{D}_\kappa(\kappa^{+n})$.

Proof : By 18 and 14. Let $\kappa = \mu^+$, let $\alpha = 0$, if $cf\ \mu \ge \aleph_n$, and α be n otherwise. So in both cases (as $\kappa > \aleph_n$), $\aleph_{\alpha+n} < \kappa$, and $\aleph_\alpha$ is regular and $cf\ \mu \ne \{\aleph_\alpha, \ldots, \aleph_{\alpha+n-1}\}$.

Let for $0 \le \ell \le n$ $W^\ell = \{\vartheta : \vartheta < \kappa, \vartheta \text{ regular }\} - \{\aleph_\alpha, \ldots, \aleph_{\alpha+\ell}\}$.

Note that $\mathcal{D}^{W^\ell}_{<\kappa}(\kappa^{+\ell})$ is a proper (fine normal) ideal as $\aleph_0 < \kappa$, $\aleph_0 \notin W$, $W \subseteq \{\mu : \mu < \kappa$ and μ is regular$\}$ also for each $\ell < n$ it is not $\kappa^{+\ell}$- saturated: by 14 as κ is a successor and $cf\ \mu \in W$ (otherwise by inspecting the definition of W, clearly $cf\ \mu \ge \aleph_\alpha$, hence $\alpha = 0$ so hence $cf\ \mu \ge \aleph_{\alpha+n}$ hence $cf\ \mu > \aleph_{\alpha+\ell}$).

Now we prove by induction on $\ell \le n$ that there is a WSC set S_ℓ for $\mathcal{D}^{W^\ell}_{<\kappa}(\kappa^{+\ell})$ with $cf(\sup a) = \aleph_{\alpha+\ell}$ for $a \in S_\ell$. For $\ell = 0$ $S = \{\delta < \kappa : cf\ \delta = \aleph_0\}$ is O.K. and for $\ell = m + 1$ use 18 with $W^m, \aleph_{\alpha+\ell}, \kappa^{+m}$ standing W, ϑ, λ. [on the problematic assumption "$\mathcal{D}^{W \cup \{\vartheta\}}_{<\kappa}(\lambda) + S_m$ is not λ^+-saturated"; this holds by 14 as $cf(\sup a) = \aleph_{\alpha+m}$ for $a \in S_m$ $\aleph_{\alpha+m} \ne cf\ \mu$ as $0 \le m < n$. We still have to show $cf(\sup a) = \aleph_\ell$ for $a \in S_\ell$, but this holds by the construction in 18]. So we get the result for $\ell = n$.

20 Lemma: Suppose

(i) $\mathcal{D}_\kappa$ is κ^+-saturated.

(ii) every $\mathcal{D}_{<\kappa}(\kappa^+)$-stationary set S is reflected i.e., for some $\alpha < \kappa^+, S \cap \mathcal{P}_{<\kappa}(\alpha)$ is $\mathcal{D}_{<\kappa}(\alpha)$-stationary.

Then there is no WSC set S for $\mathcal{D}_{<\kappa}(\kappa^+)$.

Remark: Later the assumptions were proved consistent for $\kappa = \aleph_1$ in Foreman Magidor and Shelah [FMS].

Proof : Suppose S is a counterexample, let $g_\alpha(\alpha < \kappa^+)$ be a one-to-one function from $\kappa + \alpha$ onto κ and let $S_\alpha = \{g''_\alpha(a) : a \in S, a \in \mathcal{P}_{<\kappa}(\alpha)\}$.

By (i) for some $\alpha(*) < \kappa^+$, $\kappa + \alpha(*) = \alpha(*)$ and for every $\alpha, \alpha(*) \le \alpha < \kappa^+$, and stationary $S^* \subseteq S_\alpha$ for some $\gamma < \alpha(*)$: $S^* \cap S_\gamma \ne \phi \mod \mathcal{D}_{<\kappa}(\kappa)$. Now

$S^a \stackrel{def}{=} \{a \in S : \alpha(*) \in a\}$ is $\mathcal{D}_{<\kappa}(\kappa^+)$- stationary (as $S - S^a$ is not). So by (ii) for some α, $\alpha(*) < \alpha < \kappa^+$, and $S^b = S^a \cap P_{<\kappa}(\alpha)$ is $\mathcal{D}_\kappa(\alpha)$-stationary; by the choice of $\alpha(*)$ there are $a \in S^b$, $\gamma < \alpha(*)$ such that $a \cap \gamma \in S_\gamma \subseteq S$. But $a \cap \gamma \subset a$, (not equal as $\alpha(*) \in a$ because $a \in S^a$) and we get contradiction to "S is WSC".

21 Lemma : Suppose $\lambda \rightarrow (\kappa)^{<\omega}_\kappa$, then there is a fine normal filter $\mathcal{D}$ on $P_{<\kappa}(\lambda)$ for which there is no WSC.

Proof : For every model M with universe λ and $<\kappa$ functions let $G(M) = \{A : A$ a submodel of M, and some expansion of A is generated of a sequence of length α of indiscernible $\omega \leq \alpha < \kappa\}$.

References

[FMS]
Foreman, M. Magidor, and S. Shelah, in preparation.

[BTW]
J. Baumgartner, A. Taylor, S. Wagon, Structural properties of ideal, Dissertation Math. 17 (182).

[KM] A. Kanamori, and M. Magidor, The development of large cardinals.

[M] T. K. Menas, On strong compactness and super-compactness, *Annals of Math. Logic*, 7 (1974), 327.

[Sh1]
S. Shelah, Proper Forcing, Springer Verlag Lecture Notes, 940 (1982).

[Sh2]
S. Shelah, Abstracts of A.M.S. 4 (1983), 578

[Si] J. Silver, A large cardinal in the constructible Universe, 69 *Fund. Math.* 93-100.

[So] Solovay, Strongly compact cardinals and the G.C.H. proc. of The Tarski Symp. Berkeley 1971, ed. Henlien; Symp. in Pure Math. Vol. XXV (1974), 187- 204.

[Z] B. Zwicker, $P_\kappa(\lambda)$ combinatorics, I: Stationary Coding sets, *Comtemporary Mathematics*, 31 (1984), 243-259.

Monadic Logic: Hanf Numbers[†]

Abstract

This is part of the classification developed in Baldwin Shelah [BSh]. The paper is divided into two parts. In part I we show that $(T_\infty, 2^{nd}) \not\leq (T, \text{mon})$ iff the Hanf number for the theory T in monadic logic is smaller than the Hanf number of second order logic.

For this we deal with partition relations for models of T. The main result is that if T does not have the independence property even after expanding by monadic predicates (or equivalently $(T_\infty, 2^{nd}) \not\leq (T, \text{mon})$) then: $\beth_{\omega+1}(\lambda)^+ \to_s (\lambda)^{<\omega}_T$. In Part II we analyze such T getting a decomposition theorem like that in [BSh] (but weaker) (This is needed in part I.)

Part I

§1 Preliminaries

We review here some relevant facts and definitions.

1.1. Convention:

T will be a fix complete theory, $\mathfrak{C}$ a $\bar{\kappa}$-saturated model of T, $\bar{\kappa}$ large enough (see [Sh1] I §1); M,N denote elementary submodels of $\mathfrak{C}$ of power $<\bar{\kappa}$, A,B,C subsets of such M, a,b,c,d elements of $\mathfrak{C}$, $\bar{a},\bar{b},\bar{c},\bar{d}$ finite sequences, and I,J denote linear orders. A monadic expansion of M is expansion by monadic predicates; a finite expansion is one by finitely many relations. When dealing with finite monadic expansions of $\mathfrak{C}$, we may mean a $\bar{\kappa}$-saturated one, or any such expansion. We shall not specify, because if $M \subset \mathfrak{C}$, M^+ a finite expansion of M, then we can expand $\mathfrak{C}$ to $\mathfrak{C}^+$, an

---

† I thank Rami Gromberg for many corrections.

elementary extension of M^+ which is $\bar{\kappa}$-saturated.
This paper has two parts, the major one in part I, but in order to prove an important property of decomposition of models (see claim I2.4(1)) we need a property of types which is lemma 2.3 of Part II. The sole contribution of part II to I is the proof of this lemma.
We quote from [BSh] 1.2, 1.3:

1.2. Definition:

We say $(T_\infty, 2^{nd}) \le (T, \text{mon})$ *if* in some monadic expansion of $\mathfrak{C}$, there is an infinite set on which a pairing function is defined. (a pairing function on A is a one-to-one function from $A \times A$ onto A).

1.3. Theorem:

1) If T has the independence property (see [Sh1] II §4) then $(T_\infty, 2^{nd}) \le (T, \text{mon})$. Hence, $(T_\infty, 2^{nd}) \le (T, \text{mon})$ iff some finite monadic expansion of a model of T has the independence property.

2) If in some finite monadic expansion of $\mathfrak{C}$ for some infinite sets $\{a_t : t \in I\}$, $\{b_t : t \in J\}$ and formula θ, for any $t \in I$, $s \in J$ there is d such that $(\forall u \in I)$ $(\forall v \in J)$ $[\theta(a_u, b_v, d) \longleftrightarrow t = u \wedge s = v]$ *then* $(T_\infty, 2^{nd}) \le (T, \text{mon})$.

We quote from [Sh1] VII §4:

1.4. Definition:

1) We say p is finitely satisfiable in A if every finite subset of p is realized by elements of A

2) For an ultrafilter D on ${}^I A$, and set B, we define

$$Av(D,B) = \{\varphi(\ldots,x_t,\ldots,\bar{b})_{t\in I} : \bar{b} \in B \text{ and the set}$$
$$\{\langle a_t : t \in I\rangle : \models \varphi[\ldots,a_t,\ldots,\bar{b}]_{t\in I}\} \text{ belong } D\}$$

1.5. Lemma:

1) $Av(D,B)$ is a complete type in the variables $\langle x_t : t \in I\rangle$ over B, finitely satisfiable in A; of course $B \subseteq C \Longrightarrow Av(D,B) \subseteq Av(D,C)$

2) If p is finitely satisfiable in A, p a set of formulas in the variables $\{x_t : t \in I\}$, *then* for some ultrafilter D on ${}^I A$, and some set B $p \subseteq Av(D,B)$.

3) If p is finitely satisfiable in A then p does not split over A (i.e., if $\bar{b},\bar{c}$ realize the same type over A then for no φ, $\varphi(\bar{x},\bar{b})$, $\neg\varphi(\bar{x},\bar{c}) \in p$)

4) If p is an m-type over B finitely satisfiable in A, then it can be extended to $p' \in S^m(B)$ finitely satisfiable in A,

5) If $p, q \in \bigcup_{m<\omega} S^m(C)$ are finitely satisfiable in A, $B \subseteq C$, and every m-type over A realized in C is realized in B, then $p \restriction B = q \restriction B \Longrightarrow p = q$

6) If $tp_*(C_0, A \cup B)$ is finitely satisfiable in A, and $tp_*(C_1, A \cup B \cup C_0)$ is finitely satisfiable in $A \cup C_0$ then $tp_*(C_0 \cup C_1, A \cup B)$ is finitely satisfiable in A.

1.6. Observation:

If every $p \in \bigcup_m S^m(A_0)$ is realized in A_1, (hence $A_0 \subset A_1$) $tp_*(D \cup C, A_1 \cup B)$ and $tp_*(D, A_1 \cup C)$ are finitely satisfiable in A_0 *then* $tp_*(D, A_1 \cup B \cup C)$ is finitely satisfiable in A_0

Proof: W.l.o.g. $D = \bar{d}$; by 1.5(4) there is $\bar{d}^1$ realizing $tp(\bar{d}, A_1 \cup C)$ such that $tp(\bar{d}^1, A_1 \cup B \cup C)$ is finitely satisfiable in A_0 (remember that $tp(\bar{d}, A_1 \cup C)$ is finitely satisfiable in A_0). By 1.5(6) $tp_*(C \cup \bar{d}^1, A_1 \cup B)$ is finitely satisfiable in A_0.
So $tp_*(C \cup \bar{d}^1, A_1 \cup B)$, $tp_*(C \cup \bar{d}, A_1 \cup B)$ are both finitely satisfiable in A_0, and their restriction to A_1 are equal. By 1.5(5) they are equal. Hence $tp(\bar{d}^1, A_1 \cup B \cup C) = tp(\bar{d}, A_1 \cup B \cup C)$. As $tp(\bar{d}^1, A_1 \cup B \cup C)$ is finitely satisfiable in A_0, necessarily $tp(\bar{d}, A_1 \cup B \cup C)$ is finitely satisfiable in A_0.

1.6A Remark: We can weaken the hypothesis by restricting ourselves to $p \in \bigcup_{m<\omega} S^m(A_0)$ realized in $A_1 \cup B$.

§2 A Weak Decomposition Theorem

Hypothesis: $(T_\infty, 2^{nd}) \not\leq (T, \mathrm{mon})$.

Notation: Let I, J be linear ordering

2.1. Definition:

1) We say that $\bar{A} = \langle A_t : t \in I \rangle$ is a partial decomposition of M over N iff : the A_t's are pairwise disjoint subsets of M and for every $t \in I$, $tp_*(A_t, \bigcup_{s<t} A_s \cup N)$ is finitely satisfiable in N (but not necessarily $N \subseteq M$).

2) $\bar{A}$ is a decomposition of M over N, if it is a partial decomposition of M over N and $M = \bigcup_{t \in I} A_t$.

2.2. Definition:

For partial decomposition $\langle A_t : t \in I \rangle$, $\langle B_t : t \in J \rangle$ of M over N we say $\langle A_t : t \in I \rangle \leq \langle B_t : t \in J \rangle$ if $I \subseteq J$ and for every $t \in I$, $A_t \subseteq B_t$; we say $\langle A_t : t \in I \rangle \leq^* \langle B_t : t \in J \rangle$ if $I = J$ and for every $t \in I$, $A_t \subseteq B_t$.

2.3. Claim:

1) For every $<$-increasing sequence of partial decompositions of M over N there is a least upper bound (similarly for $<^*$)

2) If $\langle A_t : t \in I \rangle$ is a partial decomposition of M over N, $I \subseteq J$, and for $t \in J - I$, we let $A_t = \phi$ *then* $\langle A_t : t \in J \rangle$ is a partial decomposition of M over N

Proof: Immediate

2.4. Claim:

1) Suppose $\langle A_t : t \in I \rangle$ is a partial decomposition of M over N and $c \in M$. Then for some $\langle B_t : t \in J \rangle \geq \langle A_t : t \in I \rangle$ (a partial decomposition of M over N), $c \in \bigcup_{t \in I} B_t$

2) If I is a well-ordering with last element then w.l.o.g. $I = J$

Proof:

1) W.l.o.g. $c \notin \bigcup_{t \in I} A_t$. Let I_1 be a maximal initial segment of I [i.e., $(\forall t \in I_1)\ (\forall s \in I)\ (s < t \rightarrow s \in I_1)$] such that $tp(c, \bigcup_{t \in I_1} A_t \cup N)$ is finitely satisfiable in N (there is such I_1, as $I_1 = \phi$ satisfies the demand, and by the finitary character of the demand). By 2.3 (2) w.l.o.g. $I_1 = \{s \in I : s < t^*\}$ for some $t^* \in I$. Now we let $J = I$, and let B_t be A_t if $t \neq t^*$, and $A_t \cup \{c\}$ if $t = t^*$. We now check Def 2.1 (1). The main non-obvious point is why for t, $t^* < t \in I$, $tp_*(B_t, \bigcup_{s<t} B_s \cup N)$ is finitely satisfiable in N. If not then for some $\bar{b} \in B_t = A_t$, $\bar{a} \in \bigcup_{s<t} B_s - \{c\} = \bigcup_{s<t} A_s$, $tp(\bar{b}, \bar{a} \cup \{c\} \cup N)$ is not finitely satisfiable in N. However we know that $tp(\bar{b}, \bar{a} \cup N)$ is finitely satisfiable in N (as $\langle A_s : s \in I \rangle$ is a partial decomposition of M over N). Also $tp(c, \bigcup_{s<t} A_s \cup N)$ is not finitely satisfiable in N (by the choice of I_1, as maximal, as $t > t^*$ and as w.l.o.g. we add t^* only if needed), hence w.l.o.g. $tp(c, \bar{a} \cup N)$ is not finitely satisfiable in N. Hence, $tp(\bar{b} \frown \langle c \rangle, \bar{a} \cup N)$ is not finitely satisfiable in N. Together, $N, \bar{a}, \bar{b}, c$ contradict Lemma II 2.3. For $t = t^*$, we should prove for $\bar{b} \in A_{t^*}$, $\bar{a} \in \bigcup_{s<t^*} B_s = \bigcup_{s<t^*} A_s$ that $tp(\bar{b} \frown \langle c \rangle, N \cup \bar{a})$ is finitely satisfieable in

N. Suppose this fails. As $\langle A_s : s \in I \rangle$ is a partial decomposition of M over N, clearly $tp(\bar{b}, N \cup \bar{a})$ is finitely satisfiable in N. By II 2.3 and the last two facts $tp(\bar{b}, (\bar{a} \frown \langle c \rangle) \cup N)$ is finitely satisfiable in N. By the choice of t^*, $tp(c, N \cup \bar{a})$ is finitely satisfiable in N. By 1.5 (6) and the last two facts, $tp(\bar{b} \frown \langle c \rangle, N \cup \bar{a})$ is finitely satisfiable in N, contradiction

2) Either there is t^* as required or $I_1 = I$, and then choose t^* as last.

2.5. Conclusion:

1) Suppose $\langle A_t : t \in I \rangle$ is a partial decomposition of M over N. Then there is a decomposition $\langle B_t : t \in J \rangle \geq \langle A_t : t \in I \rangle$ of M over N

2) If $I = (\alpha + 1, <)$ then w.l.o.g. $J = I$; also $|\{t \in J : B_t \neq \phi\}| \geq |\{t \in I : A_t \neq \phi\}|$

Proof: Immediate by 2.3(1), 2.4

Remember that (see [Sh1]) $Ded_r(\lambda)$ is the first regular cardinal μ, such that every linear order of power λ has strictly less than μ Dedekind cuts.

2.6. Lemma:

1) Suppose $\langle A_t : t \in I \rangle$ is a decomposition of M over N. Then we can find relations $P^n_{\gamma,\alpha}$ $(\alpha < \lambda_N < Ded_r(||N|| + |T|)$, $\gamma < |T|$, hence $\lambda_N \leq 2^{||N|| + |T|})$ such that:

a) $P^n_{\alpha,\gamma}$ is an n-place relation on M.

b) if $\gamma < |T|$, $n < \omega$, and $\alpha \neq \beta$, then $P^n_{\gamma,\alpha} \cap P^n_{\gamma,\beta} = \phi$, and $\bigcup_{\alpha} P^n_{\gamma,\alpha} = \bigcup_{t \in I} {}^n(A_t)$

c) for a finite sequence $\bar{b}$ from any A_t let $\alpha_\gamma(\bar{b})$, $n(\bar{b})$ be the unique n and α such that $\bar{b} \in P^n_{\gamma,\alpha}$; then if $t_1 < \cdots < t_n$, $\bar{b}_m \in A_{t_m}$ then we can compute the type of $\bar{b}_1 \frown \cdots \frown \bar{b}_n$ from $\langle n(\bar{b}_m) : m = 1, n \rangle$, and $\langle \alpha_\gamma(\bar{b}_m) : m = 1, n \rangle$, for $\gamma < |T|$

2) So as $Ded_r(|T|) \leq (2^T)^+$ we can use just $|T|$ predicates when $|N| \leq |T|$, and we waive the disjointness of the $P^n_{\gamma,\alpha}$'s.

Proof:

1) For any set A, $N \subseteq A$, and formula φ the number of $p \in S^m_\varphi(A)$ finitely satisfiable in N is $< Ded_r(||N||)$ (see [Sh2, p.202], slightly improving a result of Poizat, which suffice for (2) (alternatively use $\alpha < 2^{2^{|T|}}$))
Let N_1 be such that $N \subseteq N_1$, N_1 is $(||N||+|T|)^+$-saturated and we shall show that w.l.o.g.

(*) for each $t \in I$, $\bar{b} \in A_t$, $tp(\bar{b}, N_1 \cup \bigcup_{s<t} A_s)$ is finitely satisfiable in N

For each $t \in I$ let $A_t = \{b_v^t : v \in I(t)\}$, let $\bar{\mathbf{b}}^t = \langle b_v^t : v \in I(t) \rangle$ and by 1.5(2) we can choose an ultrafilter D_t on ${}^{I(t)}N$ such that:

$$tp(\bar{\mathbf{b}}^t, N \cup \bigcup_{s<t} A_s) = Av(D_t, N \cup \bigcup_{s<t} A_s)$$

It suffice to us that for any $t(0) < \cdots < t(n) \in I$

(*) $tp(\bar{\mathbf{b}}^{t(n)}, N_1 \cup \bigcup_{l<n} \bar{\mathbf{b}}^{t(l)}) = Av(D_{t(n)}, N_1 \cup \bigcup_{l<n} \bar{\mathbf{b}}^{t(l)})$

For any finite set $w \subseteq I$ we define $q_w = q_w(\langle x_c \cdot c \in N_1 \rangle)$, a complete type over $N \cup \bigcup_{t \in w} \bar{\mathbf{b}}^t$ by induction on $|w|$. For $w = \phi$ it is $tp_*(N_1, N)$, if $w \neq \phi$ let $w = \{t(0), \ldots, t(n)\}$, $n \geq 0$, $t(0) < \cdots < t(n)$, and we define it by

(*) if $\langle e_c : c \in N_1 \rangle$ realizes $q_{w - \{t(n)\}}$, we can find $\langle b'_v : v \in I(t(n)) \rangle$ realizing $Av(D_{t(n)}, N \cup \bigcup_{p<n} A_{t(l)} \cup \{e_c : c \in N_1\})$ and let F be an automorphism of $\mathfrak{C}$, F the identiy on $N \cup \bigcup_{l<n} A_{t(l)}$, $F(b'_v) = b_v^{t(n)}$ for $v \in I(t(n))$. Now $q_w = tp(\langle F(e_c) : c \in N_1 \rangle, N \cup \bigcup_{t \in w} A_t)$.

It is easy to prove that for $w_1 \subseteq w_2 (\subseteq I$ finite) $q_{w_1} \subseteq q_{w_2}$, (by induction on $|w_2|$) and obviously q_w is finitely satisfied. Hence $\cup\{q_w : w \subseteq I$ finite$\}$ is finitely satisfiable, hence realized by some $\langle e'_c : c \in N_1 \rangle$. We can use $\{e'_c : c \in N_1\}$ instead N_1 and then (*) holds.

Let $\{\varphi_\gamma^n(\bar{x}, \bar{y}) : \gamma < |T|\}$ be a list of the formulas $\varphi(\bar{x}, \bar{y})$, $l(\bar{x}) = n$, and $\{p_{\gamma,\alpha}^n : \alpha < \lambda_N\}$ be a list of $\{tp(\bar{b}, N_1) \restriction \varphi: \bar{b}, \varphi$ as above$\}$, lastly $\bar{b} \in P_{\gamma,\alpha}^n$ iff $\bar{b} \in \bigcup_{t \in I} {}^n(A_t)$, $n = l(\bar{b})$, and $tp(\bar{b}, N_1) \restriction \varphi_\gamma = p_{\gamma,\alpha}^n$

2) Obvious from (1).

§3 Partition relations for theories

3.1. Definition:

1) $\lambda \rightarrow (\mu)_T^n$ mean that for every model M of T of power λ, there are distinct elements a_i $(i < \mu)$ such that $\langle a_i : i < \mu \rangle$ is an n-indiscernible sequence in M.

2) $\lambda \rightarrow_s (\mu)_T^n$ means that for every model M of T and $a_i \in M (i < \lambda)$ there is $I \subseteq \lambda$, $|I| = \mu$ such that $\langle a_i : i \in I \rangle$ is an n-indiscernible sequence.

3) $\lambda \rightarrow (\mu)^{<\omega}_T$, $\lambda \rightarrow_s (\mu)^{<\omega}_T$, are defined similarly.

3.2. Discussion

This definition was suggested by R. Grossberg and the author during the winter of 1980/1, but we still know little. We can rephrase [Sh1] I 2.8 as, e.g.: if T is stable, $\lambda = \lambda^{|T|}$ then $\lambda^+ \rightarrow_s (\lambda^+)^{<\omega}_T$. We cannot hope for results on T without the strict order property (see [Sh1] II§4) or even for simple T (see [Sh2].) The reason is as follows: suppose $\lambda \not\rightarrow (\mu)^{<\omega}_2$, and let F be a function from $\{w : w \subseteq \lambda, |w| < \aleph_0\}$ to $\{0,1\}$ exemplifying it, let L consist of the predicates R_n (n place) P_n (monadic) for $n < \omega$, and let T be the model completion of $\{(\forall x)(x = x)\}$ in this language. We define an L-model M with universe $\{a_{n,i} : n < \omega, i < \lambda\}$ such that:

(i) for $w \subseteq \lambda$, $|w| = n$, $\langle a_{n,i} : i \in w \rangle \in R^M_n$ iff $F(w) = 0$.

(ii) for every n,i for some k, for every $m > k$ $a_{n,i} \in P_m$ iff m is divisible by the nth prime.

(iii) if $(\forall y_1 \cdots y_m)(\exists x)\,[\bigwedge_{l=1}^{m} x \neq y_l \wedge \varphi(x, y_1 \cdots, y_m)]$ belong to T, φ quantifier free, but R_k, P_k do not appear in φ and $a_1 \cdots, a_m \in \{a_{n,i} : n < k, i < \lambda\}$, then there is $b \in \{a_{k,i} : i < \lambda\}$ such that $\models \varphi[b, a_1, \ldots, a_m]$.

This is quite easy, M is a model of T (by T's definition and (iii),) and M exemplify $\lambda \not\rightarrow (\mu)^{<\omega}_T$. We can similarly deal with $\lambda \not\rightarrow (\mu)^n_T$.

Now T is simple, and in fact very close to T_{ind}. This leads naturally to:

3.3. Conjecture:

If T does not have the independence property, then for every μ for some λ, $\lambda \rightarrow (\mu)^{<\omega}_T$, or even $\beth_{\omega+\omega}(\mu + |T|) \rightarrow (\mu)^{<\omega}_T$.

3.4. Lemma:

Suppose $(T_\infty, 2^{nd}) \not\leq (T, \mathrm{mon})$, then

$$\beth_{\omega+1}(\lambda + |T|)^+ \rightarrow_s (\lambda)^{<\omega}_T .$$

Proof: W.l.o.g. $\lambda > |T|$, let $\mu = \beth_\omega(\lambda)$, $A = \{a_i : i < (2^\mu)^+\}$, for $i \neq j$

$a_i \neq a_j \in M$, and M is a model of T.

3.5. Fact:

At least one of the following occurs for $A = \{a_i : i < (2^\mu)^+\} \subseteq M$, $|A| = (2^\mu)^+$:

(i) There is an indiscernible sequence of length $(2^\mu)^+$ of distinct members of A (in the same length)

(ii) There is k, and $\bar{a}_i \in {}^kA (i < \mu)$ and θ such that $M \models \theta[\bar{a}_i, \bar{a}_j]$ iff $i < j$;

Proof: Repeat the proof of [Sh1] I 2.12. Let $A_i \stackrel{def}{=} \{a_j : j < i\}$
Let $S = \{\delta < (2^\mu)^+ : cf\ \delta > \mu\}$, clearly S is a stationary subset of $(2^\mu)^+$. For each $\delta \in S$ and formula φ choose if possible a subset $B_{\delta,\varphi} \subseteq A_\delta$, $B_{\delta,\varphi}$ of cardinality $< \mu$ such that: $tp_\varphi(a_\delta, A_\delta)$ does not split over $B_{\delta,\varphi}$ [i.e., if $\varphi = \varphi(x, \bar{y})$, $\bar{b}$, $\bar{c}$ sequences from A_δ of length $l(\bar{y})$ realizing the same type over $B_{\delta,\varphi}$ then $\models \varphi[a_\delta, \bar{b}] \equiv \varphi[a_\delta, \bar{c}]$]. Let $S_\varphi = \{\delta \in S: B_{\delta,\varphi}$ is defined$\}$.

Case a: For each φ for some closed unbounded $C \subseteq (2^\mu)^+$, $C \cap S = C \cap S_\varphi$

Then there is a closed unbounded $C \subseteq (2^\mu)^+$ such that for every φ, $C \cap S = C \cap S_\varphi$. For each $\delta \in C \cap S$ choose B_δ, B_δ a subset of A_δ of power μ including $\bigcup_\varphi B_{\delta,\varphi}$ such that for each φ, and $n < \omega$, every n-type over $B_{\delta,\varphi}$ realized in A_δ is realized in B_δ (possible as $|B_{\delta,\varphi}| < \mu$, μ strong limit). Now by Fodor's lemma for some stationary $S^* \subseteq C \cap S$, for all $\delta \in S^*$, B_δ $\langle B_{\delta,\varphi}: \varphi \in L(T) \rangle$, $tp(a_\delta, B_\delta)$ are the same. Continue as [Sh1] I 2.12.

Case b: For some φ, $S - S_\varphi$ is a stationary subset of $(2^\mu)^+$

So there is $\delta \in S - S_\varphi$ such that for every $B \subseteq A_\delta$, $|B| \leq \mu$, there is $\alpha < \delta$ such that a_α realizes $tp(a_\delta, B)$. So choose by induction on $i < \mu$, $\bar{b}_i, \bar{c}_i, d_i \in A_\delta$ as follows:

(α) $\bar{b}_\alpha$, $\bar{c}_\alpha$ realizes the same type over $\bigcup_{j<i} \bar{b}_j {}^\frown \bar{c}_j {}^\frown \langle d_j \rangle$

and $\models \varphi[a_\delta, \bar{b}_\alpha] \equiv \neg\varphi[a_\delta, \bar{c}_\alpha]$

(*w.l.o.g.* $\models \varphi[a_\delta, \bar{b}_\alpha] \wedge \neg\varphi[a_\delta, \bar{c}_\alpha]$)

(β) d_i realizes $tp(a_\delta, \bigcup_{j<i} (\bar{b}_j {}^\frown \bar{c}_j {}^\frown \langle d_j \rangle) \cup \bar{b}_i {}^\frown \bar{c}_i$

By the choice of δ this is possible and $\langle \bar{b}_i {}^\frown \bar{c}_i {}^\frown \langle d_i \rangle : i < \mu \rangle$ is as required.

3.6. Fact:

If $d_i \in M$ are distinct for $i < (2^\lambda)^+$.

$$B_\alpha = \{d_i : i < \alpha\}\ ,\ \ B = \{d_i : i < (2^\lambda)^+\}\ ,$$

then at least one of the following occurs:

(i) for some $\gamma < (2^\lambda)^+$ and $k < \omega$, $\{tp(\bar d, B_\gamma)$: $\bar d$ a sequence of length k from $B\}$ has power $(2^\lambda)^+$

(ii) (i) does not occur but for some $\varphi = \varphi(x, \bar y)$ for a stationary set of $\delta < (2^\lambda)^+$, $cf\ \delta > \lambda$ and $tp_\varphi(d_\delta, B_\delta)$ split over B_α for every $\alpha < \delta$

(iii) for some stationary $S \subseteq (2^\lambda)^+$, $\langle d_i : i \in S \rangle$ is an indiscernible sequence

Proof: Again as in [Sh1] I 2.12 (or 3.5 above)

Remark: In the proofs of 3.5, 3.6 we have not used the hypothesis of 3.4.

Continuation of the proof of 3.4:

Clearly if 3.5(i) holds, we finish, so w.l.o.g. 3.5(ii) holds. By Erdos-Rado theorem, for every $m, n < \omega$ there is $I = I_{n,m} \subseteq \mu$, $|I_{n,m}| = \beth_n(\lambda)$, $\{\bar a_i : i \in I_{n,m}\}$ is an m-indiscernible sequence. By the proof of [BSh] VIII 1.3, there is a formula θ^1 such that for any n there are $I_n \subseteq (2^\mu)^+$, $|I_n| = \beth_n(\lambda)^+$, and a finite monadic expansion $\mathfrak{C}^+$ of $\mathfrak{C}$ such that (for some distinct $a_i^n (i \in I_n)$):

$$(\forall i, j \in I_n)[\mathfrak{C}^+ \models \theta^1(a_i^n, a_j^n)\ \ iff\ \ i \leq j]$$

Note that a_i^n belongs to our original A. We now can deal with $\{a_i^1 : i \in I_1\}$ only. W.l.o.g. $I_1 = (2^\lambda)^+$, $\mathfrak{C} = \mathfrak{C}^+$, $a_i^1 = a_i$ and denote $B_\gamma = \{a_i : i < \gamma\}$. Applying 3.6 to $\mathfrak{C}^+$, $A^1 = \{a_i : i < (2^\lambda)^+\}$, if our conclusion fails then one of the following two cases occurs.

Case I: there are $\gamma < (2^\lambda)^+$ and $\bar b_\alpha \in A^1 (\alpha < (2^\lambda)^+)$ such that $tp(\bar b_\alpha, B_\gamma)$ are distinct (for distinct α's).

W.l.o.g. $l(\bar b_\alpha) = k$ for every α. Next we show that w.l.o.g. $k = 1$, otherwise choose an example with minimal k (possibly replacing $\mathfrak{C}$ by a finite monadic expansion). W.l.o.g. the $\bar b_\alpha$ form a Δ-system hence by k's minimality are disjoined. If $k > 1$, let $\bar b_\alpha = \bar c_\alpha \frown \langle d_\alpha \rangle$; w.l.o.g. for some $\varphi = \varphi(\bar x, y; \bar z)$, the types $tp_\varphi(\bar b_\alpha, B_\gamma)$ are distinct.

Clearly if for some α, $\{tp(d_\beta, B_\gamma \cup \bar{c}_\alpha) : \beta < (2^\lambda)^+\}$ has power $> 2^\lambda$, we get contradiction to k's minimality, hence w.l.o.g. $\alpha < \beta < (2^\lambda)^+$, $\sigma < (2^\lambda)^+$ implies $tp_\varphi(\bar{c}_\beta \frown \langle d_\beta \rangle, B_\gamma) \neq tp_\varphi(\bar{c}_\alpha \frown \langle d_\sigma \rangle, B_\gamma)$. Similarly w.l.o.g. $\alpha < \beta < (2^\lambda)^+$, $\sigma < (2^\lambda)^+$ implies $tp_\varphi(\bar{c}_\beta \frown \langle d_\beta \rangle, B_\gamma) \neq tp_\varphi(\bar{c}_\sigma \frown \langle d_\alpha \rangle, B_\gamma)$. W.l.o.g. for every β there is no $\beta' < \beta$ such that $\bar{c}_\beta \frown \langle d_{\beta'} \rangle$ satisfies this. W.l.o.g. for some monadic predicate $P, P = \{d_\beta : \beta < (2^\lambda)^+\}$, so d_β is defined from $\bar{c}_\beta$, so we can decrease k.

An alternative way to do it is as follows. Let $\bar{b}_\alpha = \langle a_{i(\alpha,0)}, \ldots, a_{i(\alpha,k-1)} \rangle$, w.l.o.g. $i(\alpha,0) < \cdots < i(\alpha,k-1)$, and as the $\bar{b}_\alpha$'s are pairwise disjoint, w.l.o.g. $\alpha < i(\alpha,k-1) < i(\beta,0)$ for $\alpha < \beta$.
We may expand $\mathfrak{C}$ by $P_m = \{a_{i(\alpha,m)} : \alpha < (2^\lambda)^+\}$, and using the order defined by θ^1 on $\{a_i : i < (2^\lambda)^+\}$ we can define the functions $a_{i(\alpha,0)} \to a_{i(\alpha,m)}$ hence can code $\bar{b}_\alpha$ by $a_{i(\alpha,0)}$.

So there are $\gamma < (2^\lambda)^+$ and $b_\alpha \in A^1$ $(\alpha < (2^\lambda)^+)$, and φ such that $tp_\varphi(b_\alpha, B_\gamma)$ are distinct for distinct α's, and w.l.o.g. γ is minimal. First assume $\varphi = \varphi(x,y)$. Also w.l.o.g. for every $\gamma_1 < \gamma < \alpha < (2^\lambda)^+$, there are $(2^\lambda)^+$ β's such that $tp_\varphi(b_\alpha, B_{\gamma_1}) = tp_\varphi(b_\beta, B_{\gamma_1})$. Hence for any n we can find $\gamma_0 < \gamma_1 < \cdots < \gamma_{2n}$, and $\alpha_\eta < (2^\lambda)^+$ for $\eta \in {}^{2n}2$ such that $\gamma_{2n} < \gamma$, $\gamma < \alpha_\eta$ and for $m \le 2n$, n, $\nu \in {}^{2n}2$:

$$tp_\varphi(b_{\alpha_\eta}, B_{\gamma_m}) = tp_\varphi(b_{\alpha_\nu}, B_{\gamma_m}) \text{ iff } \eta \restriction m = \nu \restriction m .$$

Expand $\mathfrak{C}$ by:

$$R = \{b_{\alpha_\eta} : \eta \in {}^{2n}2, \bigwedge_{m<n} (\eta(2m) = 0 \vee \eta(2m+1) = 0)\}$$

$$Q_1 = \{b_{\gamma_{2m}} : m \le n\}$$

$$Q_2 = \{b_{\gamma_{2m+1}} : m < n\}$$

$$P = B_\gamma .$$

Let (remembering θ defines the order on $\{a_i : i < (2^\lambda)^+\}$):

$$\begin{aligned} \psi(x,y) \overset{def}{=} R(x) \wedge Q_2(y) \wedge (\exists x_1, y_1)[R(x_1) \wedge Q_1(y_1) \wedge \\ \wedge (\forall y_2)[Q_2(y_2) \wedge \theta^1(y_2,y_1) \to \theta^1(y_2,y)] \wedge \\ \wedge [x, x_1 \text{ realizes the same } \varphi\text{-type over} \\ \{z \in P : \theta^1(z,y)\} \text{ but not over} \\ \{z \in P : \theta^1(z,y_1)\}] \end{aligned}$$

It is easy to see that:

$$\models \psi[b_\eta, a_{\gamma_{2m+1}}] \quad iff \quad \eta(2m+1) = 1$$

Together with compactness this shows that some finite monadic expansion of $\mathfrak{C}$ has the independence property, contradiction.

We still have to deal with the case $\varphi = \varphi(x, \bar{y})$, $l(\bar{y}) > 1$. Let $l(\bar{y}) = m$ let $<^*$ be the lexicographic order on ${}^m(B_\gamma)$, (based on e^1); so ${}^m B_\gamma = \{\bar{a}_\alpha: \alpha < \gamma_m\}$, $\bar{a}_\alpha <^* \bar{a}_\beta$ iff $\alpha < \beta < \gamma_m$. We then let $\gamma^* \leq \gamma_m$ be minimal such that $\{\{\varphi(x, \bar{a}_\beta): \beta < \gamma^*, \models \varphi(b_\alpha, \bar{a}_\beta)\}: \gamma < \alpha < (2^\lambda)^+\}$ has power $(2^\lambda)^+$. Now again necessarily γ^* is limit and we can find $\gamma_0 < \gamma_1 < \cdots < \gamma^*$ and $\gamma^* < \alpha_\eta < (2^\lambda)^+$ for $\eta \in {}^\omega 2$ which are eventually zero such that

$$\bigwedge_{\beta < \gamma_l} \varphi[b_{\alpha_\eta}, \bar{a}_\beta] \equiv \varphi[b_{\alpha_\nu}, \bar{a}_\beta] \quad iff \quad \eta \restriction l = \nu \restriction l$$

Our only problem is to code $\{\bar{a}_{\gamma_l}: l < \omega\}$ by monadic predicates, which is easy applying Ramsey theorm on the $\bar{a}_{\gamma_l}$'s and using the order on B_j.

Case II: For some finite $\bar{c} \in \mathfrak{C}$ and some $\gamma < (2^\lambda)^+$, $\{tp(\bar{b}, B_\gamma \cup \bar{c}): \bar{b} \subseteq A^1\}$ has power $(2^\lambda)^+$
Like case I.

Case III: Note case II.
We shall prove
(*) if $\bar{c} \in \mathfrak{C}$, $W \subseteq \{\delta: \delta < (2^\lambda)^+, cf\ \delta > \lambda\}$ is stationary, *then* for some closed unbounded $U \subseteq (2^\lambda)^+$, and function f, $Dom\ f = U \cap W$; $f(\alpha) < \alpha$ for $\alpha \in U \cap W$, and for each γ the sequence $\langle tp(a_\alpha, \bar{c} \cup \{a_\beta: \beta < \alpha, f(\beta) = \gamma\}): f(\alpha) = \gamma \rangle$ is increasing.

Now it suffice to prove (*). As then we define by induction on n K_n, and for $t \in K_n$, W_t, U_t, f_t, $\bar{c}_t$ such that:

(a) $K_0 = \{<0>\}$, $W_{<0>} = W \subseteq \{\delta < (2^\lambda)^+: cf\ \delta > \lambda\}$, $\bar{c}_{<0>}$ is the empty sequence.

(b) for $t \in K_n$ $\bar{c}_t \in \mathfrak{C}$ is a sequence of length n, and if $\alpha_1 < \alpha_2 \cdots < \alpha_n$ are in W_t, then

$$tp(\langle a_{\alpha_n}, a_{\alpha_{n-1}}, \ldots, a_{\alpha_2}, a_{\alpha_1} \rangle, \{a_\gamma: \gamma < \alpha_1, \gamma \in W_t\})$$

$$= tp(\bar{c}_t, \{a_\gamma: \gamma < \alpha_1, \gamma \in W_t\})$$

(c) K_n is a family of sequences of length n of ordinals $< (2^\lambda)^+$

(d) for $t \in K_n$, U_t is a closed unbounded subset of $(2^\lambda)^+$, f_t a function with domain $U_t \cap W_t$, $f_t(\alpha) < \alpha$

(e) $K_{n+1} = \{\eta\frown<\gamma> : \eta \in K_n, \ \gamma \in Rang(f_t)$ for some $t \in K_n\}$ and $W_{\eta\frown<\gamma>} = \{\alpha \in W_\eta : \alpha \in U_\eta$ and $f_\eta(\alpha) = \gamma\}$

For $n = 0$-no problem, for $n+1$: for each $W_\eta (\eta \in K_n)$ apply $(*)$ (with $\bar{c} = \bar{c}_\eta$).

Now K_0, $W_\eta, \bar{c}_\eta (\eta \in K_0)$ are defined.

If $W_\eta, \bar{c}_\eta$ are defined we can define f_η, U_η by applying $(*)$, then define $W_{\eta\frown<\gamma>}, \bar{c}_{\eta\frown<\gamma>}$ ($\gamma \in Rang(f_\eta)$) by (d). If we do this for every $\eta \in K_n$, we can define K_{n+1} by (e).

For every $\delta \in W_{<>}$, we can define by induction on $l < \omega$, $\eta_l \in K_l$, such that $\eta_l = \eta_{l+1} \restriction l$, $\delta \in W_{\eta_l}$ and $Rang\,\eta_l \subseteq \delta$ and the η_l are unique but maybe for some l, $\delta \notin U_{\eta_l}$ hence η^δ_{l+1} is not defined. Let $\varepsilon(\delta) \leq \omega$ be such that η^δ_l is defined iff $l < \varepsilon(\delta)$. If $\{\delta : \varepsilon(\delta) < \omega\}$ is stationary, we get contradiction by Fodor lemma. If $W^* = \{\delta : \varepsilon(\delta) = \omega\}$ is stationary, then $\gamma(\delta) = \sup_{l<\omega} \eta^\delta_{l+1}(l) < \delta$ for $\delta \in W^*$ (as $cf\ \delta > \lambda$) hence for some stationary $W^1 \subseteq W^*$, $\gamma(\delta)$ is constant on W^1. As $(2^\lambda)^{\aleph_0} = 2^\lambda$ w.l.o.g. $\eta^\delta_l = \eta_l$ for every $\delta \in W^1$. Now $\bigcap_{l<\omega} W_{\eta_l}$ is stationary and by (b) $\langle a_i : i \in \bigcap_{l<\omega} W_{\eta_l} \rangle$ is an indiscernible sequence.

Proof of $(*)$: For notational simplicity let $\bar{c} = \phi$. For every $\varphi = \varphi(x,\bar{y})$, and $\gamma < (2^\lambda)^+$, type $p \in S^\varphi_\gamma \overset{def}{=} \{tp_\varphi(a_i, B_\gamma) : \gamma < i < (2^\lambda)^+\}$ and natural number n we define when $Rk_\varphi(p) \geq n$:

For $n = 0$-always.

For $n = 2m+1$, $Rk(p) \geq n$ *iff* there is β, $\gamma < \beta < (2^\lambda)^+$ and distinct $p_1, p_2 \in S^\varphi_\beta$ extending p with $Rk_\varphi(p_1), Rk_\varphi(p_2) \geq 2m$.

For $n = 2m+2$, $Rk(p) \geq n$ *iff* for every β, $\gamma < \beta < (2^\lambda)^+$ there is $p_1 \in S^\varphi_\beta$ extending p with $Rk_\varphi(p_1) \geq 2m+1$.

If there are p, φ such that $Rk_\varphi(p) \geq n$ for every $n < \omega$, the proof is as in case I. Suppose not, then for every $p \in \bigcup_\gamma S^\varphi_\gamma$ let $Rk_\varphi(p)$ be the maximal n such that $Rk_\varphi(p) \geq n$. Clearly

$(*)$ $p_1 \leq p_2$ (both in $\bigcup_\beta S^\varphi_\beta$) implies $Rk_\varphi(p_1) \geq R_\varphi(p_2)$

Now for every $\delta \in W_0 = \{i < (2^\lambda)^+ : cf\ i > \lambda\}$, and α, there is $\gamma(\delta, p) < \delta$ such that:

$$\gamma(\delta,\varphi) \leq \gamma < \delta \Longrightarrow Rk_\varphi(tp_\varphi(a_\delta, B_{\gamma(\delta,\varphi)})) = Rk_\varphi(tp_\varphi(a_\delta, B_\gamma))$$

Let $\gamma(\delta) = \bigcup_\varphi \gamma(\delta,\varphi)$ so $\gamma(\delta) < \delta$. As we can use several f 's (by coding) we can restrict ourselves to some stationary $W_1 \subseteq W_0$, such that for some γ^* $(\forall \delta \in W_1)\ [\gamma(\delta) = \gamma^*]$.

As not case II similarly w.l.o.g. for some p $(\forall \delta \in W_1)\ [tp(a_\delta, B_{\gamma^*}) = p]$.

Clearly $Rk_\varphi(p \restriction \varphi)$ is not even, hence is odd, (for every φ). Suppose $\gamma^* < \delta_1 < \delta_2$ in W_1, $tp(a_{\delta_1}, B_{\delta_2}) \not\subset tp(a_{\delta_2}, B_{\delta_2})$, then for some φ and $\alpha < \delta_1, \alpha > \gamma^*$ and both $tp_\varphi(a_{\delta_1}, B_\alpha) \neq tp_\varphi(a_{\delta_2}, B_\alpha)$ have the same rank ($Rk_\varphi(-)$) as p, contradiction.

§4 From indiscernibles to finitely satisfiable and Hanf numbers

4.1. Lemma:

Suppose $\langle a_t : t \in I \rangle$ is an indiscernible sequence (I infinite). Then we can find a model N of power T such that for every $t \in I$, $tp(a_t, N \cup \{a_s : s < t\})$ is finitely satisfiable in N.

Proof: Let $I \subseteq J$, $t(n) \in J - I$, $(\forall t \in I)\ [t < t(n+1) < t(n)]$.

Let $\{a_t : t \in I\} \subseteq M \subseteq \mathfrak{C}$, and let M^* be an expansion of M by Skolem functions (so M^* is an L^*-model, $L \subseteq L^*$). By Ramsey theorem and the compactness theorem, there is a model M^+ of the theory of M^*, and $b_t \in M^+$ $(t \in J)$ such that:

(*) for every $\varphi(x_1, \ldots, x_n) \in L^*$, and $s_1 < \cdots < s_n \in J$ if

$M^+ \models \varphi[b_{s_1}, \ldots, b_{s_n}]$ *then* for some $t_1 < \cdots < t_n \in I$,

$M^* \models \varphi[a_{t_1}, \ldots, a_{t_n}]$.

Clearly for every $s_1 < \cdots < s_n \in J$, $t_1 < \cdots < t_n \in I$ the L-types of $\langle b_{s_1}, \ldots, b_{s_n} \rangle$ in M^+ and $\langle a_{t_1}, \ldots, a_{t_n} \rangle$ in M are equal, hence w.l.o.g. the L-reduct of M^+ is an elementary submodel of $\mathfrak{C}$ and $a_t = b_t$ for $t \in I$. Lastly let $N \subseteq \mathfrak{C}$ be the model whose universe is the Skolem hull of $\{b_{t(n)} : n < \omega\}$ in M^+, and $a_t \stackrel{def}{=} b_t$ also for $t \in J - I$.

So let $t \in I$ and we should prove that $tp_L(a_t, N \cup \{a_s : s < t, s \in I\})$ is finitely satisfiable in N. Let $\bar{d} \in N$, $t_0 < t_1 < \cdots < t_n = t \in I$, $\varphi \in L$, $\mathfrak{C} \models \varphi[b_{t_n}, b_{t_{n-1}}, \ldots, b_{t_0}, \bar{d}]$ so for some L^*-term $\bar{\tau}$, and $k < \omega$,

$\bar{d} = \bar{\tau}(b_{t(0)}, \ldots, b_{t(k)})$. As $\langle b_t : t \in J \rangle$ is indiscernible in M^+, and $M^+ \models \varphi[b_{t_n}, b_{t_{n-1}}, \ldots, b_{t_0}, \bar{\tau}(b_{t(0)}, \ldots, b_{t(k)})]$ clearly

$$M^+ \models \varphi[b_{t(k+1)}, b_{t_{n-1}}, \ldots, b_{t_0}, \bar{d}] \ .$$

As $b_{t(k+1)} \in N$, we finish.

4.2. Conclusion:

If $\lambda \rightarrow (\mu)^{<\omega}_{T}$, M a model of power λ, *then* for some N of power $|T|$, M has a decomposition $\langle A_i : i < \alpha \rangle$ over N, $A_i \neq \phi$, $\alpha \in \{\mu, \mu+1\}$

Proof: Immediate by 2.5, 4.1.

Remember $\mathcal{L}^{\delta}_{\infty,\lambda}$ is the set of sentences of $\mathcal{L}_{\infty,\lambda}$ with quantifier depth $< \delta$.

4.3. Theorem:

Suppose $(T_\infty, 2^{nd}) \not\leq (T, \mathrm{mon})$.

1) For limit ordinal δ and every λ the Hanf numbers of the logic $\mathcal{L}^{\delta}_{\infty,\lambda}$, μ_1 for models of T expanded by $\leq |T|$ monadic predicates, and μ_2 for linear well ordering expanded by $\leq |T|$ monadic predicates, *satisfies* $\beth_{\omega^2}(\mu_1) = \beth_{\omega^2}(\mu_2)$

2) The Lowenheim and Hanf number of $\mathcal{L}^{\delta}_{\infty,\lambda}$, for well ordering expanded by $\leq |T|$ monadic predicates, are equal; so if λ, α are definable in second order logic, then those numbers are smaller than the Hanf number of 2nd order logic.

Proof: 1) By 2.6, 4.2 this is reduced to a problem on monadic theory of sum of models, for complete proof see [Sh4]. However if $(\forall \alpha)(\alpha < \delta \rightarrow \alpha + \alpha < \delta)$, $\beth_\delta > |T|$ there are no problems.

2) See [BSh].

Now by 4.3 and 3.4:

4.4. Conclusion:

For T as above.

1) The Hanf number of $L_{\omega,\omega}(\mathrm{mon})$ for models of T is strictly smaller than the Hanf number of second order logic.

2) Even in $L_{\lambda,\lambda}$ we cannot interpret a pairing function on arbitrarily large sets in models of T.

Part II

Hypothesis: $(T_\infty, 2^{nd}) \not\leq (T,\text{mon})$

§1 On a rude equivalence relation

1.1. Context:

Let M_0 be a fixed model $(\subseteq \mathfrak{C}) M_0 \subseteq M_1 \subseteq \mathfrak{C}$, and in M_1 every type over M_0 (with $< \omega$ variables) is realized. The case $||M_0|| = |T|$, $||M_1|| \leq 2^{|T|}$ will suffice. We let $\mathcal{B}$ be an elementary extension of M_0, which is the model we want to analyze: and we assume $tp_*(\mathcal{B}, M_1)$ is finitely satisfiable in M_0 (and $\mathcal{B} \subseteq \mathfrak{C}$).

We usually suppress members of M_0 when used as individual constants.

We further let I be a κ-saturated dense linear order, $\kappa > 2^{|T|}$, and we can find elementary maping $f_t (t \in I)$ such that $Dom\ f_t = \mathcal{B}$, $f_t \restriction M_0 =$ the identity, and for some ultrafilter D on $^{\mathcal{B}}M_0$, $tp_*(f_t(\mathcal{B}), M_1 \cup \bigcup_{s<t} f_s(\mathcal{B}))$ is $Av(D, M_1 \cup \bigcup_{s<t} f_s(\mathcal{B}))$ (see for definition I 1.4, 1.5).

We denote by $\mathcal{B}_t$ the image of $\mathcal{B}$ by f_t.

For $a \in \mathcal{B}$ let $a_t = f_t(a)$, $\langle a_1, \ldots, a_n \rangle_t = \langle f_t(a_1), \ldots, f_t(a_n) \rangle$, $0 \in I$, $f_0 =$ the identify.

1.1A Remark:

Except in 2.1, 2.3, we use just the indiscernibility of the $\mathcal{B}_t$'s.

1.2. Definition:

1) On $\mathcal{B} = \mathcal{B}_0$, we define a relation E_0:
aE_0b *iff* in some monadic finite expansion of $\mathfrak{C}$ the set $\{<a_t, b_t> : t \in I\}$ is first order definable.

2) For $a \in \mathcal{B}$, $Od(a)$ hold if in some monadic finite expansion of $\mathfrak{C}$ the set $\{<a_t, a_s> : t \in I, s \in I, t < s\}$ is first order definable.

1.3. Claim:

1) E_0 is an equivalence relation

2) aE_0b implies $Od(a) \longleftrightarrow Od(b)$

Proof: Easy

1.3. Claim

If $\bar{a}^k \subseteq b_k/E_0 \subset \mathcal{B}(k=1,n)$ and $b_k/E_0 \neq b_m/E_0$ for $k \neq m$ then:

(i) $tp(\bar{a}^1_{t_1} \frown \bar{a}^2_{t_2} \frown \cdots \frown \bar{a}^n_{t_n}, M_1)$ is the same for all $t_1, \ldots, t_n \in I$

(ii) $tp(\bar{a}^n, M_1 \cup \bigcup_{k=1}^{n-1} \bar{a}^k)$ is finitely satisfiable in M_0

(iii) if $a^n = \bar{b} \frown \bar{c}$, $tp(\bar{c}, M_1 \cup \bar{b})$ is finitely satisfiable in M_0, then $tp(\bar{c}, M_1 \cup \bar{b} \bigcup_{k=1}^{n-1} \bar{a}_k)$ is finitely satisfiable in M_0.

Proof: Clearly (ii) follows from (i) (just choose $t_n > t_1, \ldots, t_{n-1}$ in (i)] and also (iii) follows by I 1.6 from (ii).

So let us prove (i), and we prove it by induction on n and then on $k \leq n$, restricting ourselves to $\langle t_1, \ldots, t_n \rangle$ such that $|\{t_1, \ldots, t_n\}| \geq n-k$ (for $k = n$ we get the conclusions)

Suppose we have prove it for $n' < n$ and for $n' = n, k' < k$.

1.3A Fact:

By replacing $\mathfrak{C}$ by a monadic finite expansion we can replace $\bar{a}^m$ by a singleton $<a^m>$. Replacing $\mathfrak{C}$ by a finite monadic expansion $\mathfrak{C}^+$ does not preserve the properties of $M_0, M_1, \langle \mathcal{B}_s : s \in I \rangle$. However we can w.l.o.g. assume that $\langle \mathcal{B}_s : s \in I \rangle$ is indiscernible over M_1 in $\mathfrak{C}^+$. We could here also use $L(\mathfrak{C}^+)$-formulas only of the form $\varphi(\cdots, x_n \cdots, F_k(x_n) \cdots)$ where $\varphi \in L(\mathfrak{C})$, F_k are definable in $\mathfrak{C}^+$ and maps each $\mathcal{B}_s$ into itself and commute with the functions f_s.

1.4. Notation:

For non-decreasing sequences $\langle s_1, \ldots, s_n \rangle$, $\langle t_1, \ldots, t_n \rangle$ from I, we say that $\langle s_1, \ldots, s_n \rangle$ is *closed to* $\langle t_1, \ldots, t_n \rangle$ if

either (α) $t_1 < \cdots < t_n$, $s_m = t_{m+1}$ $s_{m+1} = t_m$, $s_i = t_i$ for $i \neq m, m+1$, for some m, $1 \leq m \leq n$

or (β) for some $1 \leq l < m \leq n$

$$t_1 \leq \cdots \leq t_{l-1} < t_l = t_{l+1} = \cdots = t_m < t_{m+1} \leq \cdots \leq t_n, t_m < s_m < t_{m+1},$$

and $(\forall i)$ $[1 \leq i \leq n \wedge i \neq m \rightarrow s_i = t_i]$.

We shall prove:

1.5. Fact:

If $\langle s_1, \ldots, s_n \rangle$ is closed to $\langle t_1, \ldots, t_m \rangle$, both non-decreasing sequences from I, $|\{t_i : i = 1, n\}| = n-k$, *then* $tp(\langle a^1_{t_1}, \ldots, a^n_{t_n} \rangle, M_1) = tp(\langle a^1_{s_1}, \ldots, a^n_{s_n} \rangle, M_1)$

This suffice for proving 1.3 as any equivalence relation E on

$$\{\langle t_1, \ldots, t_n \rangle : t_i \in I,\ |\{t_i : i = 1,\ n\}| \geq n-k\}$$

satisfying the following has just one class:

(a) if $\bar{s}$ is closed to $\bar{t}$ both non decreasing then $\bar{s}\ E\ \bar{t}$

(b) if $\langle s_1, \ldots, s_n \rangle\ E\ \langle s_{n+1}, \ldots, s_{2n} \rangle$ and $(\forall i,j \in [1,2n])\ [s_i < s_j \equiv t_j < t_j]$ *then* $\langle t_1, \ldots, t_n \rangle\ E\ \langle t_{n+1}, \ldots, t_{2n} \rangle$.

Proof of the Fact 1.5:

Note that 1.4(α) occurs only when $k = 0$, and 1.4(β) occurs only when $k > 0$

Case A: $k = 0$.

So there is a formula φ with parameters from $M_1 \cup \{a^1_{t_1}, \ldots, a^{i-1}_{t_{i-1}}, a^{i+2}_{t_{i+2}}, \ldots, a^n_{t_n}\}$, such that $\models \varphi[a^i_{t_i}, a^{i+1}_{t_{i+1}}]$ but $\models \neg\varphi[a^i_{t_{i+1}}, a^{i+1}_{t_i}]$. So clearly (by the indiscernibility of $\langle \mathcal{B}_t : t \in I \rangle$ over M_1) there is a formula φ with parameters from $\mathfrak{C}$ such that for any $s < t$ in I $\models \varphi[a^i_s, a^{i+1}_t] \wedge \neg\varphi[a^i_t, a^{i+1}_s]$ and w.l.o.g. $\models \varphi[a^i_s, a^{i+1}_s]$.

Adding monadic predicates $P^i = \{a^i_t : t \in I\}$, $P^{i+1} = \{a^{i+1}_t : t \in I\}$, we easily find that:

$$\theta(x,y) = \varphi(x,y) \wedge P^i(x) \wedge P^{i+1}(y) \wedge (\forall z)\,[P^i(z) \wedge x <^i z \rightarrow \neg\varphi(z,y)]$$

define $\{< a^i_s, a^{i+1}_s > : s \in I\}$, where

$$x <^i z \overset{def}{\equiv} (\forall y)\,[P^{i+1}(y) \wedge \varphi(z,y) \rightarrow \varphi(x,y)] \wedge x \neq z \wedge P^i(x) \wedge P^i(z)\,.$$

Now θ contradict the non E_0-equivalence of a^i, a^{i+1}.

Case B: $k > 0$

So there is a formula φ with parameters from $M_1 \cup \{a^1_{t_1}, \ldots, a^{l-1}_{t_{l-1}}, a^{m+1}_{t_{m+1}}, \ldots, a^n_{t_n}\}$ such that:

(a) $\models \varphi[a^l_{t_l}, \ldots, a^{m-1}_{t_{m-1}}, a^m_{s_m}]$

(b) $\models \neg\varphi[a^l_{t_l}, \ldots, a^{m-1}_{t_{m-1}}, a^m_{t_m}]$

by the induction hypothesis on k, from (a) it follows

(c) for any $v_l, \ldots, v_m \in \{t \in I : t_{l-1} < t < t_{m+1}\}$,

not all of them equal $\models \varphi[a^l_{v_l}, \ldots, a^m_{v_m}]$

By (b), as $t_l = \cdots = t_m$,

(d) for any $v \in \{t \in I : t < t < t_{m+1}\}$

$\models \neg\varphi[a^l_v, \ldots, a^m_v]$

Using the indiscernibility of $\langle \mathcal{B}_t : t \in I \rangle$ over M_1 there is a formula φ' (with parameters from $\mathfrak{C}$) such that (c), (d) holds for any $v_l, \ldots, v_m \in I$ not all equal, and for any $v \in I$ respectively. Expanding $\mathfrak{C}$ by $P^i = \{a^i_t : t \in I\}$, we find that the formula

$$\theta(x,y) = P^l(x) \wedge P^{l+1}(y) \wedge (\exists z_{l+2}, \ldots, z_m)[\bigwedge_{i=l+2}^{m} P^i(z_i) \wedge \neg\varphi(x,y,z_{l+2}, \ldots, z_n)]$$

define the set $\{< a^l_t, a^{l+1}_t > : t \in I\}$ of pairs, contradicting the non E_0-equivalence of a^l, a^{l+1}.

§2 Extending a pair of finitely satisfiable

We continue to use the context of §1 (of part II)

2.1. Claim:

If $\bar{a}, \bar{b} \in \mathcal{B}$ *then* $tp(\bar{a}\frown\bar{b}, M_1) = tp(\bar{a}_s\frown\bar{b}_t, M_1)$ for some (every) $s < t \in I$ *iff* $tp(\bar{b}, M_1 \cup \bar{a})$ is finitely satisfiable in M_0

Proof: Easy

2.2. Lemma:

There are no $s < t \in I$, $\bar{a}, \bar{b} \in \mathcal{B}$ and $c \in \mathfrak{C}$ and formula φ with parameters from M_1, such that:

(a) $\models \varphi[c, \bar{a}_s, \bar{b}_t]$

(b) $\models \neg\varphi[c, \bar{a}_s, \bar{b}_{t_1}]$ for every $t_1 > t$ (in I)

(c) $\models \neg\varphi[c, a_{s_1}, \bar{b}_t]$ for every $s_1 < s$ (in I)

(d) $\bar{a}\frown\bar{b}$ is included in one E_0-equivalence class .

Proof: By (d) and 1.3A, replacing $\mathfrak{C}$ by a monadic finite expansion w.l.o.g. $\bar{a} = \langle a \rangle$, $\bar{b} = \langle b \rangle$. By Ramsey theorem and compactness we can assume that if $\langle v_1, \ldots, v_m \rangle$, $\langle u_1, \ldots, u_m \rangle$, are increasing sequences from I, $(\exists k)(v_k = u_k = s)$, $(\exists k)(v_k = u_k = t)$ then

$$tp_*(\langle \mathcal{B}_{v_1}, \ldots, \mathcal{B}_{v_m} \rangle, M_1 \cup \{c\}) =$$

$$tp_*(\langle \mathcal{B}_{u_1}, \ldots, \mathcal{B}_{u_m} \rangle, M_1 \cup \{c\}).$$

By II. 1.3A, w.l.o.g. $\mathfrak{C}$ has predicates for $\{a_t : t \in I\}$, $\{b_t : t \in I\}$, and $\{\langle a_t, b_t \rangle : t \in I\}$. We shall try to use c for coding $\{s, t\}$ (i.e., $\{a_s, b_t\}$), which contradict $(T_\infty, 2^{nd}) \not\leq (T, \text{mon})$ (see I. 1.3(2)).

Case A: not $Od(a)$

Subcase A1: For any $v \in I$, $s < v < t$, $\models \varphi[c, a_v, b_t]$.

Then we can fix t, and define $\{\langle a_v, b_u \rangle : v < u < t\}$ as in the proof of 1.5 Case A and then define $\{\langle a_v, a_u \rangle : v < u \in I\}$, contradicting not $Od\ (a)$.

Subcase A2: Not A1 but for any $v \in I$, if $v > t$, then $\models \varphi[c, a_v, b_t]$

Similar contradiction: fix s, and using the function $\{\langle a_v, b_v \rangle : v \in I\}$ define $\{\langle b_v, b_u \rangle : s < v < u \rangle\}$.

Subcase A3: For $v \in I$, if $s < v < t$ then $\models \varphi[c, a_s, b_v]$ like subcase A1 (interchanging a and b)

Subcase A4: Note A3 but if $v \in I$, $v < s$ then $\models \varphi[c, a_s, b_v]$ like A2 (interchanging a and b)

Subcase A5: Not A1-A4

Here c code the pair $\langle a_s, b_t \rangle$: a_s is unique for t such that $s \neq t$ and $\varphi(c, a_s, b_t)$ (by not A1, A2). By symmetry (i.e., as not A3, A4) t is unique for s, by the indiscernibility we have over c and as I is dense this shows that c determine $\langle s, t \rangle$, so we get the contradiction to the hypothesis of Part II.

Case B: $Od(a)$

Let $\Theta(x, y, z)$ says all the relevant things on $\langle a, b, c \rangle$: $x \in \{a_v : v \in I\}$, $y \in \{b_v : v \in I\}$, $\varphi(x, y, z)$, $\neg\varphi(z, x', y)$ where $x' < x$ [i.e., $(\exists v < u)$ $(x' = a_v \wedge x = a_u)$] and $\neg\varphi(z, x, y')$ where $y' < y$ [i.e., $(\exists v < u)$ $(y = b_v \wedge y' = b_u)$] and the amount of $\varphi(z, -, -)$-indiscernibility of $\langle \langle a_v, b_v \rangle : v \in I \rangle$ over $\{c\}$ which holds.

Clearly $\models \theta[a_s, b_t, c]$

It suffices to prove that

(*) If $\theta[a_{s(k)}, b_{t(k)}, c]$ for $k = 1, 2$ then $s(1) = s(2)$, $t(1) = t(2)$.

By symmetry we can assume $t(1) < t(2)$ (if $t(2) < t(1)$ interchange the order, if $t(2) = t(1)$ neccessarily $s(1) \neq s(2)$ and invert the order). Below u, v denote elements of I.

Suppose $s(2) < u < v$, we can find $u_1, v_1, t(1) < u_1 < v_1$ such that $s(2) < u_1$, $u < t(2) \iff u_1 < t(2)$, $u = t(2) \iff u_1 = t(2)$, $v < t(2) \iff v_1 < t(2)$, and $v = t(2) \iff v_1 = t(2)$.

As $\models \theta[a_{s(2)}, b_{t(2)}, c]$, it follows that

(i) $\varphi(c, a_{u_1}, b_{v_1}) \equiv \varphi(c, a_u, b_v)$

Now choose $u_2 > v_2 > t(2)$, as $\models \theta[a_{s(1)}, b_{t(1)}, c]$, clearly

(ii) $\varphi(c, a_{u_1}, b_{v_1}) \equiv \varphi(c, a_{u_2}, b_{v_2})$

By transitivitiy of $\equiv$

(iii) the truth value of $\varphi(c, a_u, b_v)$ is the same for all $v > u > s(2)$.

Now (iii) is a property of c and $s(2)$, and it fails for any $s' < s(2)$ as $\models \varphi[a_{s(2)}, b_{t(2)}, c]$ but $\models \neg\varphi[a_{s(2)}, b_v, c]$ when $v > t(2)$; so $a_{s(2)}$ is definable from c, and then we can easily define $b_{t(2)}$, and so get the desired contraction.

2.3. Lemma:

If $\bar{a}, \bar{b}, c \in \mathfrak{C}$, $tp(\bar{b}, M_0 \cup \bar{a})$ is finitely satisfiable in M_0 *then:*

$tp(\bar{b} \frown \langle c \rangle, M_0 \cup \bar{a})$ is finitely satisfiable in M_0 or

$tp(\bar{b}, M_0 \cup \bar{a} \frown \langle c \rangle)$ is finitely satisfiable in M_0

Proof: Suppose $\bar{a}, \bar{b}, c$ form a counterexample. W.l.o.g. $\mathcal{B}$ is $||M_0||^+$-saturated. Choose $\bar{a}' \in \mathcal{B}$ realizing $tp(\bar{a}, M_0)$, then choose $\bar{b}'$ such that $tp(\bar{a}' \frown \bar{b}', M_0) = tp(\bar{a} \frown \bar{b}, M_0)$. Then choose $\bar{b}''$ realizing $tp(\bar{b}', M_0 \cup \bar{a}')$ such that $tp(\bar{b}'', M_1 \cup \bar{a}')$ is finitely satisfiable in M_0; now $tp(\bar{a}' \frown \bar{b}'', M_1)$ is finitely satisfiable in M_0, so we could have chosen $\mathcal{B}$, D such that $\bar{a}' \frown \bar{b}'' \subseteq \mathcal{B}$.

Now choose $c' \in \mathcal{B}$ such that $tp(\bar{a}' \frown \bar{b}'' \frown \langle c' \rangle, M_0) = tp(\bar{a} \frown \bar{b} \frown \langle c' \rangle, M_0)$; hypothesis 2.2 (d) may fail for $\bar{a}', \bar{b}'', c'$, but by 1.3 (iii) we get it by replacing $\bar{a}', \bar{b}''$ by $\bar{a}' \cap (c'/E_0)$, $\bar{b}'' \cap (c/E_0)$.

We can choose c'', such that $c''\frown\bar{a}'_s\frown\bar{b}''_t$, $c\frown\bar{a}\frown\bar{b}$ realizes the same type over M_0, and $tp_*(\{c''\} \cup\cup \{\mathcal{B}_v : s \leq v \leq t\}, M_1 \cup\cup \{\mathcal{B}_v : v < s\})$ is finitely satisfiable in M_0. We can furthermore assume as in the proof of I. 2.6 that for $v > t$ $tp_*(\mathcal{B}_v, \bigcup_{u<v} \mathcal{B}_u \cup \{c''\} \cup M_1)$ is finitely satisfiable in M_0, so $tp_*(\bigcup_{v>t} \mathcal{B}_v, M_1 \cup \bigcup_{u \leq t} \mathcal{B}_u \cup \{c'\})$ is finitely satisfiable in M_1. Now $\bar{a}'_s$, $\bar{b}''_t$, c'' satisfies (a) (b) (c) (d) of 2.2 if $\bar{a}$, $\bar{b}$, c where a counterexample to 2.2, where $s < t \in I$. So by 2.2 we have proved 2.3.

References

[BSh] J. Baldwin and S. Shelah. Second order quantifiers and the complexity of theories, Proc. of the 1980/1 model theory year in Jerusalem, *Notre Dame J. of Formal Logic*, 1985.

[GSh] Y. Gurevich and S. Shelah. Monadic Logic and the next world. Proc. of the 1980/1 model theory year in Jerusalem; *Israel J. Math*, 1985.

[Sh1] S. Shelah. Classification theory, North Holland Publ. Co. 1978.

[Sh2] S. Shelah. Simple unstable theories. *Annals of Math Logic* 19(1980) 177-204.

[Sh3] S. Shelah. On the monadic theory of order. *Annals of Math* 102 (1975) 379-419.

[Sh4] S. Shelah, More on monadic theories, in preparation.

More on Stationary Coding

We here continue our investigations in [Sh1] on stationary coding sets (introduced and investigated by Zwicker [Z]) making some improvements and additions.

The various claims are not so connected. They include:

A If $\kappa = \kappa^{\aleph_0}$, $\lambda = \lambda^{\kappa}$ then there is a (κ^+,λ^+)-stationary coding (see 23)

B If $\lambda = \lambda^{\aleph_0}$ is regular, $S \subseteq \{\delta < \lambda^+ : cf\ \delta = \aleph_0\}$ is stationary but does not reflect *then* there is an $(\aleph_1,\lambda^+)$-stationary coding (see 24, 25)

C If $\lambda = \lambda^{\aleph_0}$ then $\diamondsuit\ (\mathcal{D}_{\aleph_1}(\lambda^+))$ (see 28); for more on diamonds see 13, 14, 15.

D We note that Martin Maximum implies that "there is no $(\aleph_1,\lambda)$-weak stationary coding for every λ" and we show that statement for $\lambda = \aleph_2$ when $2^{\aleph_0} \geq \aleph_3$, (see 3). We note also that for κ first inaccessible, strong stationary coding may not exist (see 4).

E We also give an elementary presentation of "a normal fine filter on λ (or $\mathcal{P}_{<\kappa}(\lambda)$) concentrating on the wrong cofinality is not λ^+-saturated" (see 6,7,8). $\diamondsuit\ (\mathcal{D})$ has an even stronger conclusion (see 17).

F On strong stationary coding see 18.

1. Notation:

1) If $<^* \restriction a$ well order the set a let $otp\ (a, <^*)$ be the order type. If a is a set of ordinals, $<^*$ the usual order then we write $otp\ (a)$.

Let *ord* be the class of ordinal.

2) $H_{<\kappa}(\alpha)$ is

$\{ a : |a| < \kappa$ and for every n and $x_1, \ldots, x_n$,

if $x_1 \in x_2 \in x_3 \cdots \in x_n \in a$,

then x_1 is an ordinal $< \alpha$ or

a set of power $< \kappa \}$

$H_{<\kappa}(0)$ is written $H(\kappa)$.

3) Observe that $|H_{<\kappa}(\alpha)| = |2+\alpha|^{<\kappa}$ when κ is regular.

4) For $\kappa \leq \lambda$ let $\mathcal{B} = \mathcal{B}_{\kappa,\lambda}$ be a subset of $H_{<\kappa}(\lambda)$ of power λ, such that for some M^*, $M^* <_{es} (H((2^\lambda)^+), \in)$, $\kappa \in M^*$, $||M^*|| = \lambda$, $\lambda \in M^*$, $\lambda \subseteq M^*$ and $\mathcal{B} = M^* \cap H_{<\kappa}(\lambda)$, hence

(*i*) if $\lambda^{<\kappa} = \lambda$ then $\mathcal{B} = H_{<\kappa}(\lambda)$

(*ii*) if $\lambda^{<\kappa} > \lambda$ but there is $\mathcal{B} \subseteq H_{<\kappa}(\lambda)$,

$|\mathcal{B}| = \lambda$, $(\forall a \in \mathcal{P}_{<\kappa}(\lambda)) (\exists b \in \mathcal{B}) (a \subseteq b)$

then $\mathcal{B}$ satisfies this

5) Let $cd_{\kappa,\lambda}$ be a one-to-one function from $\mathcal{B}_{\kappa,\lambda}$ onto λ, and let $dcd_{\kappa,\lambda}$ be its inverse

Let $dcd``(a) = \{dcd(x) : x \in a\}$

6) Let $\mathcal{D}_\theta$ (θ an uncountable regular cardinal), be the filter generated by the closed unbounded subsets of θ, $\mathcal{D}^{cb}_\theta$ is the filter of co-bounded subsets of θ.

7) For $f, g : I \to ord$, $f <_D g$ means $\{t \in I : f(t) < g(t)\} \in D$, $f/\mathcal{D} < g/\mathcal{D}$ has the same meaning

8) If $\mathcal{D}$ is an $\aleph_1$-complete filter on a set I, $f: I \to ord$ then the $\mathcal{D}$-rank of f is denoted by $Rk(f, \mathcal{D})$, is an ordinal. We define it by defining by induction on α when $Rk(f, \mathcal{D}) = \alpha$:
$Rk(f, \mathcal{D}) = \alpha$ iff $\alpha = \cup\{\beta+1 : \beta < \alpha$, and for some $g/\mathcal{D} < f/\mathcal{D}$, $Rk(g, \mathcal{D}) = \alpha.\}$
9) If $Dom\ f = \theta$ a regular uncountable cardinal, let $Rk(f) = Rk(f, \mathcal{D}^{cb}_{\theta})$.

10) For $\mathcal{D}$ a fine normal filter on $\mathcal{P}_{<\kappa}(A)$, $B \subseteq A$ let

$$\mathcal{D} \restriction B = \left\{ \{a \cap B : a \in I\} : I \in \mathcal{D} \right\}$$

$\mathcal{D} \restriction B$ is a fine normal filter on $\mathcal{P}_{<\kappa}(B)$

2. Lemma:

1) The following are equivalent for a regular uncountable κ and stationary $T \subseteq \kappa$:
(i) there are function $g_\alpha (\alpha < \kappa^+)$, g from κ to κ, such that $(\forall i < \kappa)$ $g(i) < (\aleph_0 + |i|)^+$ and $g_\alpha/\mathcal{D}_\kappa < g_\beta/\mathcal{D}_\kappa$ for $\alpha < \beta < \kappa^+$ and $g/\mathcal{D}_\kappa \not\leq g_\alpha/\mathcal{D}_\kappa$
(v) *for any* cardinal μ such that $(\forall \delta \in T)\ [cf\ \delta > \mu \wedge |\delta|^\mu < \kappa]$, cardinal $\lambda > \kappa$ and subsets $P_i \subseteq \lambda$ $(i < \mu)$ *there are* functions $g_i : \kappa \to \kappa$ $(i < \mu)$ such that the following set is stationary (i.e., $\neq \phi\ mod\ \mathcal{D}_{<\kappa}(\lambda)$)
$\{a \in \mathcal{P}_{<\kappa}(\lambda)$: $a \cap \kappa$ is an ordinal and for $i < \mu$ the order type of $a \cap P_i$ is $g_i(a \cap \kappa)$, and if δ is an accumulation point of a, $cf\ \delta \neq cf\ (a \cap \kappa)$ then $\delta \in a\}$

2) Assume (i) of 1) holds (for κ, T), $\lambda = \kappa^{+\alpha}$, $|\alpha|^+ < \kappa$ and $(\forall \gamma < \kappa)$ $[|\gamma|^{|\alpha|} < \kappa]$. If T is a set of inaccessibles (not necessarily strong limit) *then* there is a (κ,λ) $-$stationary coding.

2.A Remark:

Lemma 2 (1) says that in [Sh1] 12, 12A we can add condition (v) to the four equivalent conditions. Lemma 2 (2) says we can strengthen [Sh1] 13 (which uses the same assumption and deduce the existence of a (κ,λ)-weak stationary coding (with no additional condition on T)).

Proof:

1) We use [Sh1] 12A which has the same proof of [Sh1]12. Now (v) here implies (iv) there trivially. The proof there of (ii)$\Longrightarrow$ (iv) gives (ii)$\Longrightarrow$ (v).

2) Like the proof of [Sh1]13.

3. Fact:

1) If $2^{\aleph_0} > \aleph_2$ *then* there is a stationary $S \subseteq S_{\leq\aleph_0}(\aleph_2)$ which does not reflect, i.e., $S \neq \phi \bmod \mathcal{D}_{\leq\aleph_0}(\aleph_2)$ but for every $\alpha < \aleph_2$ (but $\geq \aleph_1$), $S \cap S_{\leq\aleph_0}(\alpha) = \phi \bmod \mathcal{D}_{\leq\aleph_0}(\alpha)$

2) If $S \subseteq \{a : a \subset \kappa^+,\ a \cap \kappa \text{ an ordinal},\ |a| < \kappa\} \subseteq S_{<\kappa}(\kappa^+)$ is a stationary set which does not reflect κ regular uncountable, *then* for some $C \in \mathcal{D}_{\leq\kappa_0}(\kappa^+)$, $C \cap S$ is a weak (κ,κ^+)-stationary coding for (κ,κ^+)

Proof:

1) Let for an ordinal i, h_i be a one to one function from $|i|$ onto i. In $V' \overset{\text{def}}{=} L[\langle h_i : i < \omega_2\rangle]$ there are at most $\aleph_2^V$ countable subsets of ω_2^V (and $\aleph_1^V = \aleph_1^{V'}$, $\aleph_2^V = \aleph_2^{V'}$ $[V' \models "a \in S_{\leq\aleph_0}(\aleph_2)" \Longrightarrow V \models "a \in S_{\leq\aleph_0}(\aleph_2)"]$). But it is known that every $C \in \mathcal{D}_{\leq\aleph_0}(\aleph_2)$ has power $\leq 2^{\aleph_0} > \aleph_2$. So $S \overset{\text{def}}{=} \{a : a \in S_{\leq\aleph_1}(\aleph_2),\ a \notin V'\}$ is $\neq \phi \bmod \mathcal{D}_{\leq\aleph_0}(\aleph_0)$. But for every $\alpha < \aleph_2$ using h_α there is $C_\alpha \in \mathcal{D}_{\leq\aleph_0}(\alpha)$, $C_\alpha \subseteq V'$ (each member of C_α has the form $\{h_\alpha(i) : i < \delta\}$ for some $\delta < \omega_1$). So S does not reflect.

2) Let $S \subseteq S_{<\kappa}(\kappa^+)$ be $\neq \phi \bmod \mathcal{D}_{<\kappa}(\kappa^+)$, but $S \cap S_{<\kappa}(\alpha) = \phi \bmod \mathcal{D}_{<\kappa}(\alpha)$ when $\kappa \leq \alpha < \kappa^+$. Let h_β be a one to one function from $|\beta|$ onto β. When $\kappa \leq \alpha < \kappa^+$ let $\alpha = \bigcup_{i<\kappa} a_i^\alpha$, a_i^α increasing continuous in i, $a_i^\alpha \notin S$. (Possible by the choice of S). Let $C_\alpha = \{i < \kappa : h_\alpha$ maps i onto $a_i^\alpha\}$ so C_α is a club of κ. Let $g_\alpha : \kappa \to \kappa$ be defined by $g_\alpha(i) = Min(C_\alpha - i)$.

Let $C^* = \{a \in S_{<\kappa}(\kappa^+) : a$ is closed under h_α, h_α^{-1} and g_α and $a \cap \kappa$ is an ordinal$\}$

Obviously $C^* \in \mathcal{D}_{\leq\aleph_0}(\aleph_0)$. So $S \cap C^*$ is $\neq \phi \bmod \mathcal{D}_{\leq\aleph_0}(\kappa^+)$.

Suppose $(*)$ $a, b \in S \cap C^*$, $a \subseteq b$, $a \cap w_1 = b \cap w_1$, $a \neq b$, and we shall get a contradiction.

Let $\delta = a \cap \kappa = b \cap \kappa$. If $\alpha \in a \cap b$, $\alpha \geq \kappa$, then $a \cap \alpha = \{h_\alpha(i) : i < \delta\} = b \cap \kappa$. We know $b - \neq \phi$, let $\beta = Min\,(b - a)$; by the previous sentence $a \subseteq \beta_1$ hence $a = b \cap \beta$. Now as b is closed by g_β, clearly $\delta \in C_\beta$, hence (using h_α and the definition of C_β) $a = a_\delta^\beta$, so $a \notin S$, contradiction.

So $(*)$ is impossible hence $S \cap C^*$ is a weak (κ,κ^+)-stationary coding.

Remark: The proof is similar to some proofs in [FMS].

4. Fact:

It is consistent that e.g. the first inaccessible cardinal λ, is a strong limit and for no (regular uncountable) $\kappa < \lambda$, a strong (κ, λ)-stationary coding exists (assuming the consistency of suitable large cardinals)

Proof: Woodin constructs a model of set theory in which the first inaccessible λ is strong limit and $\diamondsuit_\lambda$ fail. By [Sh1] 7*A* for $\kappa < \lambda$, strong (κ,λ)-stationary coding does not exists.

Why 7*A* holds? By the known (folk?) proof that club implies diamond i.e.

4.A Fact: (= 7A of [Sh1])

If there is a strong (κ,λ)-stationary coding, $\kappa < \lambda$, $\lambda = \lambda^{<\lambda} > 2^{<\kappa}$ then $\diamondsuit_{\{\delta < \lambda : cf\, \delta < \kappa\}}$

Proof: As $\lambda = \lambda^{<\lambda}$ let $\{A_i : i < \lambda\}$ be a list of all bounded subset of κ. Let $\{a_\delta : \delta \in S\}$ be a strong (κ, λ)-stationary coding, for some stationary $S \subseteq \{\delta < \lambda : cf\, \delta < \kappa\} \subseteq \lambda$, $\delta = \sup a_\delta$ and $|a_\delta| < \kappa$. Let $P_\delta = \{\bigcup_{i \in b} A_i : b \subseteq a\}$, so for $\delta \in S$, P_δ is a family of $\leq 2^{<\kappa}$ subsets of δ. Now we shall prove that $\langle P_\delta : \delta \in S \rangle$ satisfies

$(*)$ for $X \subseteq A$, $\{\delta < \lambda : X \cap \delta \in P_\delta\}$ is a stationary subset of λ.

For let $h: \lambda \to \lambda$ be defined by

$$h(i) = \text{Min}\,\{j : A_j \cap i = X \cap i\}$$

So for stationarily many δs, a_δ is closed under h hence $X \cap \delta = \bigcup_{i \in a_\delta} (X \cap i) = \bigcup_{i \in a_\delta} A_{h(i)} = \bigcup \{A_\gamma : \gamma \in a, \ \gamma \in \text{Rang}\,(h \restriction a)\} \in P_\delta$. As $2^{<\kappa} < \lambda$ we are finished by a Theorem of Kunen.

5. Lemma:

1) It is consistent (in fact follows from the axiom from Foreman Magidor and Shelah [FMS] Martin Maximum) *that:* for no $\lambda > \aleph_1$ is there an $(\aleph_1, \lambda)$-weak stationary coding

2) It suffice to assume that $\mathcal{D}_{\omega_1}$ is $\aleph_2$-saturated, and for every stationary $S \subseteq \mathcal{P}_{<\aleph_1}(\lambda)$, $\{A \in \mathcal{P}_{<\aleph_2}(\lambda)$: $S \cap \mathcal{P}_{<\aleph_1}(A) \neq \phi$ $mod\ \mathcal{D}_{<\aleph_1}(A)\} \neq \phi$ $mod\ \mathcal{D}_{<\aleph_2}(\lambda)$.

Proof:

1) We prove 1) by 2), the assumptions of 2) holds by [FMS], and for 2) we may repeat [Sh1] 20.

Alternatively assume S is a weak $(\aleph_1,\lambda)$-stationary coding, let I_1 be the family of $T \subseteq \omega_1$ such that: there is an increasing continuous sequence $\langle a_i : i < \omega_1 \rangle$ of countable subsets of λ satisfying:
$\{i < \omega_1$: if $i \in T$ then $(\exists\, b \in S)\ [i = a_i \cap \omega_1 \subseteq b \subseteq a_i]\}$ contains a club C.
For $T \in I_1$ let $\langle a_i(T) : i < \omega_1 \rangle$, $C(T)$ be witnesses. Now I_1 is a normal ideal on ω_1, hence modulo the non-stationary ideal on ω_1 has a maximal member T^* (as $\mathcal{D}_{\omega_1}$ is $\aleph_2$-saturated).
<u>*If* $T^* = \omega_1$</u> (or just contains a club), then

$$S^* = \{b \in \mathcal{P}_{<\aleph_1}(\lambda) : (\exists i)\ [b \cap \bigcup_{j<\omega_1} a_j(T^*) = a_i(T^*) \wedge i \in C(T^*)]$$

$$\text{and } b \not\subseteq \bigcup_{j<\omega_1} a_j(T^*)\ \}$$

is a club of $\mathcal{P}_{<\aleph_1}(\lambda)$, and any member of $S^* \cap S$ contradict the assumption "S is a weak $(\aleph_1,\lambda)$-stationary coding", but such an element exists.

<u>*If* $\omega_1 - T^*$ is stationary,</u> $S_1 = \{b \in S : b \cap [\bigcup_{j<\omega_1} a_j(T^*)] = a_i(T^*)$ for some $i \notin T^*\}$ cannot be stationary otherwise by the second hypothesis of 5(2) we get contradiction to the maximalety of T^*. So for some $C_1 \in \mathcal{D}_{<\aleph_1}(\lambda)$, $C \cap S_1 = \phi$
Let $C_2 = \{b \in \mathcal{P}_{<\aleph_1}(\lambda)$: $b \not\subseteq \bigcup_{j<\omega_1} a_j(T^*)$, and

$b \cap \cap[\bigcup_{j<\omega_1} a_j(T^*]$ is $a_i(T^*)$ for some $i < \omega_1\}$.

Clearly $C_2 \in \mathcal{D}_{<\aleph_1}(\lambda)$. Hence $C_1 \cap C_2 \in \mathcal{D}_{<\aleph_1}(\lambda)$ hence there is $b \in C_1 \cap C_2 \cap S$. As $b \in C_1$ we know $b \notin S_1$, and as $b \in C_2$ for some $i < \omega_1$ $b \cap [\bigcup_{j<\omega_1} a_j(T^*)] = a_i(T^*)$, $b \neq a_i(T^*)$. This implies as $b \notin S_1$ by the definition of S_1 that $i \in T^*$, hence there is $a_i \in S_1$ $a_i(T^*) \cap \omega_1 \subseteq a \subseteq a_i(T^*)$. As by the choice of b and i $b \cap \omega_1 \subseteq a_i(T^*) \subseteq b$, $a_i(T^*) \neq b$ we get a, b contradicting "S is a weak $(\aleph_1,\lambda)$-stationary coding".

* * *

We give an elementary (i.e. with no forcing) presentation of the proof of [Sh1] 14.

6. Theorem:

If $\mathcal{D}$ is a fine normal filter on $I = \{a \subseteq \lambda : cf(\sup a) \neq cf\, |a|\}$, and λ is regular *then* there are functions f_i for $i < \lambda^+$ such that: $\mathrm{Dom}\, f_i = I$, $f_i(a) \in a$ and for $i \neq j$, $\{a \in I : f_i(a) = f_j(a)\} = \phi \bmod \mathcal{D}$

Proof: We can find $A_i (i < \lambda^+)$ such that:

(*) A_i is a subset of λ, unbounded in λ, and for $j < i$, $A_i \cap A_j$ is bounded in λ

[just let $A_i (i < \lambda)$ be pairwise disjoint subsets of λ of power λ, and then define A_i $(\lambda \leq i < \lambda^+)$ by induction on i: for each i let $\{j : j < i\}$ be $\{j_\alpha : \alpha < \lambda\}$, and let $A_i = \{\gamma^i_\beta : \beta < \lambda\}$ where $\gamma^i_\beta = \mathrm{Min}\, (A_{j_\beta} - \bigcup_{\alpha < \beta} A_{j_\alpha})$,) it exists as $|A_{j_\beta} \cap A_{j_\alpha}| < \lambda$ for $\alpha < \beta$].

Let for $i < \lambda^+$, $g_i : i \to \lambda$ be such that $\{A_j - g_i(j) : j < i\}$ are pairwise disjoint. Let f_i be a strictly increasing function from λ onto A_i (for $i < \lambda^+$) hence $f_i(\alpha) \geq \alpha$. So $C_i = \{a : a \text{ is closed under } f_i\}$ belongs to $\mathcal{D}$. For each $a \in I$ let $a = \{x^a_\alpha : \alpha < |a|\}$.

Now for each $a \in C_i$, $a \cap A_i$ is unbounded is a, (by the definition of C_i) so for some $\alpha_i(a) < |a|$, $A_i \cap \{x^a_\alpha : \alpha < \alpha_i(a)\}$ is unbounded in a (as $cf(\sup a) \neq cf\, |a|$).

Next for $i < \lambda^+$ let h_i be a one-to-one function from λ onto $\lambda \cup \{j : j < i\}$ and define by induction on i:

$$C_i^1 = \{a \subseteq i \cup \lambda : \begin{array}{l} a \text{ closed under } h_i\, ,\ h_i^{-1}\, ,\ a \cap \lambda \in I \\ a \cap \lambda \text{ closed under } f_i\, ,\ f_i^{-1}\, , \\ a \text{ closed under } g_j\, ,\ (j \in a \text{ or } j = i) \\ \text{and for } j \in a\, ,\ a \cap (j \cup \lambda) \in C_j^1\, \} \end{array}$$

Clearly $C_i^1 \restriction \lambda = \{a \cap \lambda : a \in C_i^1\}$ is in $\mathcal{D}$, and for each $a \in I$ there is at most one $a' \in C_i^1$ satisfying $a' \cap \lambda = a$, namely $h_i``(a)$.

Now we define for $i < \lambda^+$ a function d_i with domain I.

$$d_i(a) = \begin{cases} \langle \alpha_i(a), otp(\{j \in h_i``(a) : \alpha_j(a) = \alpha_i(a)\} \rangle, & \text{if } h_i``(a) \cap \lambda = a \\ & h_i``(a) \in C_i^1 \\ \text{Min } a & \text{otherwise} \end{cases}$$

Now we shall finish by showing:

A: for $i_1 \neq i_2$, $\{a \in I : d_{i_1}(a) = d_{i_2}(a)\} = \phi \bmod \mathcal{D}$

B: for $a \in I$, $\{d_i(a) : i < \lambda^+\}$ has cardinality $\leq a$

Why this suffice? As for each $a \in I$ we can find a one-to-one function e_a from $\{d_i(a) : i < \lambda^+\}$ into a and now use the λ^+ functions $\langle e_a(d_i(a)) : i < \lambda^+ \rangle$

Proof of A: *W.l.o.g.* $i_1 < i_2$ and $\lambda \leq i_1$ for notational simplicity. Clearly $R = \{a \in I : h_{i_2}``(a) \in C_{i_2}^1, \ i_1 \in h_{i_2}``(a)$ (hence $h_{i_1}``(a) = h_{i_2}``(a) \cap i_1 \in C_{i_1}^1$

belongs to $\mathcal{D}$. Let a be in it, and $d_{i_1}(a) = d_{i_2}(a)$. Clearly $d_{i_1}(a) \neq$ *Min* a hence by the first coordinale in $d_i(a)$, $\alpha_{i_1}(a) = \alpha_{i_2}(a)$. Now $\{\xi \in h_{i_1}``(a) : \alpha_\xi(a) = \alpha_{i_1}(a)\}$ is an initial segment of $\{\xi \in h_{i_2}``(a) : \alpha_\xi(a) = \alpha_{i_2}(a)\}$ (as $a \in R$) and a proper one (as i_1 belong to the latter but not the former). As the ordinals are well ordered, their order types are not equal. That means that the second coordinate in the $d_{i_1}(a)$, $d_{i_2}(a)$ are distinct. So $d_{i_1}(a) \neq d_i(a)$ is true for $i_1 \neq i_2$, $a \in R$, as required.

Proof of B: As the number of possible $\alpha_i(a)$ is $\leq |a|$, and the number of order types of well orderings of power $< |a|$ is $|a|$ it suffice to prove:

(*) for $i < \lambda^+$, $a \in C_i^1$, the set $u = \{j \in a : \alpha_j(a \cap \lambda) = \alpha_i(a \cap \lambda)\}$ has power $< |a|$

Why (*) holds? Because for $j \in u$ the set

$$A_j \cap \{x_\alpha^a : \alpha < \alpha_i(a \cap \lambda)\}$$

is unbound in $a \cap \lambda$

but $A_j \cap g_i(j)$ is bounded in $a \cap \lambda$ (as a is closed under g_i)

hence

$$r_j \stackrel{def}{=} (A_j \setminus g_i(j)) \cap \{x^a_\alpha : \alpha < \alpha_i(a\cap\lambda)\}$$

is an unbounded subset of $a \cap \lambda$, hence non empty.

But $\langle r_j : j \in a,\ \alpha_j(a\cap\lambda) = \alpha_i(a\cap\lambda)\rangle$ is a sequence of pairwise disjoint subsets of $\{x^a_\alpha : \alpha < \alpha_i(a\cap\lambda)\}$ (by the choice of g_i). As they are non empty their number is $\leq |\{x^a_\alpha : \alpha < \alpha_i(a\cap\lambda)\}| < |a|$.

7. Claim:

Let $\mathcal{D}$ be a fine normal filter on $I \subseteq \mathcal{P}_{<\kappa}(\lambda)$, λ singular and $(\forall a \in I)$ $(|a| \geq cf\ \lambda \wedge \quad cf\ |a| \neq cf\ \lambda \quad \wedge\ cf\ \lambda = \sup(cf\ \lambda \cap a))$ and $Rk(|a|, \mathcal{D}^{cb}_{cf\lambda}) \leq |a|^+$

Then there are functions f_i for $i < \lambda^+$, Dom $f_i = I$, $(\forall a \in I)[f_i(a) \in a]$ and for $i \neq j$ $\{a \in I : f_i(a) = f_j(a)\} = \phi \bmod \mathcal{D}$

Proof: Let $\sigma = cf\ \lambda$, $\lambda = \sum_{\xi<\sigma} \lambda_\xi$, each λ_ξ regular, $\sum_{\xi<\zeta} \lambda_\xi < \lambda_\zeta < \lambda$ for $\zeta < \sigma$. We can find for $i < \lambda^+$ functions A_i from σ to λ, $\sum_{\xi<\zeta} \lambda_\xi < A_i(\zeta) < \lambda_\zeta$ such that for $i < j < \lambda^+$ there is $\xi < \sigma$ such that

$$\xi \leq \zeta < \sigma \Longrightarrow A_i(\zeta) < A_j(\zeta)$$

Let again $a = \{x^a_\alpha : \alpha < |a|\}$, so for each $i < \lambda^+$, $a \in I$ if Range A_i is unbounded in a then for some $\alpha_i(a) < a$, (Range A_i) $\cap \{x^a_\alpha : \alpha < \alpha_i(a)\}$ is unbounded in a (and $\alpha_i(a) =$ Min a otherwise).

Now for $i < \lambda^+$ we define a function d_i with domain I (h_i - a one-to-one function from λ onto $i \cup \lambda$):

$$d_i(a) = \begin{cases} \langle \alpha_i(a),\ otp\{j \in h_i``(a) : \alpha_j(a) = \alpha_i(a)\}\rangle & \text{if } a = h_i``(a) \cap \lambda, \\ & (\forall \zeta \in (a \cap cf\ \lambda))A(\zeta) \in a \\ & \text{and } (\forall j \in a) \\ & a = h_j``(a) \cap \lambda \\ \text{Min } a & \text{otherwise} \end{cases}$$

We finish as in 6.

7A Remark:

1) Really we use $Rk(|a|, \mathcal{D}^{cb}_\sigma) \leq |a|^+$ (where $\sigma = cf\ \lambda$) just to get, that for every $\zeta < |a|$ for some $\xi_\zeta < |a|^+$

(*) there are no $f_i : \sigma \to \zeta$ for $i < \xi_\zeta$, $[i<j \Longrightarrow f_i <_{D^{cb}_\sigma} f_j]$

We should observe that for $a \in I$, $a \cap \sigma$ has order type σ.

Note that if for each $\zeta < |a|$ there is such ξ_ζ then $\xi(*) = \bigcup_{\zeta < |a|} \xi_\zeta$ is $< |a|^+$ and work for all ζs.

Similar remark apply to 8.

8. Claim:

Suppose $\kappa \leq \sigma = cf\ \lambda < \lambda$,
$I \subseteq \{a \in \mathcal{P}_{<\kappa}(\lambda)$: $cf\ |a| \neq cf(\sup(a \cap \sigma))$, and $Rk(|a|, \mathcal{D}^{cb}_{cf(\sup(a \cap \sigma))}) \leq |a|^+$ when $cf(\sup a) > \aleph_0$ and $|a|^{\aleph_0} = |a|$ when $cf(\sup a) = \aleph_0\}$,
and $\mathcal{D}$ a normal fine filter on I.

Then there are for $i < \lambda^+$ functions $f_i : I \to \lambda$, $f_i(a) \in a$ and for $i \neq j$ $\{a \in I : f_i(a) = f_j(a)\} = \phi \bmod \mathcal{D}$.

Proof: Let A_i, λ_ζ be as in the proof of 7, $a = \{x^a_\alpha : \alpha < |a|\}$. Let h_i be a one-to-one function from λ onto $\lambda \cup \{j : j < i\}$. For each i the set $C^1_i \overset{def}{=} \{a \in I$: a is closed under A_i, and (Range A_i) $\cap a$ is unbounded in a, $h_i``(a) \cap \lambda = a$ and $a \in C^1_j$ for $j \in h_i``(a)$ and $cf(\sup a) = cf(\sup(a \cap \sigma))\}$ belongs to $\mathcal{D}$, and for $a \in C^1_i$ let $\alpha_i(a) < |a|$ be minimal such that (Range A_i) $\cap \{x^a_\alpha : \alpha < \alpha_i(a)\}$ is unbounded in a. We then let

$$d_i(a) = \begin{cases} \langle \alpha_i(a),\ otp\{j : j \in h_i''(a),\ \alpha_j(a) = \alpha_i(a)\} \rangle & \text{if } a \in C^1_i, \\ \text{Min } a & \text{otherwise} \end{cases}$$

and we proceed as in the proof of 6, 7 (and see 7A).

9. Definition:

1) For $\kappa < \lambda$, κ regular, and a model N with universe $|N|$ which is an ordinal $< \kappa$, two place relation R^N_1, R^N_2, a three place relation R^N_3 and a partial one place function F^N (if one of them is not memtioned this means it is empty), let (see notation 1(5)):
$T_{\kappa,\lambda}(N) = \{a \in \mathcal{P}_{<\kappa}(\lambda)$: $dcd``(a) \cap \lambda = a$,

and there are b_s (for $s \in |N|$) such that:

(i) $b_s \subseteq a$, $a = \bigcup_{s \in N} b_s$, $b_s \in dcd``(a)$ (equivalently $cd_{\kappa,\lambda}(b_s) \in a$)

(ii) $sR_1^N t$ implies $b_s \subseteq b_t$

(iii) $sR_2 t$ implies $cd_{\kappa,\lambda}(b_s) \in b_t$

(iv) for each t, $cd\{<\alpha, cd_{\kappa,\lambda}(b_s)> : \alpha \in N, R_3^N(\alpha, s, t)\} \in b_t$

(v) for $t \in \mathrm{Dom} F^N$, $|b_t| \leq F(t)$

2) For K a family of models N, $T_{\kappa,\lambda}(K) = \bigcup_{N \in K} T_{\kappa,\lambda}(N)$

3) $N_\theta^0 = (\theta)$ (so R_1, R_2, R_3, F are empty)

$N_\theta^1 = (\theta, <)$ (so R_2, R_3, F are empty)

$N_\theta^2 = (N, <, <)$ (so R_3, F are empty)

$N_\theta^3 = (\theta, <, <, R_3)$ where $R_3 = \{<\alpha, \alpha, \gamma> : \alpha < \gamma < \theta\}$ (so F is empty)

* * *

We now show that [Sh1] 13 (and 12) is applicable sometimes. (see 2, 2A above for what they say). This is when $\kappa = \lambda$ in 10.

10. Claim:

Suppose $\kappa = \mu^+ \leq \lambda$, θ regular, $\aleph_0 < \theta < \mu$, and $Rk(\mu^+, \mathcal{D}_\theta^{cb}) = \mu^+$. Then there is a function g from $T = T_{\kappa,\lambda}(N_\theta^1)$ to κ such that for every well ordering $<^*$ of λ

$$\{a \in \mathcal{P}_{<\kappa}(\lambda) : otp(a, <^*) < g(a)\} \supseteq T \bmod \mathcal{D}_\kappa(\lambda)$$

11. Remark:

1) We can use other N's, but then have to change accordingly the filter by which we define the rank.

2) In [Sh2] various sufficient conditions for $Rk(\mu^+, \mathcal{D}^{cb}_\theta) = \mu^+$ are given: (When $cf\,\mu \neq \theta$):

$$(\forall \sigma < \mu)\,[\sigma^\theta \leq \mu]$$

and

$$\text{"}\mu > 2^\theta \text{ and } \mu \leq (\sup\{\sigma : \sigma^\theta \leq \mu\})\text{"}$$

4) As for $a \in T$ $\{\beta : \beta < g(a)\}$ has power μ, and $C' = \{a \in \mathcal{P}_{<\kappa}(\lambda): |a| = \mu\} \in \mathcal{D}_{<\kappa}(\lambda)$, we can deduce that:

If the conclusion of 10 holds for T then there are functions $g_i : T \to \lambda$ (for $i < \lambda^+$) $g(a) \in a$ such that for $i \neq j$ $\{a \in T: g_i(a) = g_j(a)\} = \phi \bmod \mathcal{D}_{<\kappa}(\lambda)$

Proof of 10: For each well ordering $<^*$ of λ let

$$C[<^*] = \{a \in \mathcal{P}_{<\kappa}(\lambda): \text{for each } i \in a,\ cd_{\kappa,\lambda}(i) \cap \lambda \subset a \text{ and } otp(cd_{\kappa,\lambda}(i) \cap \lambda, <^*) < otp(a)\}$$

It is clearly closed unbounded, i.e., belongs to $\mathcal{D}_{<\kappa}(\lambda)$. Now if $a \in T(N^1_\theta) \cap C[<^*]$, let $\langle b_\alpha : \alpha < \theta \rangle$ witness "$a \in T$" (i.e., $i_\alpha \in a$, $i_\alpha = cd_{\kappa,\lambda}(b_\alpha) \in \mathcal{B}_{\kappa,\lambda}$, $a = \bigcup_{\alpha<\theta} b_\alpha$, b_α is increasing in α), so $otp(b_{i_\alpha}, <^*) < otp(a)$ for each α. So clearly it suffices to prove:

12. Fact:

If θ is regular cardinal, $\aleph_0 < \theta < \mu$, $\theta \neq cf\,\mu$ and $Rk(\mu^+, \mathcal{D}^{cb}_\theta) = \mu^+$ *then* for every $\xi < \mu^+$ there is $\zeta_\xi < \mu^+$ such that: if $\zeta = \bigcup_{i<\theta} A_i$, A_i increasing, then for some $i < \theta$ $otp(A_i) \geq \xi$

Proof:

Suppose $otp(A_i) < \xi$ for $i < \theta$, A_i increasing, and $\zeta = \bigcup_{i<\theta} A_i$. Define for $\gamma < \zeta$ a function $h_\gamma : \theta \to \xi$ by: $h_\gamma(i) = otp(A_i \cap \gamma)$. So each h_γ is a function from θ to ordinals, and for $\beta < \gamma$ $(\forall i < \theta)$ $[h_\beta(i) \leq h_\gamma(i)]$, moreover for some $j < \theta$ $\beta \in A_j$ hence $(\forall i)$ $[j < i < \theta \to h_\beta(i) < h_\gamma(i)]$. This clearly implies $Rk(\xi, \mathcal{D}^{cb}_\theta) \geq \zeta$ but $Rk(\xi, \mathcal{D}^{cb}_\theta) < \mu^+$.

13. Definition

For $\kappa \leq \lambda$, κ regular, $\mathcal{D}$ a normal fine filter on $I \subseteq \mathcal{P}_{<\kappa}(\lambda)$,

1) $\diamondsuit(\mathcal{D})$ means that there are $\langle A_a : a \in I \rangle$, $A_a \subseteq a$, such that for every $A \subseteq \lambda$, $\{a \in I : A \cap a = A_a\} \neq \phi \bmod \mathcal{D}$

2) $\diamondsuit^*(\mathcal{D})$ means that there are $\langle \mathcal{P}_a : a \in I \rangle$, $\mathcal{P}_a$ a family of $\leq |a|$ subsets of a, such that for every $A \subseteq \lambda$ $\{a \in I : A \cap a \in \mathcal{P}_a\} \in \mathcal{D}$

3) We replace $\mathcal{D}$ by I when $\mathcal{D}$ is the filter generated by the family of closed unbounded subsets of I. We write $I, \mathcal{D}$ instead of $\mathcal{D} + I$,

14. Remark:

We implicitly assume $I \neq \phi \bmod \mathcal{D}_{<\kappa}(\lambda)$,

15. Fact:

1) For $I \subseteq J \subseteq \mathcal{P}_{<\kappa}(\lambda)$, $\mathcal{D}_1 \subseteq \mathcal{D}_2$ normal fine filter on $\mathcal{P}_{<\kappa}(\lambda)$,

i) $\diamondsuit^*(\mathcal{D}_1+J) \Longrightarrow \diamondsuit^*(\mathcal{D}_2+I)$

ii) $\diamondsuit^*(\mathcal{D}_1+J) \Longrightarrow \diamondsuit(\mathcal{D}_1)$

iii) $\diamondsuit(\mathcal{D}_2+I) \Longrightarrow \diamondsuit(\mathcal{D}_1+J)$

iv) $\diamondsuit^*(\mathcal{D}_1+J) \Longrightarrow \diamondsuit(\mathcal{D}_2+I)$

(remember $\mathcal{D}_{<\kappa}(\lambda)+I \subseteq \mathcal{D}$ for any fine normal filter on I)

2) Suppose $\kappa <\cdot \lambda = \lambda^{<\kappa}$,

$T = \{a$: for some θ, $a \in T_{\kappa,\lambda}(N^0_\theta)$, $|a|^\theta = |a|$

or $a \in T_{\kappa,\lambda}(N^1_\theta)$, and $cf\,|a| \neq \theta$ $(\forall \sigma < |a|)\sigma^\theta \leq |a|$

or $(\exists \chi,\sigma,\alpha)$ $(2^\chi \leq \lambda \wedge \lambda = \chi^{+\alpha} \wedge |a|^{<\sigma} = |a| \wedge (\forall \gamma < \alpha)$ $[cf\,(a \cap \chi^{+(\gamma+1)} < \sigma] \wedge \alpha < \sigma)\}$

Suppose further $T \neq \phi \bmod \mathcal{D}_{<\kappa}(\lambda)$. Then $\diamondsuit^*(T, \mathcal{D}_\kappa(\lambda))$

Proof: By straightforward generalization of the proof for the case $\lambda = \kappa$, due to Kunen for 1, (i.e., 1(ii), the rest being trivial) Gregory and Shelah for 2) (see e.g. [Sh3]). I.e., for 1)(ii), suppose $\langle \mathcal{P}_a : a \in \mathcal{P}_{<\kappa}(\lambda) \rangle$ exemplifies $\diamondsuit^*(\mathcal{D}_1 + J)$. Let $\mathcal{P}_a = \{A^a_i : i \in a\}$. Let $<\ ,\ >$ be a pairing function on λ, and for each $i < \lambda$, $a \in \mathcal{P}_{<\kappa}(\lambda)$ let

$$B^i_a = \{\alpha : \alpha \in a\ ,\ <\alpha,i> \ \in A^a_i\}$$

So $B^i_a \subseteq a_i$; is $\langle B^i_a : a \in \mathcal{P}_{<\kappa}(\lambda) \rangle$ a $\Diamond(\mathcal{D}_1)$-sequence for some i? If yes we finish, if not let $B^i \subseteq \lambda$ exemplify this i.e.,

$$C^i = \{a \in \mathcal{P}_{<\kappa}(\lambda) :\ B^i \cap a \neq B^i_a\} \ \in \mathcal{D}_1$$

Hence

$$C = \{a \in \mathcal{P}_{<\kappa}(\lambda) :\ (\forall i \in a)\ a \in C^i\ ,\ \text{and } a \text{ is closed under } <\ ,\ >\} \ \in \mathcal{D}$$

and let

$$A = \{<\alpha,i> \ : \alpha \in B^i\}\ .$$

So for some $a \in C$, $A \cap a \in \mathcal{P}_a$ hence for some $i \in A$, $A \cap a = A^a_i$ hence $B^i \cap a = B^i_a$ contradiction.

16. Remark:

We can enlarge T in 15(2) to:
the set of $a \in \mathcal{P}_{<\kappa}(\lambda)$ satisfying:
(*) there is a family H of $\leq |a|$ functions from a to a such that: *for any* $h : a \to a$, *for some* $b \subseteq a$, $h \restriction b \in H$ and $a \subseteq \bigcup_{i \in b} dcd_{\kappa,\lambda}(i)$

* * *

Now 15(2) can be combined with (15(ii) and):

17. Observation:

If $\mathcal{D}$ is a fine normal filter on $\mathcal{P}_{<\kappa}(\lambda)$, and $\Diamond(\mathcal{D})$ holds, *then:* there are $J_\alpha \subseteq \mathcal{P}_{<\kappa}(\lambda)$ for $\alpha < 2^\lambda$ such that:

$$J_\alpha \neq \phi \bmod \mathcal{D}\ ,\quad J_\alpha \cap J_\beta = \phi \bmod \mathcal{D} \ \text{ for } \ \alpha \neq \beta$$

18. Conclusion:

Suppose $\lambda = \lambda^{<\kappa}$, $\theta < \kappa$, κ regular, λ regular, and there is a strong (κ,λ)-stationary coding set S^* such that $(\forall a \in S^*)$ $[cf(\sup a) = \theta]$ and $\diamondsuit(\mathcal{D}_\kappa(\lambda)+S^*)$. Then there are $S_\alpha \subseteq \{\delta < \lambda: cf\ \delta = \theta\}$ for $\alpha < 2^\lambda$, each stationary, the intersection of any two non-stationary (any normal filter $\mathcal{D}$ on λ will satisfy this if $\{\sup a : a \in S^*\} \neq \phi \bmod \mathcal{D}$ and $\diamondsuit(\mathcal{D}'+S^*)$ where $\mathcal{D}' = \mathcal{D}+\{\{a : \sup a \in A\} : A \in \mathcal{D}\}$).

19. Conclusion:

If $\theta < \kappa \leq \lambda$, $\kappa = \mu^+$, $\mu^\theta = \mu$, *then* for $T = T_{\kappa,\lambda}(N^0_\theta)$, $T \neq \phi \bmod \mathcal{D}_\kappa(\lambda)$ and $\diamondsuit(T, \mathcal{D}_\kappa(\lambda))$.

Remark: This is closely related to [Sh6], [Sh7], (see particularly last section of [Sh7]) which continues [Sh4] VIII 2.6.

Proof: By 15(2).

20. Lemma:

1) Suppose $\theta < \kappa \leq \chi \leq \lambda$, $T \subseteq P_{<\chi^+}(\lambda)$, $T \neq \phi \bmod \mathcal{D}_{\chi^+}(\lambda)$, $\diamondsuit(T, \mathcal{D}_{\chi^+}(\lambda))$ and for each $a \in T$, $\chi \subseteq a$ and:

$$(i)\quad (\exists b \subseteq a)[|b| < \kappa \wedge a = \bigcup_{\alpha \in b} dcd_{\chi^+,\lambda}(\alpha)]$$

Then we can find $T_1 \subseteq P_{<\kappa}(\lambda)$, $T_1 \neq \phi \bmod \mathcal{D}_\kappa(\lambda)$ such that $\diamondsuit(T_1, \mathcal{D}_\kappa(\lambda))$ holds.

2) Suppose in addition that for $a \in T$:

$$(ii)\quad (\forall c \subseteq a)[|c| < \kappa \rightarrow cd_{\chi^+,\lambda}(c) \in a]$$

Then we can demand $T_1 \subseteq T_{\kappa,\chi}(N^3_\theta)$

Proof: 1) As in the proof of claim 7 in [Sh1].
As $\diamondsuit(T_1, \mathcal{D}_{\chi^+}(\lambda))$, we can find $\langle M_a : a \in T \rangle$ such that M_a is a model with universe a and countably many (finitary) functions, and for every model M with universe λ and countably many functions $\{a : M_a = M \restriction a\} \neq \phi \bmod \mathcal{D}_{\chi^+}(\lambda)$

For $a \in T$ we can find $b_a \subseteq a$, $|b_a| < \kappa$ such that b_a is closed under the functions of M_a and $a \subseteq \bigcup_{\alpha \in b_a} dcd_{\chi^+,\lambda}(\alpha)$. By the last condition, and as $[\alpha \in a \Longrightarrow dcd_{\chi^+,\lambda}(\alpha) \subseteq a]$ clearly $[a_1 \neq a_2 \Longrightarrow b_{a_1} \neq b_{a_2}]$. We define $N_{b_a} = M_a \restriction b_a$, and let $T_1 = \{b_a : a \in T\}$. So $N_b (b \in T_1)$ is well defined. Now
(i) $T_1 \subseteq \mathcal{P}_{<\kappa}(\chi)$,
(ii) $T_1 \neq \phi \, mod \, \mathcal{D}_\kappa(\chi)$ [if M is a model with universe λ and countably many functions, for some $a \in T$ $M_a = M \restriction a$, so b_a is closed under the functions of M and $b_a \in T_1$
(iii) For every model M with universe λ and countably many functions, for some $b \in T_1$, $N_b = M \restriction b$. [same proof as in (ii)]. Hence $\Diamond \, (T_1, \mathcal{D}_\kappa(\lambda))$ holds.
2) Easy from the proof of 1), choosing b_a in $T_{\kappa,\lambda}(N^3_\theta)$

21. Lemma:

Suppose $\mathcal{D}_1$ is a fine normal filter on $\mathcal{P}_{<\kappa}(\lambda_1)$, $\kappa \leq \lambda_1 < \lambda$. Let $\mathcal{D}$ be the normal fine filter on $\mathcal{P}_{<\kappa}(\lambda)$ generated by $\left\{\{a \in \mathcal{P}_{<\kappa}(\lambda) : a \cap \lambda_1 \in S\} : S \in \mathcal{D}_1\right\}$. Suppose further that $T_1 \subseteq \mathcal{P}_{<\kappa}(\lambda_1)$,

$T_1 \neq \phi \, mod \, \mathcal{D}_1$, $\Diamond \, (T_1, \mathcal{D}_1)$ and T_1 is a (κ,λ_1)-weak stationary coding.

Lastly suppose $NSi(\kappa,\lambda)$ holds (see [Sh1] Def.8) or at least: for some algebra M will universe λ and countably many functions, M has no isomorphic but distinct subalgebras $M_1 \subseteq M_2$, $M_1 \cap \lambda_1 = M_2 \cap \lambda_1 \in T$

Then there is a (κ,λ)-weak stationary coding set T, for which $\Diamond \, (T, \mathcal{D})$ holds.

Proof: Just like 10 of [Sh1].

Remark: We can combine 21 or 22 with 23 or 24, getting existence for many cardinals.

22. Lemma:

Suppose in the previous lemma, κ is a strongly Mahlo cardinal, T is a (κ,λ_1)-stationary coding. Suppose further that if $b \subset a$ are in T then for every subset c of a of power $\leq |b|$, $cd_{\kappa,\lambda}(c) \in a$. Then $\mathcal{P}_{<\kappa}(\lambda)$ has a (κ,λ)-

stationary coding.

23. Lemma:

1) Suppose $\aleph_0 < \kappa < \lambda$, κ is regular, $\lambda^{<\kappa} = \lambda$ and $(\forall \sigma < \kappa)\, \sigma^{\aleph_0} < \kappa$, (hence $2^{\aleph_0} < \kappa$).

Then there is a (κ,λ^+)-stationary coding set T.

2) Also we can have $\diamondsuit\ (T, \mathcal{D}_\kappa(\lambda))$

3) Suppose that $\lambda^{<\kappa} \le \lambda^+$, $\mathcal{D}_1$ is a normal fine filter on $\mathcal{P}_{<\kappa}(\lambda)$, $T^* \in \mathcal{D}_1$, T^* has cardinality λ, and

(a) $(\forall a \in T^*)\,(\forall b \subseteq a)\,[\,|b| \le \aleph_0 \rightarrow cd_{\kappa,\lambda}(b) \in a]$

Let $\mathcal{D}$ be the minimal normal fine filter on $\mathcal{D}_{<\kappa}(\lambda^+)$ such that $\mathcal{D} \restriction \lambda = \mathcal{D}_1$. *Then* for some $\mathcal{D}$-stationary T, $(\mathcal{D} + T) \restriction \lambda = \mathcal{D}_1$, and T is a stationary coding set.

4) For 3) if $\lambda = \lambda^{\aleph_1}$, $\lambda^{<\kappa} \le \lambda^+$ and for some $T_0 \subseteq \mathcal{P}_{<\kappa}(\lambda)$

$$|T_0| = \lambda \wedge (\forall a \in \mathcal{P}_{<\kappa}(\lambda))(\exists b \in T_0)\,[a \subseteq b]$$

then $\mathcal{D}_{<\kappa}(\lambda) + T$ is as required where

$T = \{a \in \mathcal{P}_{<\kappa}(\lambda)$: there are $b_i \in T_0$ $(i < \omega_1)$ increasing $a = \bigcup_{i<\omega_1} b_i$,

$$a = \lambda \cap dcd_{\kappa,\lambda}``(a),\ cd_{\kappa,\lambda}(b_i) \in a\}$$

Proof:

1) Let $\mathcal{P}_{<\kappa}(\lambda) = \{b_i:\ i < i(*)\}$, $i(*) \le \lambda^+$, and let for $i < i(*)$ $S_i \subseteq S^* = \{\delta < \lambda^+: cf\ \delta = \aleph_0\}$ be pairwise disjoint stationary subsets of λ^+, $S^* = \bigcup_i S_i$. For $\delta \in \bigcup_{i<i(*)} S_i$ let $i(\delta)$ be the unique i such that $\delta \in S_i$.

Let f, g be such that: f, g two place functions from λ^+ to λ^+, for $i < \lambda^+$, $i = \{j : j < i\} = \{f(i,j): j < |i|\}$ and for $j < |i| < \lambda^+$ $g(i, f(i,j)) = j$.

23.A. Observation:

If $a \in \mathcal{P}_{<\kappa}(\lambda^+)$ is closed under f and g, $w \subseteq a$ is unbounded in a and $a \cap \lambda = b_i$ *then* a is totally determined by w and i, and we write $a = a_i[w]$.

Let for $\delta \in S_i$

$$T^i_\delta = \{a \in \mathcal{P}_{<\kappa}(\lambda^+) : \sup a = \delta,\ a \cap \lambda = b_i\ ,\ a \text{ closed under f and g, and for any bound countable } w \subset a\ ,\ \text{with } \sup w \in S^*,\ cd_{\kappa,\lambda^+}(a_{i(\sup w)}[w]) \in a\}$$

$$T^i = \bigcup_{\delta \in S_i} T^i_\delta$$

$$T = \bigcup_{i < i(*)} T^i$$

23.B. Observation:

If $c \subset d, d \neq c$ and $c, d \in T$ then $cd_{\kappa,\lambda^+}(c) \in d$

Proof: Let $d \cap \lambda = b_i$, $c \cap \lambda = b_j$, $w \subset c$ a countable subset of c with $\sup w = \sup c$ (w exists as for each $a \in T$, $cf(\sup a) = \aleph_0$).) As $c \in T$, $c \cap \lambda = bj$ necessarily $\sup w \in S_j$. If $i = j$ then $d \cap \lambda = c \cap \lambda$ and w is an unbounded subset of both so $d = c = a_i[w]$ contradiction. So assume $i \neq j$, so necessarily $\sup w \neq \sup a$ hence $\sup w < \sup a$ hence $a_{i(\sup w)}[w] = c$ but as $d \in T$ by the definition of the T^i_δ's we know that $a_{i(\sup w)}(w)) \in d$. So $cd_{\kappa,\lambda^+}(c) \in d$.

23.C. Observation: $T \neq \phi \mod \mathcal{D}_\kappa(\lambda)$

Proof: By Rubin and Shelah [RS]. (see proof of 24 after 24A)

Continuation of the proof of 23.

The observations above finishes the proof of 23(1).

2) We let $\{(b_i, M_i): i < i(*)\}$ list all pairs (b, M), where $b \in \mathcal{P}_{<\kappa}(\lambda)$, $M = (\alpha^M, A^M)$, $\alpha^M < \kappa$, $A^M \subseteq \alpha$. We use $\{b_i : i < i(*)\}$ as above and for $a \in T$, $\sup a \in T_i$, let $A_a = \{\xi \in A : otp(a \cap \xi) \in A^{M_i}\}$. Now $\langle A_a : a \in T \rangle$ is a wittness for $\diamondsuit(T, \mathcal{D}_\kappa(\lambda))$.

3) Same proof.

4) Left to the reader.

24. Lemma:

Suppose $\aleph_0 < \kappa \leq \lambda$, κ regular, $S^* \subseteq \{\delta < \lambda^+ : cf\ \delta = \aleph_0\}$, and $\mathcal{D}$ is a normal fine filter on $\mathcal{P}_{<\kappa}(\lambda)$ such that:

(i) $\lambda^+ = (\lambda^+)^{\prime -}$

(ii) there is $Y^* \in \mathcal{D}$ of power λ

(iii) if $\lambda < \alpha < \lambda^+$, and $\mathcal{D}_\alpha$ is the unique normal fine filter on α such that $\mathcal{D}_\alpha \restriction \lambda = \mathcal{D}$ then:

$$\{a \in \mathcal{P}_{<\kappa}(\alpha) : \text{there is } \delta \in S^* \cap \alpha - a \text{ such that } \delta = \sup(\delta \cap a)\} = \phi \bmod D_\alpha$$

(iv) $2^{<\kappa} \leq \lambda$

Let $\mathcal{D}_1$ be the minimal normal fine filter on

$$\mathcal{P}_{<\kappa}(\lambda^+) \text{ such that } \mathcal{D}_1 \restriction \lambda = \mathcal{D}$$

Then there is $T \subseteq \mathcal{P}_{<\kappa}(\lambda^+)$, such that T is a (κ,λ^+)-stationary coding, $(\mathcal{D}_1 + T) \restriction \lambda = \mathcal{D}$ and $\diamondsuit\ (T,\mathcal{D}_1)$

Proof: Let $\{(b_i,M_i): i < i(*)\}$ (where $i(*) \in \{\lambda,\lambda^+\}$) list the pairs (b,M), $b \in Y^*$, $M = (\alpha^M,A^M)$, $\alpha^M < \kappa$, $A^M \subseteq \alpha^M$ (by (i) this is possible). Let $S_i \subseteq S^*$ (for $i < i(*)$) be pairwise disjoint stationary subsets of λ^+, $S^* = \bigcup_{i<i(*)} S_i$. For $\delta \in S^*$ let $i(\delta)$ be the unique $i < i(*)$ such that $\delta \in S_i$. Let f, g be two-place functions on λ^+ such that for $i < \lambda^+$ $i = \{f(i,j): j < |i|\}$ and for $j < |i|$ $g(i,f(i,j)) = j$. Let $C_0 = \{a \in \mathcal{P}_{<\kappa}(\lambda^+)$: a closed under f and g and $x+1\}$

For $w \subseteq \lambda^+$ countable with $\sup w \in S^*$ let set $[w]$ be the closure of $w \cup b_{i(\sup w)}$ under f and g. For $i < i(*)$, $\delta \in S_i$ let

$T^i_\delta \overset{def}{=} \{a \in \mathcal{P}_{<\kappa}(\lambda^+) : \sup a = \delta,\ a \cap \lambda = b_{i(\delta)}$

a is a closed under f and g ,

and for any bounded countable

$w \subseteq a$: if $\sup w \in S^*$ (and

set $[w] \cap \lambda = b_{i(\sup w)}$) *then* $cd_{\kappa,\lambda}$ (set $[w]) \in a$ }

$$T^i \overset{def}{=} \bigcup_{\delta \in S_i} T^i_\delta$$

For $a \in T^i$ let h_a be the unique order preserving function from a onto the ordinal $otp(a)$ (= the order type of a). Let $A_a = \{j \in a : h_a(j) \in A^{M_i}\}$, so A_a is a subset of a.

$$T \overset{def}{=} \bigcup_{i<i(*)} T^i$$

24A. Observation: If $c \subseteq d$, $c \neq d$ both are in T then $cd_{\kappa,\lambda^+}(c) \in d$
As in the previous proof (i.e., see 23A).

Now let M be an algebra with universe λ^+ and countably many functions including f, g and $A \subseteq \lambda^+$, and let $Y \subseteq \mathcal{P}_{<\kappa}(\lambda)$, $Y \neq \phi \bmod \mathcal{D}$. We shall find $a \in T$, $a \cap \lambda \in Y$ and a is a subalgebra of M such that $A \cap a = A_a$. This will prove $T \neq \phi \bmod \mathcal{D}_1$, $(\mathcal{D}_1 + T) \restriction \lambda = \mathcal{D}$ and $\diamondsuit\ (T, \mathcal{D}_\kappa(\lambda))$.

We imitate Rubin and Shelah [RSh]: We define a game $\mathcal{G}$ which lasts ω moves. In the nth move player I chooses $a_n \in \mathcal{P}_{<\kappa}(\lambda)$ and then player II chooses an ordinal α_n, which satisfies:

(I) (i) a_n is a subalgebra of M

(ii) $a_n \cap \lambda \in Y$

(iii) $a_n \cap \alpha_{n-1} = a_{n-1}$ when $n > 0$

(iv) there is no $\delta \in (\sup a_n) \cap S^* - a_n$, $\delta = \sup (a_n \cap \delta)$

(II) (i) $\alpha_n > \sup a_n$, $\alpha_n > \lambda$ and when $n > 0$, $\alpha_n > \alpha_{n-1}$

The game is determined being closed. If player I has a winning strategy, a_0 his first move, let $b_0 = a_0$ and simulate a play $\langle a_n, \alpha_n : n < \omega \rangle$ in which player I uses his winning strategy and $\cup\, \alpha_n \in S_i$. Now $a \overset{def}{=} \bigcup_{n<\omega} a_n$ is in T

and is a subalgebra of M. What about $A_a = A \cap a$? For each $\alpha < \kappa$, $B \subseteq \alpha$ we define a game $\mathcal{G}(\alpha,B)$, similar to $\mathcal{G}$, but player I also choose in his n^{th} move an order preserving $h_n : a_n \to \alpha$, $\bigcup_{m<n} h_m \subseteq h_n$ and $(\forall \alpha \in a_n)$ $(\alpha \in A \equiv h_n(\alpha) \in B)$. If for some α, B player I has a winning strategy, we have no problem. If not then (as the games are closed hence determined) player II has a winning strategy $F_{\alpha,B}$ for $\mathcal{G}(\alpha,B)$ for each $\alpha < \kappa$, $B \subseteq \alpha$. Now we define a strategy for player II in G:

$F(a_0, \ldots, a_n) = \bigcup \{F_{\alpha,B}(a_0, h_0, a_n, h_1, \ldots, a_n, h_m)+1$: for $l \leq n$, h_l a function from a_c into α, $\alpha < \kappa$, $B \subseteq \alpha\}$

Clearly this gives a legal move for player II, and in the end we can define $\alpha = otp(\bigcup_{m<\omega} a_n)$, $B = \{otp(\xi \cap \bigcup_m a_n)$: $\xi \in A \cap \bigcup_{n<\omega} a_n\}$, and define $h_m : a_m \to \alpha$ by $h_m(\gamma) = otp(\gamma \cap C_m)$ and get contradiction.

So it is enough to prove that player I wins $\mathcal{G}$, or equivalently that player II has no winning strategy. So suppose F is a winning strategy. Now by assumption (ii) of 24 w.l.o.g. $|Y| = \lambda$ and (by 24 (iv)) $\{a \cap \kappa : a \in Y\}| \leq \lambda$. Now let for $\zeta < \kappa\omega$ M_ζ be an elementary submodel of $H((2^{\lambda^+})^+, \in)$ to which $S^*, \mathcal{D}, M, F, Y$ belongs, $\{i : i < \lambda\} \subseteq M_\zeta$, $\langle M_\xi : \xi \leq \zeta \rangle \in M_{\zeta+1}$, $||M_\zeta|| = \lambda$. Let $\beta_\zeta = \sup(M_\zeta \cap \lambda^+) = \mathrm{Min}(\lambda^+ - |M_\zeta|)$, and let $\beta = \bigcup_\zeta \beta_\zeta$ So M_ζ is increasing. Choose $a \subseteq (\bigcup_{\zeta < \kappa\omega} M_\zeta) \cap \lambda^+$, $a \cap \lambda \in Y$ and $a \cap \{\beta_{\kappa m+\xi}$: $\xi < \kappa\} = \{\beta_{\kappa m+\xi}$: $\xi \in a \cap \kappa\}$, a is closed under f, g, and there is no $\delta \in S^* \cap \beta - a$, $\delta = \sup(a \cap \delta)$. (This demand "$a \cap \lambda \in Y$" restrict ourselves to a positive set $mod\ \mathcal{D}_\beta$, the rest to a member of $\mathcal{D}_\beta$ (the last demand by (iii) of 24) so there is such a.)

As $a \cap \lambda \in Y$, clearly for each ζ $a \cap \lambda \in M_\zeta$, and as $a \cap \{\beta_{\kappa m+\xi}$: $\xi < \kappa\} = \{\beta_{\kappa m+\xi}$: $\xi < \kappa$, $\xi \in a\}$, and $a \cap \kappa \in M_\zeta$, (by the restriction on Y) and $f, g \in M$, and $\langle M_\zeta : \zeta < \kappa m + (\sup \kappa \cap a) \rangle \in M_{\kappa m+1}$ (as for $\sup(\kappa \cap a) < a$) clearly we get $a \cap M_{\kappa(m+1)} \in M_{\kappa(m+1)}$. Now we can simulate a play of the game in which player II uses his winning strategy F, whereas player I choose $a_n = a \cap M_{\kappa(n+1)}$. By what we say above $F(a_0, \ldots, a_n) \in M_{\kappa(n+1)}$ hence $F(a_0, \ldots, a_n) < \beta_{\kappa(n+1)}$, so actually player I wins the play, contradiction.

25. Conclusion:

Suppose κ is regular $> \aleph_0$, $\lambda = \lambda^{<\kappa}$, and $S^* \subseteq \{\delta < \lambda^+ : cf\ \delta = \aleph_0\}$ is stationary, but for no $\delta < \lambda^+$ of cofinality κ is $S^* \cap \delta$ stationary in δ. *Then*, there is a (κ, λ^+)-stationary coding $T \subseteq T_{\kappa, \lambda^+}(N^2_\omega)$ and even $\diamondsuit(T, \mathcal{D}_\kappa(\lambda))$ holds.

26. Remark:

1) When does such a S^* exist? It follows from the existence of square on $\{\delta < \lambda^+ : cf\ \delta < \kappa\}$, which $\neg 0^\#$ implies holds when $\kappa < \lambda$ (and even for many $\kappa = \lambda$'s (see Magidor's work).

2) We can weaken the non-reflection as in 7 of [Sh1].

27. Claim:

In 24 if we do not require $\diamondsuit\ (T, \mathcal{D}_1)$ then we can omit (i) and (iv).

We can deduce from the proof of 24 also:

28. Lemma:

1) $\diamondsuit\ (\mathcal{D}_{<\aleph_1}(\lambda^+))$ when $\lambda = \lambda^{\aleph_0}$

2) If $\mathcal{D}$ is normal fine filter on $\mathcal{P}_{<\aleph_1}(\lambda^+)$, $\mathcal{D}_1$ is the minimal normal fine filter on $\mathcal{P}_{<\aleph_1}(\lambda^+)$ such that $\mathcal{D}_1 \restriction \lambda = \mathcal{D}$ and $\lambda = \lambda^{\aleph_0}$ *then* $\diamondsuit\ (\mathcal{D}_1)$

References

[FMR] M. Foreman, M. Magidor and S. Shelah, Martin Maximum, Saturated ideals and non-regular ultrafilters, *Annals of Math.*

[RSh] M. Rubin and S. Shelah. Combinatorial problems on trees: Partitions, Δ-systems and large free subsets, *Annals of Pure and Applied Logic* in press.

[Sh1] S. Shelah, "The existence of coding sets," ~~A Springer Lecture notes volume.~~ this

[Sh2] S. Shelah, A note on cardinal exponentiation, *J. of Symb. Logic*, 45(1980) 56-66.

[Sh3] S. Shelah, On successor of singular cardinals, *Proc. of the A.S.L. meeting in Mons*, Aug. 1978, Logic Colloquium 78 ed. M. Boffa, D. van Dalen and K. McAloon, *Studies in Logic and the Foundation of Math.* vol.97, North Holland Publishing Co., Amsterdam, 1979, 357-380.

[Sh4] S. Shelah, Classification Theory, North Holland Publishing Co., Amsterdam, 1978.

[Sh5] S. Shelah, Existence of rigid-like families of abelian p-groups, model theory and Algebra, *A memorial tribute to A. Robinson*, ed. Saraceno and Weisphenning, Lecture notes in *Math 498*, Springer-Verlag 1975, 385-402.

[Sh6] S. Shelah, A combinatorial principle and endomorphism rings of p-groups I., *Proc. of the 1980/1 Jerusalem model theory year*, *Israel J. Math.*

[Sh7] S. Shelah, A combinatorial principle and endomorphism of abelian groups II, *Proc. of the conference on Abelian groups*, Iudine, Italy, April 84.

[Sh8] S. Shelah, Proper forcing, *Springer lecture notes*, 940, 1982.

[Z] B. Zwicker, $P_\kappa(\lambda)$ combinatories I: Stationary coding *Contemporary Mathematics*. 31 (1984) 243-259.

On normal ideals and Boolean Algebras

In [Sh 1] 3.1 we prove: If $\mathcal{B}$ is a Boolean algebra of power $\kappa^+, \kappa = \kappa^{<\kappa}$, and $\mathcal{B}$ satisfies the κ-chain condition then $\mathcal{B} - \{0\}$ is the union of κ ultrafilters (why not "$\mathcal{B}$ of power λ^{++}"? see [Sh 3] mainly 2.4, p.245). We here replace "κ-chain condition" by a weaker condition we introduce here (κ-SD, (see Definition 1), which says that for almost all $\mathcal{B} \subseteq \mathcal{B}$ of power κ, $\mathcal{B} \lessdot \mathcal{B}$ (for the right interpretation of almost).

The other theorem (6) is that $2^{\aleph_0} < 2^{\aleph_1}$ implies $\mathcal{D}_{\omega_1}$ (the club filter on ω_1), cannot be $\aleph_2$- dense. We then observe we cannot improve this to $[2^{\aleph_0} < 2^{\aleph_0} \Longrightarrow \mathcal{D}_{\omega_1}$ not $\aleph_2$-saturated] as by Forman Magidor Shelah [FMS], a universe V, $V \models$"$\mathcal{D}_{\omega_1}$ is $\aleph_2$-saturated understructibly under c.c.c. forcing" was obtained and discuss the large cardinal needed. For proving Theorem 6 we use normal filters connected with variants of the weak diamonds (see Devlin Shelah [DS], Shelah [Sh 2]) and prove a more general such theorem. Compare with a recent result of Woodin: from ADR + "ϑ regular" he gets the consistency of "$\mathcal{D}_{\omega_1} + X$ is $\aleph_1$-dense" for some stationary $X \subseteq \omega_1$. The conception of this work is closely connected with Forman Magidor and Shelah [FMS], and also Shelah and Woodin [SW], and [Sh 5]; it was done subsequently to most of [FMS].

Notation: $\mathcal{P}(\lambda) = \{A: A \subseteq \lambda\}$, it is a Boolean algebra and we sometimes say λ instead of $\mathcal{P}(\lambda)$. $\mathcal{B}$ denotes a Boolean algebra; the filter $E \subseteq \mathcal{B}$ generated is $<E>_{\mathcal{B}} = \{x \in \mathcal{B}$: there are $n < \omega$, $x_1 \in E, \ldots, x_n \in E$ such that $\bigcap_{\ell=1}^{n} x_\ell \leq x\}$, it is proper if $0 \notin <E>_{\mathcal{B}}$; an ultrafilter is a maximal proper filter. Let $\mathcal{B}_1 \lessdot \mathcal{B}_2$ means $\mathcal{B}_1$ is a subalgebra of $\mathcal{B}_2$, and every maximal antichain of $\mathcal{B}_1$ is a maximal antichain of $\mathcal{B}_2$, or what is equivalent: for every $x \in \mathcal{B}_2, x \neq 0$ there is $y \in \mathcal{B}_1, y \neq 0$ such that $(\forall z \in \mathcal{B}_1)[0 < z \leq y \rightarrow x \cap z \neq 0]$. Let $\mathcal{B}_1 <\!\!\!\circ^* \mathcal{B}_2$ means that $\mathcal{B}_1$ is subalgebra of $\mathcal{B}_2$ and $\{x \in \mathcal{B}_2 : \{y \in \mathcal{B}_1 : y \cap x = 0\}$ is dense in $\mathcal{B}_1\}$ is dense below no $z \in \mathcal{B}_1, z \neq 0$.

For a regular $\lambda > \aleph_0$ let $\mathcal{D}_\lambda$ be the filter (on $\mathcal{P}(\lambda)$) generated by the closed unbounded subsets of λ. For I an ideal of $\mathcal{B}$ let $\mathcal{B}/I$ be the quotient algebra, similarly we define $\mathcal{B}/\mathcal{D}$, $\mathcal{D}$ a (proper) filter on $\mathcal{B}$.

1 Definition : Let $\mathcal{B}$ be a Boolean algebra of cardinality κ^+, $\mathcal{B} = \bigcup_{\alpha<\kappa^+} \mathcal{B}_\alpha$, $\mathcal{B}_\alpha$ increasing continuous, each $\mathcal{B}_\alpha$ of cardinality $\leq \kappa$. We say $\mathcal{B}$ is κ-SD if $\{\alpha :$ if $cf\ \alpha = cf\ \kappa$ then $\mathcal{B}_\alpha <\!\!\!\circ\, \mathcal{B}\}$ belong to $\mathcal{D}_{\kappa^+}$. We say $\mathcal{B}$ is almost κ-SD if $\{\alpha : cf\ \alpha = cf\ \kappa$ and $\mathcal{B}_\alpha <\!\!\!\circ\, \mathcal{B}\} \neq \phi \ mod\ \mathcal{D}_{\kappa^+}$. We say $\mathcal{B}$ is almost κ-WSD if for some stationary $S \subseteq \{\alpha\ cf\ \alpha = cf\ \kappa\}$, for every $i < j, [i \in S, j \in S \Longrightarrow \mathcal{B}_i <\!\!\!\circ^* \mathcal{B}_j]$. We say $\mathcal{B}$ is κ-WSD if we can choose an S above such that $S \cup \{\alpha : cf\ \alpha \neq cf\ \kappa\} \in \mathcal{D}_\lambda$.

1A Remark: 1) We can define naturally κ-SD, κ-WSD for $\mathcal{B}$ of cardinality $> \kappa^+$, see the proof of Theorem 2 and Claim 3.

2) if $\kappa = \kappa^{<\kappa}$, $\mathcal{B}$ satisfies the κ-chain condition, $\mathcal{B}$ has cardiality κ^+ then $\mathcal{B}$ is κ-SD.

2. Theorem : If $\mathcal{B}$ is κ-SD, $\kappa = \kappa^{<\kappa}$ then $\mathcal{B} - \{0\}$ is the union of κ, ultrafilters.

Proof : Let $\mathcal{B} = \bigcup_{\alpha<\kappa^+} \mathcal{B}_\alpha$, $\mathcal{B}_i$ increasing continuous, $\mathcal{B}_i$ of cardinality $\leq \kappa$.

As $\kappa = \kappa^{<\kappa}$, and as we can replace $\mathcal{B}$ by any extension satisfying the same conditions, w.l.o.g. $\mathcal{B}$ is closed under unions of $< \kappa$ elements.

Let $S = \{i < \kappa^+ : i = 0, i$ is a successor ordinal or i is a limit ordinal with cofinality $\kappa\}$.

By renaming the $\mathcal{B}_i$ we can assume;

(α) if $i \in S$ then $\mathcal{B}_i \lessdot \mathcal{B}$ and $\mathcal{B}_i$ is $(<\kappa)$-complete, i.e. if $\alpha < \kappa$, $a_\gamma \in \mathcal{B}_i$ for $\gamma < \alpha$, then $\bigcup_{\gamma,\alpha} a_\gamma \in \mathcal{B}_i$ (where $\bigcup_{\gamma<\alpha} a_\gamma$ is taken in $\mathcal{B}$).

Let $\chi = (2^{\kappa^+})^+$ and w.l.o.g. $\mathcal{B}_i \in H(\chi)$. Now for each $y \in \mathcal{B}, y \neq 0$ we define by induction on $n < \omega$, an elementary submodel N^y_n of $(H(\chi), \in)$ such that :

(i) $y \in N^y_n$, $\langle \mathcal{B}_i : i < \kappa^+ \rangle \in N^y_n$.

(ii) N^y_n has cardinality $< \kappa$ but $N^y_n \cap \kappa$ is an ordinal.

(iii) $N^y_n \prec N^y_{n+1}$ and $N^y_n \in N^y_{n+1}$ (remember $N^y_n \in H(\chi)$).

Now for every $z, y \in \mathcal{B}, y \neq 0$, natural number n and ordinal $\alpha \in S \cap N^y_n$ we define

$$G^n_\alpha(z;y) = \bigcup \{a \in \mathcal{B}_\alpha : a \in N^y_n \text{ and } (\forall b \in \mathcal{B}_\alpha)[0 < b \leq a \rightarrow b \cap z \neq 0]\}.$$

Let $y \in \mathcal{B}, m < \omega$ we define by induction on $n, m \leq n < \omega$ a set $P^{n,m}_y$ of terms $\tau = \tau(t)$:

$$P^{m,m}_y = \{t\}$$

$$P^{n+1,m}_y = \{G^n_\alpha(\bigcap_{\ell=1}^{k} \tau_\ell, y) : \alpha \in S \cap N^y_n,\ k < \omega \text{ and for } \ell = 1, \ldots, k,\ \tau_\ell \in P^{n,m}_y\}$$

2A Fact: For $\tau(t) \in P^{n,m}_y$ and $z \in N^y_m$, $\tau(z)$ is define naturally and it belongs to N^y_n, and if $\tau(t) = G^{n-1}_\alpha(\cdots)$ then $\tau(z) \in \mathcal{B}_\alpha$.

2B Fact; 1) For any $y \in \mathcal{B}$, $m \leq n < \omega$, $z \in N^y_n \cap \mathcal{B}$, $z \neq 0$ and $\tau \in P^{n,m}_y$ the element $\tau(z)$ is not zero.

2) *if* $m \leq n$, $k < \omega$, $\tau_\ell(t) \in P^{n,m}_y$ and for $\ell < k$, $z_\ell \in N^y_m \cap \mathcal{B}$, $z_\ell \neq 0$, and $\bigcap_{\ell<k} z_\ell \neq 0$ *then* $\bigcap_{\ell<k} \tau_\ell(z_\ell) \neq 0$.

Proof ; Clearly 1) follows from 2). We prove 2) by induction on n.

When $n = m$, necessarily $\tau_\ell(t) = t$ and there is no problem.

When $n > m$, let $\tau_\ell(t) = G^{n-1}_{\alpha_\ell}(\bigcap_{i<i(\ell)} \tau_{\ell,i}(t),y)$ (where $\alpha_\ell \in N^y_{n-1} \cap S$) so $\tau_{\ell,i}(t) \in P^{n-1,m}_y$. Let $z_{\ell,i} = \tau_{\ell,i}(z_\ell)$, so $z_{\ell,i} \in N^{n-1}_y$, (by Fact 2A) and by the induction hypothesis on n, $z \stackrel{def}{=} \bigcap_{\substack{i<i(\ell)\\ \ell<k}} z_{\ell,i} \neq 0$ and clearly $z \in N^y_{n-1} \cap \mathcal{B}$. Clearly $G^{n-1}_{\alpha_\ell}(z,y) \leq G^{n-1}_{\alpha_\ell}(\bigcap_{i<i(\ell)} \tau_{\ell,i}(z_\ell),y)$ for each ℓ. So it suffices to prove that $\bigcap_{\ell<k} G^{n-1}_{\alpha_\ell}(z,y)$. W.l.o.g. $\alpha_0 > \alpha_1 > \cdots > \alpha_{k-1}$, and we define by induction on $\ell \leq k$, an element s_ℓ of $\mathcal{B} \cap N^{n-1}_y$ as follows:

(a) $s_0 = z$,

(b) $s_{\ell+1} \in \mathcal{B}_{\alpha_\ell} \cap N^y_{n-1}$ is such that;

$$(\forall b \in \mathcal{B}_{\alpha_\ell})[0 < b \leq s_{\ell+1} \rightarrow b \cap s_\ell \neq 0]$$

We can find such $s_{\ell+1} \in \mathcal{B}_{\alpha_\ell}$ as $\mathcal{B}_{\alpha_\ell} \lessdot \mathcal{B}$, and we can choose it in N^{n-1}_y as s_ℓ, α_ℓ and $\langle \mathcal{B}_\alpha : \alpha < \kappa^+ \rangle$ belong to N^{n-1}_y, and N^{n-1}_y is an elementary submodel of $(H(\chi),\in)$.

We can prove that when $i \leq j < k$, $(\forall b \in \mathcal{B}_{\alpha_j})[0 < b \leq s_j \rightarrow b \cap \bigcap_{\ell=i}^{j} s_\ell \neq 0]$. This is done by induction on j; when $j = i$ this is trivial. When $j > i$, let $b \in \mathcal{B}_{\alpha_j}$, $0 < b \leq s_j$, by the choice of s_j, $b \cap s_{j-1} \neq 0$, so $0 < b \cap s_{j-1} \leq s_{j-1}$ and clearly $b \cap s_{j-1} \in \mathcal{B}_{\alpha_{j-1}}$, so by the induction hypothesis on j, $(b \cap s_{j-1}) \cap \bigcap_{\ell=1}^{j-1} s_\ell \neq 0$ but $b \leq s_j$ so $b \cap \bigcap_{\ell=i}^{j} s_\ell \neq 0$.

Hence $\bigcap_{\ell<k} s_\ell \neq 0$, and also (when $0 \leq i < k$) that $(\forall b \in \mathcal{B}_{\alpha_j})[0 < b \leq s_j \rightarrow b \cap s_i \neq 0]$, now for $i = 0$ $s_i = z$, hence by definition of $G^{n-1}_{\alpha_j}(z,y)$, clearly $s_j \leq G^{n-1}_{\alpha_j}(z,y)$. So $0 \neq \bigcap_{\ell<k} s_\ell \leq \bigcap_{\ell<k} G^{n-1}_{\alpha_\ell}(z,y)$, so we have proved the induction step for $n > m$, hence Fact 2B:

2C Fact; If $\alpha \in \bigcup_{n<\omega} N^y_n$, $\alpha \in S$, $y \in \mathcal{B}$, $y \neq 0$, $\mathcal{D}$ an ultrafilter on $\mathcal{B}_\alpha$, and

$\Gamma = \{\tau(y) : \tau \in P^{n,m}_y$ for some $m \leq n < \omega\}$ and $\Gamma \cap \mathcal{B}_\alpha \subset \mathcal{D}$

then $\mathcal{D} \cup \{\Gamma \cap \mathcal{B}_{\alpha+1}\}$) generates a proper filter.

Proof : Immediate, because :

2D Fact: When $m \leq n < \omega$, $\{\tau(y) : \tau \in P_y^{n,m}\} \subset \{\tau(y) : y \in P_y^{n,0}\}$.

Proof: This can be proved by induction on n: for $n = m > 0$ choose $\alpha_0 > \cdots > \alpha_{m-1}$ in $S \cap N_0^y$ such that $y \in \mathcal{B}_{\alpha_{m-1}}$ and define $\tau_\ell \in P_y^{\ell,0}$ by induction on $\ell \leq m$: $\tau_0 = \tau_1$, $\tau_{\ell+1} = G_{\alpha_\ell}^{\ell}(\tau_\ell, y)$; the other cases are trivial.

Continuation of the proof of Theorem 2:

Let E^y be any ultrafilter of $\mathcal{B} \cap (\bigcup_{n<\omega} N_y^n)$ which includes $\{\tau(y) : \tau \in P_y^{n,m}$ for some $m \leq n < \omega\}$; by Fact 2B,2D it is proper. The rest of the proof is as in [Sh 1] 3.1. By Engelking and Karlowicz [EK] there are functions $f_\xi : \kappa \to \kappa$ (for $\xi < \kappa^+$) such that for every distinct $\xi_\beta (\beta < \beta_0 < \kappa)$ and $\gamma_\beta < \kappa$ $(\beta < \beta_0)$ for some $\varepsilon < \kappa$, $\bigwedge_{\beta<\beta_0} f_\xi(\varepsilon) = \gamma_\beta$. Let $g_\beta : \kappa^+ \to \kappa$ be defined by: $g_\beta(\xi) = f_\xi(\beta)$.

Let $\mathcal{B}_{\xi+1}$ be generated by $\mathcal{B}_\xi \cup \{y_\beta^\xi : \beta < \kappa\}$ (and w.l.o.g. $\mathcal{B}_0 = \{0,1\}$, and w.l.o.g. $\langle\langle y_\beta^\xi, \xi, \beta\rangle : \xi < \kappa^+, \beta < \kappa\rangle$ belongs to every N_0^y). Let $\langle Y_\gamma^\xi : \gamma < \gamma\rangle$ list all subsets of $\{y_\beta^\xi : \beta < \kappa\}$ of cardinality $< \kappa$. We define by induction on $\xi < \kappa^+$ for each $\beta < \kappa$ an ultrafilter $\mathcal{D}_\beta^\xi$ of $\mathcal{B}_\beta$ such that:

(A) $\mathcal{D}_\beta^\xi$ is increasing continuous in ξ.

(B) if $\mathcal{D}_\beta^\xi \cup Y_{g_\xi(\beta)}^\xi$ generates a proper filter then $\mathcal{D}_\beta^\xi \cup Y_{g_\xi(\beta)}^\xi \subset \mathcal{D}_\beta^{\xi+1}$.

Clearly this can be done and each $\mathcal{D}_\beta = \mathcal{D}_\beta^{\kappa^+}$ is a (proper) ultrafilter of $\mathcal{B}$. Now if $y \in \mathcal{B}$, $y \neq 0$ then for each $\xi \in S \cap (\bigcup_{n<\omega} N_n^y)$ $(E_y \cap \{y_\alpha^\xi : \alpha < \kappa\}) \cup (E_y \cap \mathcal{B}_\xi)$ generates $E_y \cap \mathcal{B}_{\xi+1}$, [as $\mathcal{B}_\xi \cup \{y_\alpha^\xi : \alpha < \kappa\}$ generates $\mathcal{B}_{\xi+1}$, $\mathcal{B}_\xi \in N_n^y$, $\{y_\beta^\xi : \beta < \kappa\} \in N_n^y$, and $\mathcal{B}_\alpha \in N_n^y$ for every n such that $\alpha \in N_n^y$], so there is $\beta < \kappa$ such that for every $\xi \in \bigcup_{n<\omega} N_n^y$, $g_\beta(\xi) = \gamma_\xi$, and by Fact 2C, $E_y \subset \mathcal{D}_\beta$.

3 Claim; 1) In Theorem 2 we can replace κ^+ by 2^κ (its proof is written so that the changes are minimal, but the set $\{y_\beta^\xi : \beta < \kappa\}$ should still have cardinality κ.

2) In Theorem 2 (and Claim 3(1)) we really get that for every $Y \subseteq \mathcal{B}$ of cardinality $< \kappa$ which generates a proper filter, for some $\beta < \kappa$, $Y \subseteq \mathcal{D}_\beta$ (define $N_n^Y, \mathcal{P}_Y^{\eta,m}$ for any such Y, now Fact 2A, 2B have the same proof, and Fact 2C should be modified by having $\Gamma = \{\tau(y) : y \in Y, \ \ \tau \in \mathcal{P}_Y^{\eta,m}, m \le n < \omega\}$.

4. Remark: We can go beyond 2^κ, see [Sh 4], Lemma 4.

5. Observation: Suppose $\lambda > \aleph_0$ is regular, $2^\lambda = \lambda^+$, I an ideal on λ, $\mathcal{B} = \mathcal{P}(\lambda)/I$. Suppose $\mathcal{B} = \bigcup_{i<\lambda^+} \mathcal{B}_i$, $\mathcal{B}_i$ increasing continuous. $\mathcal{B}_i$ of power $\le \lambda$. Suppose further $S_{\mathcal{B}} = \{\xi < \lambda^+ : cf \ \xi = \lambda, \mathcal{B}_\xi <_\circ \mathcal{B}\}$ is stationary. Then some forcing notion Q of power λ^+, forcing by it does not add new subsets of λ, (so all relevant properties of I, are preserved), and in V^Q, $S_{\mathcal{B}} \cup \{\xi < \lambda^+ : cf \ \xi < \lambda\}$ contains a closed unbounded set.

This help us to show the consistency of "$\mathcal{P}(\lambda)/I$ is the union of λ ultrafilters" for a suitable ideal I.

Proof : The well known $Q = \{f : f$ and increasing continuous function from some $\alpha+1 < \lambda^+$ to λ^+, $[\beta \le \alpha$ and $cf(\alpha) = \lambda \Longrightarrow f(\alpha) \in S_{\mathcal{B}}]\}$.

* * *

6. Theorem : If $2^{\aleph_0} < 2^{\aleph_1}$ then $\mathcal{D}_{\omega_1}$ is not $\aleph_1$-dense (which means the Boolean algebra $\mathcal{P}(\omega_1)/\mathcal{D}_{\omega_1}$ is not $\aleph_1$-dense.)

This will follow from Conclusion 14.

7. Definition ; A Boolean algebra $\mathcal{B}$ is λ-dense if there is $B \subseteq \mathcal{B}$, $|B| \le \lambda$ which is dense i.e., $(\forall x \in \mathcal{B})[x \ne 0 \to (\exists y \in B)(0 < y \le x)]$.

Note in this connection the following two observations.

8. Observation: By [FMS] we can obtain a universe of set theory [starting with a model of ZFC + 'κ is supercompact') in which $\mathcal{D}_{\omega_1}$ is $\aleph_2$-saturated and this is preserved by forcing satisfying the $\aleph_1$-chain condition, so if we add e.g. $\beth_{\omega_1}$ Cohen reals, still $\mathcal{D}_{\omega_1}$ is $\aleph_2$-saturate but $2^{\aleph_0} = \beth_{\omega_1} < \beth_{\omega_1+1} = 2^{\aleph_1}$.

We may be interested in using smaller large cardinals:

8A. Observation: 1) It is consistent with ZFC that $2^{\aleph_0} < 2^{\aleph_1}$ but $\mathcal{D}_{\omega_1}$ is $\aleph_2$-saturated if we assume the consistency of ZFC + "κ is a suitable hypermeasurable as in [SW]."

2) If in V, $\mathcal{D}$ is a normal filter on ω_1, and $\mathcal{D}$ is $\aleph_2$-saturated.

Q is the forcing of adding λ-Cohen reals, *then* in V^Q;

a) $\mathcal{D}' = \{A \in V^Q : A \subseteq \omega_1$ and $(\exists B \in \mathcal{D})\ B \subseteq A\}$ is $\aleph_2$-saturated normal filter [so $\mathcal{D} = (\mathcal{D}_{\omega_1})^V \Longrightarrow \mathcal{D}' = (\mathcal{D}_{\omega_1})^{V^Q}$].

b) $(2^{\aleph_0})^{V^Q} = (\lambda + \aleph_0)^{\aleph_0}$ (the second term is computed in V).

c) $(2^{\aleph_1})^{V^Q} = (\lambda + \aleph_1)^{\aleph_1}$ (the second term is computed in V.)

Proof : 1) By 2), starting with a universe of set theory in which $\mathcal{D}_{\omega_1}$ is $\aleph_2$-saturated, from Shelah and Woodin [SW] .

Note that if in V, $\beth_{\omega_1+1}(\kappa) > \beth_{\omega_1}(\kappa)^{+\alpha}$, κ is supercompact, and P a forcing notion of cardinality κ, such that in V^P, $\kappa = \aleph_2, \mathcal{D}_{\omega_1}$ $\aleph_2$-saturated; choose in (2) $\lambda = \beth_{\omega_1}(\kappa)$, then in $V^{P*Q}, (2^{\aleph_0})^{+\alpha} < 2^{\aleph_1}$.

2) Straightforward.

Suppose $Q = \{f : f$ a finite function from λ to $\{0,1\}\}$, and $q \in Q$, $q \Vdash_Q$ " $\langle \underset{\sim}{S}_\alpha : \alpha < \omega_2 \rangle$ is a counterexample: Let for $\alpha < \omega_2$, $S^0_\alpha = \{\delta < \omega_1$: there is q', $q \le q' \in Q$, $q' \Vdash$" $\delta \in \underset{\sim}{S}_\alpha$ "$\}$, and for $\delta \in S^0_\alpha$ choose $q^\alpha_\delta \in Q$, $q \le q^\alpha_\delta$, $q^\alpha_\delta \Vdash$ " $\delta \in \underset{\sim}{S}_\alpha$ ", (so $\langle \langle q^\alpha_\delta : \delta \in S^0_\alpha \rangle : \alpha < \omega_2 \rangle$ is in V) Clearly $S^0_\alpha \neq \phi \mod \mathcal{D}$ hence for each $\alpha < \omega_2$ for some $k_\alpha < \omega$, $S'_\alpha = \{\delta \in S^0_\alpha :$ Dom q^α_δ has cardinality $k_\alpha\} \neq \phi \mod \mathcal{D}$ hence for some k , $W = \{\alpha < \omega_2 : k_\alpha < k\}$ has cardinality $\aleph_2$. Let m be a natural number such that $m \to (3)^2_{2^{k^2}}$.

As $\mathcal{D}$ is $\aleph_2$-saturated there are distinct $\alpha_1, \ldots, \alpha_m \in W$ suc that $S \overset{def}{=} \bigcap_{\ell=1}^{m} S^1_{\alpha_\ell} \neq \phi \mod \mathcal{D}$ For every $\delta \in S$ for some distinct

$\ell(1),\ell(2) \in \{1,\ldots,m\}$, $q_\delta^{a_{\ell(1)}}, q_\delta^{a_{\ell(2)}}$, are compatible. Hence there are distinct $\ell(1),\ell(2) \in \{1,\ldots,m\}$ such that $\{\delta \in \omega_1 : \delta \in S$, and $q^{a_{\ell(1)}}, q_\delta^{a_{\ell(2)}}$ are compatible$\} \neq \phi \bmod \mathcal{D}$ Now it is easy to show that for some $q', q \subseteq q' \in Q$, $q' \Vdash \{\delta \in S : q_\delta^{a_{\ell(1)}} \cup q_\delta^{a_{\ell(2)}} \in \underset{\sim}{G}_Q\} \neq \bmod \mathcal{D}$ contradiction.

Remark: The inaccessible f needed in 8A(8) is $\{\kappa : \kappa$ strongly inaccessible with $Pr_2(\kappa)\}$ is stationary is not in the weak compactness ideal) "$\mathcal{D}_{\omega_1}$ is indestructible by $\aleph_1$-c.c. forcing big hyperinaccessible like in"

9. Observation: If $\mathcal{D}$ is a normal filter on a regular $\mu > \aleph_0, 2^\mu = \mu^+$ then the following are equivalent:

(a) $\mathcal{D}$ is μ-dense.

(b) there are normal filters $\mathcal{D}_i$ $(i < \mu)$, $\mathcal{D} \subseteq \mathcal{D}_i$, and $[A \neq \phi \bmod \mathcal{D} \Longrightarrow A \in \bigcup_{i<\mu} \mathcal{D}_i]$.

(c) for every $A_i \subseteq \lambda$, $A_i \neq \phi \bmod \mathcal{D}$ for $i < \mu^+$, *there is* $S \subseteq \mu^+, |S| = \mu^+$, such that for any distinct $i(\alpha) \in S$ $(\alpha < \lambda)$ the diagonal intersection of $A_{i(\alpha)} (\alpha < \lambda)$ (i.e. $\{\gamma < \lambda : \gamma \in \bigcap_{\alpha<\gamma} A_{i(\alpha)}\}$) is $\neq \phi \bmod \mathcal{D}$

Proof : (a) $\Longrightarrow$ (b). Suppose $\{A_i/\mathcal{D} : i < \mu\}$ is a dense subset of $\mathcal{P}(\lambda)/\mathcal{D}$ Let (for $i < \mu$), $\mathcal{D}_i \overset{\text{def}}{=} \mathcal{D} + A_i = \{X \subseteq \lambda : X \cup (\lambda - A_i) \in \mathcal{D}\}$, then the $\mathcal{D}_i$'s exemplify that (b) holds.

(b) $\Longrightarrow$ (c): Let $\mathcal{D}_i (i < \mu)$ exemplify (b), and let $A_i \subseteq \mu$, $A_i \neq \phi \bmod \mathcal{D}$ for $i < \mu^+$. For each $i < \mu^+$ for some $\gamma(i) < \mu^+$, $A_i \in \mathcal{D}_{\gamma(i)}$. So for some γ $S = \{i : \gamma(i) = \gamma\}$ has power μ^+. Clearly $\{\gamma(i) : i \in S\}$ is as required.

(c) $\Longrightarrow$ (a): Assume (a) fails. Let $\{A \subseteq \mu : A \neq \phi \bmod \mathcal{D}\}$ be listed as $\{A_\alpha : \alpha < \mu^+\}$. As for $\xi < \mu^+$ $\{A_\alpha : \alpha < \xi\}$ cannot exemplify "$\mathcal{D}$ is μ-dense" there is $\alpha(\xi) < \mu^+$ such that for no $\beta < \xi$, $A_{\alpha(\xi)} \subseteq A_\beta \bmod \mathcal{D}$ By (c) there is $S \subseteq \mu^+$ of cardinality μ^+ such that for any $\alpha_i \in S$ $(i < \mu^+)$, $\{\gamma < \lambda : \gamma \in A_{\xi(\alpha_i)}$ for every $i < \gamma\} \neq \phi \bmod \mathcal{D}$ Let for $\zeta < \mu^+$, B_ζ be the diagonal intersection of $\{A_{\alpha(\xi)} : \xi < \zeta\}$. Note that B_ζ is not uniquely determined as a set (it depends on the enumeration of ζ) but *mod* $\mathcal{D}$ (and even *mod* $\mathcal{D}_\lambda$) it is uniquely determined. Clearly $\zeta_1 < \zeta_2 \Longrightarrow B_{\zeta_1} \supseteq B_{\zeta_2}$ *mod* $\mathcal{D}$. Now necessarily for some $\zeta(*)$ for every $\zeta \geq \zeta(*)$ (but $< \mu^+$), $B_\zeta = B_{\zeta(*)}$ *mod* $\mathcal{D}$ as otherwise there is an increasing sequence $\zeta(i)$ for $i < \mu^+$, such that $B_{\zeta(i+1)} \neq B_{\zeta(i)}$ *mod* $\mathcal{D}$ so $\{B_{\zeta(i+1)} - B_{\zeta(i)} : i < \mu^+\}$ show $\mathcal{D}$ is not μ^+-saturated and clearly contradict (c) which we are assuming.

Now as $B_{\zeta(*)} \neq \phi$ *mod* $\mathcal{D}$ for some $\gamma(*) < \mu^+$, $B_{\zeta(*)} = A_{\gamma(*)}$. Choose $\beta < \mu^+$, $\beta > \gamma(*)$, $\beta > \zeta(*)$. So by the choice of $\zeta(*)$ $B_{\beta+1} = B_{\zeta(*)}$ *mod* $\mathcal{D}$ but by the choice of $B_{\beta+1}$, $B_{\beta+1} \subseteq A_{\xi(\beta)}$ *mod* $\mathcal{D}$ hence $B_{\zeta(*)} \subseteq A_{\xi(\beta)}$ *mod* $\mathcal{D}$ but $B_{\zeta(*)} = A_{\gamma(*)}$ so $A_{\gamma(*)} \subseteq A_{\xi(\beta)}$ *mod* $\mathcal{D}$. But remember the choice of $\xi(\beta)$, as $\beta > \gamma(*)$ it implies $A_{\gamma(*)} \not\subseteq A_{\xi(\beta)}$ *mod* $\mathcal{D}$. Contradiction.

10. Definition: 1) For a regular uncountable λ and $\mu < 2^\lambda$ let

(a) Dom $(\lambda,\mu) = \{f : f$ a function with domain ${}^{\omega >}\alpha - \{\Lambda\}$ for some ordinal $\alpha < \lambda, f(\eta) < \mu$, for $\eta \in {}^{\omega \geq}\alpha - \{\Lambda\}$, where Λ is the empty sequence.

(b) Dom $^+(\lambda,\mu) = \{f : f$ a function from ${}^{\omega >}\lambda - \{\Lambda\}$ to $\mu\}$.

(c) Let $I_{\lambda,\mu}$ be the set of $A \subseteq \lambda$ such that :

for some function F from Dom (λ,μ) to $\{0,1\}$, for every $h : A \to \{0,1\}$ there is $f \in$ Dom $^+(\lambda,\mu)$ such that for some $C \in \mathcal{D}_\lambda$ $(\forall \delta \in A \cap C)\,[h(\delta) = F(f \restriction \delta)]$.

2) For λ,μ as above and function F from Dom (λ,μ) to $\{0,1\}$ let $I^F_{\lambda,\mu}$ be the set of $A \subseteq \lambda$ such that ; for every $B \subseteq A$, there is $f \in$ Dom (λ,μ) such that for some $C \in \mathcal{D}_\lambda$

$$(\forall \delta \in C)[\delta \in B \quad iff \quad F(f \restriction \delta) = 1]$$

3) For λ, μ, F as above let $J^F_{\lambda,\mu}$ be the normal ideal on λ which $I^F_{\lambda,\mu}$ generates.

Remark: This is close by related with the weak diamond, see Devlin and Shelah [SD] and Shelah [Sh , Ch. XIV , §1].

11. Lemma : 1) $I_{\lambda,\mu}$ is a normal ideal on λ (but it may be $\mathcal{P}(\lambda)$) and we could have in the definition of Dom (λ,μ) replace ${}^{\omega >}\alpha$ by α .

2) If $\kappa < \lambda$, $2^\kappa = 2^{<\lambda}$, $\mu = \mu^{<\lambda} < 2^\lambda$, $\mu < \lambda^{+\lambda}$ (i.e. $\mu < \aleph_{\alpha+\lambda}$ where $\lambda = \aleph_\alpha$) (or even a weaker restriction) then $\lambda \notin I_{\lambda,\mu}$.

3) $I^F_{\lambda,\mu} \subseteq J^F_{\lambda,\mu} \subseteq I_{\lambda,\mu}$, and $I_{\lambda,\mu} = \bigcup \{I^F_{\lambda,\mu} : F$ a function from Dom (λ,μ) to $\{0,1\}\}$.

4) For every function F : Dom $(\lambda,\mu) \to \{0,1\}$, there is a function F^* : Dom $(\lambda,\mu) \to \{0,1\}$ such that

$$J^{F^*}_{\lambda,\mu} = I^{F^*}_{\lambda,\mu} = J^F_{\lambda,\mu}$$

5) For any function F : Dom $(\lambda,\mu) \to \{0,1\}$, for every $C \in \mathcal{D}_\lambda$, $\lambda - C \in I^F_{\lambda,\mu}$.

Proof : Part 1) is straightforward. For 2) see [Sh 2 , Ch. XIV §1]. Now (3), (5) are trivial and for (4), note that in Definition 10(2) we demand $(\forall \delta \in C)[\delta \in B \Longrightarrow F(f \restriction \delta) = 1]$ and not just $(\forall \delta \in C \cap A)[\delta \in B \Longleftrightarrow F(f \restriction \delta) = 1]$.

12. Lemma : Suppose λ is regular and uncountable, $\mu < 2^\lambda$, and $\lambda \notin I_{\lambda,\mu}$.

Then for no F is $J^F_{\lambda,\mu}$ μ-dense, λ^+-saturated.

Proof : Suppose F is a counterexample and let $\{A_i / J^F_{\lambda,\mu} : i < \mu\}$ be a dense subset of $\mathcal{P}(\lambda)/J^F_{\lambda,\mu}$. We now define a function H from Dom $(\lambda,\mu) = \bigcup\{f : f$ a function from some ${}^{\omega>}\delta - \{\Lambda\}$ into μ where $\delta < \lambda\}$ to $\{0,1\}$.

Suppose $\delta < \lambda$ is limit, $f : ({}^{\omega>}\delta - \{\Lambda\}) \to \mu$, for $\nu \in {}^{\omega>}\delta$ let f_ν be the function from ${}^{\omega>}\delta - \{\Lambda\}$ to $\{0,1\}$ defined by $f_\nu(\eta) = f(\nu{}^\frown\eta)$. We define $H(f)$ by cases:

Case I: For some $\alpha,\beta < \delta$, $F(f_{<0,\alpha,\beta>}) = 1$.

Then we let $H(f)$, be $F(f_{<1,\alpha,\beta>})$ for the minimal such α,β (lexicographically).

Case II: Not Case I, but for some $\alpha < \delta$, $\delta \in A_{<2,\alpha>}$.

Then $H(f) = f(<3,\alpha>)$ for the minimal such α.

Case III: Not Case I nor II.

Then $H(f) = 0$.

If $f : {}^{\omega>}\alpha - \{\Lambda\} \to \mu$, α not limit, let $H(f) = 0$.

Now we get contradiction by Fact 12A below (as $\lambda \notin I_{\lambda,\mu}$, $I_{\lambda,\mu}$ is normal and $J^H_{\lambda,\mu} \subseteq I_{\lambda,\mu}$).

12A Fact: $\lambda \in I^H_{\lambda,\mu}$.

Let $B \subseteq \lambda$ and we shall find $f \in \mathrm{Dom}^+(\lambda,\mu)$ such that for some $C \in \mathcal{D}_\lambda$, $(\forall \delta \in C)[\delta \in B$ iff $H(f \restriction \delta) = 1]$.

Let $\mathcal{P} \subseteq \{A_i : i < \mu\}$ be a maximal subset satisfying:

(a) for every $a \neq b \in \mathcal{P}$, $a \cap b \in J^F_{\lambda,\mu}$ (i.e. $\mathcal{P}$ is $J^F_{\lambda,\mu}$- disjoint.)

(b) for every $a \in \mathcal{P}$, $a \subseteq B \bmod J^F_{\lambda,\mu}$ or $a \cap B = \phi \bmod J^F_{\lambda,\mu}$.

As F is a counterexample, $\mathcal{P}(\lambda)/J^F_{\lambda,\mu}$ is λ^+-saturated hence $|\mathcal{P}| \leq \lambda$, so let $\mathcal{P} = \{A_{i(\alpha)} : \alpha < \alpha(*)\}$, $\alpha(*) \leq \lambda$. We shall assume $\alpha(*) = \lambda$ (the other case is easier). Let B^* be the diagonal union of the $A_{i(\alpha)}$ i.e. $\{\beta < \lambda : \beta \in \bigcup_{\alpha<\beta} A_{i(\alpha)}\}$, so clearly $a_0 \overset{def}{=} \lambda - B^* \in J^F_{\lambda,\mu}$. For each $\alpha < \lambda$ let $a_{1+\alpha}$ be $A_{i(\alpha)} - B$ if $A_{i(\alpha)} \subseteq B \bmod J^F_{\lambda,\mu}$ and $A_{i(\alpha)} \cap B$ if $A_{i(\alpha)} \cap B = \phi \bmod J^F_{\lambda,\mu}$. So in any case $a_\alpha \in J^F_{\lambda,\mu}$, so there are sets $a_{\alpha,\beta} \in I^F_{\lambda,\mu}$, (for $\beta < \lambda$) such that $a_\alpha = \{\gamma < \lambda : \gamma \in \bigcup_{\beta<\gamma} a_{\alpha,1+\beta}\}$. As $a_{\alpha,\beta} \in I^F_{\lambda,\mu}$ there are functions $f^0_{\alpha,\beta}, f^1_{\alpha,\beta}$ from ${}^{\omega>}\lambda - \{\Lambda\}$ to μ, such that for some $C_{\alpha,\beta} \in \mathcal{D}_\lambda$:

$$(\forall \delta \in C_{\alpha,\beta})[\delta \in a_{\alpha,\beta} \cap B \iff F(f^1_{\alpha,\beta} \restriction \delta) = 1]$$

$$(\forall \delta \in C_{\alpha,\beta})[\delta \in a_{\alpha,\beta} \iff F(f^0_{\alpha,\beta} \restriction \delta) = 1]$$

Now we can define $f^* : ({}^{\omega>}\lambda - \{\Lambda\}) \to \mu$

$$f^*(<0,\alpha,\beta> {}^\frown \eta) = f^0_{\alpha,\beta}(\eta)$$

$$f^*(<1,\alpha,\beta>) = f^1_{\alpha,\beta}(\eta)$$

$$f^*(<2,\alpha>) = 1 \qquad \text{if } \delta \in A_{i(\alpha)},$$

$$f^*(<3,\alpha>) = 1 \qquad \text{if } \delta \in A_{i(\alpha)} \subseteq B \ mod \ I_{\lambda,\mu}$$

$$f^*(\eta) = 0 \qquad otherwise.$$

It is easy to check that $\{\delta : H(f^* \restriction \delta) = 1 \Longleftrightarrow \delta \in B\}$ belong to $\mathcal{D}_\lambda$. As B was any subset of λ this shows $\lambda \in I^H_{\lambda,\mu}$ but $I^H_{\lambda,\mu} \subseteq I_{\lambda,\mu}$, $\lambda \notin I_{\lambda,\mu}$, contradiction.

13. Conclusion: Suppose λ is regular uncountable and $\lambda \notin I_{\lambda,\mu}$ (see 11(1)). Then $\mathcal{D}_\lambda$ is not μ-dense, λ^+-saturated.

Proof: As $\mathcal{D}_\lambda$ is λ^+-saturated, and $I_{\lambda,\mu}$ a normal ideal on λ, it is known that for every appropriate F, for some $Y(F) \subseteq \lambda$ $Y(F) \neq \phi \ mod \ \mathcal{D}_\lambda$ and $J^F_{\lambda,\mu} = \{A \subseteq \lambda: (Y(F)-A) \cup (\lambda - Y(F)) \in \mathcal{D}_\lambda\}$ and so $J^F_{\lambda,\mu}$ is μ-dense λ^+-saturated too contradicting 12.

14. Conclusion: If $\lambda = \kappa^+, 2^\lambda > 2^\kappa$, $\mu = \mu^{<\lambda} < Min\{2^\lambda, \lambda^{+\lambda}\} < 2^\lambda$ then $\mathcal{D}_\lambda$ cannot be λ^+-saturated, μ-dense.

Proof : By 13 and 11(2) (so we could get a little more).

References

[BHM]
J. E. Baumgartner, A. Hajnal and A.Mate. Weak saturation propertes of ideals. *Infinite and Finite Sets*. Proc. of a Symp. for Erdos 60th Birthday, Budapest 1973. Colloq. Math. Soc. Jano Bolayi 10 ed. . Hajnal, R. Rado and T. Sos, North-Holland Publ. Co. Vol 11 (1975), 137-158.

[DS] K. Devlin and S. Shelah, A weak form of the diamond follows from $2^{\aleph_0} < 2^{\aleph_1}$, *Israel J. Math.* 29 (1978), 239-247.

[FMR]
M. Forman, M. Magidor and S. Shelah, In preparation.

[KE] R. Engelking and M. Karlowicz. Some theorems of set theory and their topological consequences. *Fund Math.* 57 (1965), 275-285.

[Sh1]
S. Shelah, Remarks on Boolean algebras, *Algebra Universalis*, 11 (1980), 77-84.

[Sh2]
———, Proper Forcing, *Springer Lecture Notes*, 940 (1982).

[Sh3]
———, On saturation for a predicate, *Notre Dame J. of Formal Logic*, 22 (1981), 301-307.

[Sh4]
———, A note on κ-freeness, *A Springer Lecture Notes*, volume here.

[Sh5]
———, From supercompacts to special normal ideals on small cardinal, in preparation.

[SW] S. Shelah, and H. Woodin, Hypermeasurability cardinals implies every projective set is Lebesgue measurable, In preparation.

A note on κ-freeness of abelian groups

Introduction: Lately Foreman proved that the assertion $(*)_n$ below follows from some axioms (speaking on the $\aleph_n$, seemingly of consistency strength like the determinancy axioms), and $(*)_2$ consistent if some large cardinal axioms ($\approx$ there is a huge cardinal) are consistent.

$(*)_n$ every $\aleph_n$ free abelian group of power $\aleph_n$ is the union $\aleph_1$ free subgroups.

Let in this note a group mean an abelian group.

We consider mainly some variants (which his proofs easily gives); give some sufficient conditions in ZFC, and find the consistency strength for $n=2$ which is Mahlo, and prove the consistency of $(*)_n$ using super compact cardinals.

1. Definition : 1) $P(\lambda,\kappa) \overset{def}{=}$ if G is λ-free of power λ, $G = \bigcup_{i<\lambda} G_i, |G_i| < \lambda, G_i$ increasing continuous *then* $\{i: G/G_i$ not κ-free$\}$ is not stationary.

2) Let $P^+(\lambda,\kappa)$ mean that every λ-free group of power λ is κ-freely represented (see 2(4)).

2. Definition : 1) A group G is (μ,κ)-coverable if we can find $H_\alpha(\alpha < \mu)$, free pure subgroups of G, such that: for every $A \subseteq G$ of power $< \kappa$, for some α, $A \subseteq H_\alpha$.

2) We define "weakly (μ,κ)-coverable" similarly if omitting the "purity".

3) G is (μ,κ)-freely represented if it has a (μ,κ) free representation i.e.

$\langle G_i : i < i(*) \rangle$, G_i increasing continuous, $G_0 = \{0\}$, $G_{i(*)} = G$, and G_{i+1}/G_i is κ-free of power $\leq \mu$. (κ-free means: every subgroup of rank $< \kappa$ is free, so κ may be finite).

3. Lemma : If G_i is increasing continuous, $G_{i(*)} = G$, $G_0 = \{0\}$, $\mu \leq \lambda$ *then* G is (μ,κ)-coverable provided that:

(*) there are sequences $\langle H_{i,\xi} : \xi < \mu \rangle$ of pure subgroups of G_{i+1} such that

(a) $H_{i,\xi} + G_i / G_i$ is a free and a pure subgroup of G_{i+1}/G_i.

(b) if $A \subset G, |A| < \kappa$ then for some set $S \subset i(*)$, and ξ such that:

(α) $A \subset \sum_{i \in S} H_{i,\xi}$ and

(β) $(\forall i < j)([i \in S \wedge j \in S \rightarrow H_{j,\xi} \cap G_{i+1} \subset H_{i,\xi} + G_i]$ and

(γ) $(\forall j \in S)$ $(\forall i < j)$ $[H_{j,\xi} \cap (G_{i+1} - G_i) \neq \phi \rightarrow i \in S]$.

Proof: Define $K_{i,\xi}$ by induction on $i < i(*)$:
$K_{0,\xi} = \{0\}$, $K_{\delta,\xi} = \bigcup_{i<\delta} K_{i,\xi}$,
$K_{i+1,\xi}$ is $K_{i,\xi} + H_{i,\xi}$ if $H_{i,\xi} \cap G_i \subset K_{i,\xi}$, and $K_{i,\xi}$ otherwise.

Easily by (a) of (*) $K_{i(*),\xi}$ is a pure subgroup of G. Now we should prove: for every $A \subset G$, if $|A| < \kappa$, then $(\exists \xi < \mu) A \subset K_{\lambda,\xi}$. Let S,ξ be as in (b) of (*) (for the set A). We prove by induction on $i \in S$ that:

(i) $H_{i,\xi} \cap G_i \subset K_{i,\xi}$

(ii) $H_{i,\xi} \subset K_{i+1,\xi}$

For $i = 0$, everything is trivial as $G_i = \{0\}$; when we arrive to i, if (i), fail, choose $j \leq i$ minimal such that $H_{i,\xi} \cap G_j \not\subset K_{j,\xi}$, necessarily j is successor, so $H_{i,\xi} \cap (G_j - G_{j-1}) \neq \phi$ so by (*) (b) (γ) $(j-1) \in S$. By the minimality of j, $H_{i,\xi} \cap G_{j-1} \subset K_{j-1,\xi}$ and as by (β) of (b) $H_{i,\xi} \cap G_{(j-1)+1} \subset H_{j-1,\xi} + G_{j-1}$, by the choice of $K_{j,\xi} = K_{(j-1)1,\xi}$, $H_{i,\xi} \cap G_j \subset K_{j,\xi}$, contradicting the choice of j. Now we can prove (ii).

So $\{K_{\lambda,\xi}:\xi < \mu\}$ exemplify G is (μ,κ)-coverable.

4. Lemma: If G is (μ,κ)-freely representable, $\kappa > \aleph_0$, and $\mathcal{P}_{<\kappa}(\mu)$ has a stationary subset S of power μ^1 (see below) *then* G is (μ^1,κ)-coverable.

Proof: We use Lemma 3.

Let $\langle G_i : i \leq i(*)\rangle$ be a (μ,κ)-free representation of G, and let $L_i \subseteq G_{i+1}$ be such that $G_{i+1} = G_i + L_i$, (L_i a pure subgroup of G_{i+1}) and $|L_i| = \mu$ (but maybe $G_i \cap L_i \neq \{0\}$). Let g_i be a one-to-one mapping from μ onto L_i.

Let $S = \{s_\xi : \xi < \mu^1\}$, be an enumeration of S in increasing order, and $H_{i,\xi}$ be the subgroup of G_{i+1} (of L_i in fact) generate by $\{g_i(x): x \in s_\xi\}$. We can finish as:

$\otimes \langle H_{i,\xi}:\xi < \mu\rangle$ $(i<i(*))$ satisfies $(*)$

(apply second sentence of 5(2)). Remember:

5. Definition : 1) $\mathcal{P}_{<\kappa}(A) = \{s : s \subseteq A, |s| < \kappa\}$.

2) $S \subseteq \mathcal{P}_{<\kappa}(A)$ is stationary, if for every $< \kappa$ (finitary) functions from A to A, some $s \in S$ is closed under all of them.

Note: if $B \supseteq A$, $\gamma < \kappa$, f_i is an n_i -place function from B to B for $i < \gamma$, and from each $\alpha \in B, g_\alpha$ is a one-to-one map from A onto some $B_\alpha \subseteq B$, then for some $s \in \mathcal{P}_{<\kappa}(B)$ closed under the f_i's, $s \cap A \in S$ and for every $\alpha \in s$, $s \cap B_\alpha = \{g_\alpha(x): x \in (s \cap A)\}$.

6. Fact: 1) If κ is regular, $\mu = \kappa^{+n}$ then $\mathcal{P}_{<\kappa}(\mu)$ has a stationary subset of power μ.

2) $\mathcal{P}_{<\kappa}(\mu)$ has a stationary set of power $\mu^{<\kappa}$.

7. Lemma : If $\kappa \leq \mu$, $\mathcal{P}_{<\kappa}(\mu)$ has no stationary subset of power μ *then* $0^\#$ exist, there is an inner model with a measurable cardinal B, etc.

Proof: By [Sh 3] Ch. XIII.

8. Lemma: Suppose $||G|| = \lambda, G$ has a (μ,κ)-free representation, $\kappa = \aleph_0$,

$2^\mu \geq \lambda$. Then G is (μ,κ)-coverable.

Proof : Let $\langle G_i : i \leq i(*)\rangle$ be a (μ,κ)-representation of G, $L_i \subseteq G_{i+1}, |L_i| = \mu$, $G_{i+1} = G_i + L_i$, L_i a pure subgroup of G_{i+1}. Let $\langle H^0_{i,\xi} : \xi < \mu\rangle$ be a list of all pure subgroups of L_i of finite rank.

Let $g_i : \mu \to \mu$ $(i < i(*)$ be functions such that for every distinct $i_1, \ldots, i_m (m < \omega)$ and $\xi_1, \ldots, \xi_m < \mu$ for some $\alpha < \mu$ $g_{i_\ell}(\alpha) = \xi_\ell$ for $\ell = 1,m$ (exists by Engelking and Karlowiz [EK] as w.l.o.g. $|i(*)| \leq \|G\| \leq 2^\mu$.)

Let $H^0_{i,\alpha} = H^0_{i,g_i(\alpha)}$.

Now apply 3 to $\langle\langle H_{i,\alpha} : \alpha < \mu\rangle : i < i(*)\rangle$.

9. Lemma : Suppose $\|G\| = \lambda \leq 2^\mu$, G has a (μ,κ)-free representation, $\kappa < \aleph_0$. Then G is (μ,κ)-coverable.

Proof : Like 8 but $G_{i+1} = L_i$ and we restrict ourselves to $H = H^0_{i,\xi}$ disjoint to G_i (more exactly, $H^0_{i,\xi} \cap G_i = \{0\}$).

We prove by induction on i, that for $A \subseteq G_i, |A| < \kappa$, the (*) (b) of Lemma (3) holds. For $i = 0,i$ limit - no problem. For $i+1$: let $A = \{a_\ell : \ell < |A|\}$, w.l.o.g. a_ℓ belong to the pure closure of $\langle G_i, a_0, \ldots, a_{\ell-1}\rangle$ iff $\ell \geq m$. We first define by induction on $\ell < m$, $b_\ell \in G_{i+1}$, $c_\ell \in G_i$.

(i) $\{b_0 + G_i, \ldots, b_\ell + G_i\}$ is independent, and generates a pure subgroup of G_{i+1}/G_i (of course $b_\ell + G_i$ is not torsion).

(ii) $a_\ell \in \langle b_0, \ldots, b_{\ell-1}, b_\ell, c_\ell\rangle_G$ (= the subgroup generated by then).

As $m \leq |A| < k$ in the ℓ-stage, $G_{i+1}/\langle G_i, a_0, \ldots, a_{\ell-1}\rangle$ is $(\kappa-\ell)$-free, so there is a maximal integer n_ℓ dividing $a_\ell + \langle G_i, b_0, \ldots, b_{\ell-1}\rangle$, and let b_ℓ be such that $n_\ell b_\ell - a_\ell \in \langle G_i b_0, \ldots, b_{\ell-1}\rangle$. So for some $n_{\ell,0}, \ldots, n_{\ell,\ell-1}$; $a_\ell - n_\ell b_\ell + n_{\ell,0} b_0 + \cdots + n_{\ell,\ell-1} b_{\ell-1} \in G_i$, and call it c_ℓ.

Now we define for $m \leq \ell < |A|, c_\ell$ such that

(iii) $a_\ell \in \langle c_0, \ldots, c_\ell\rangle$

Arriving to ℓ, for some $m_\ell \neq 0$ $m_\ell a_\ell \in \langle G_i, a_0, \ldots, a_{\ell-1} \rangle$, hence $m_\ell a_\ell + G_i \in \langle a_0 + G_i, \ldots, a_{\ell-1} + G_i \rangle \subseteq \langle b_0 + G_i, \ldots, b_{m-1} + G_i \rangle$, but the latter is pure so or some $n_{\ell,0}, \ldots, n_{\ell,m-1}$, $a_\ell + G_i \in \langle b_0 + G_i, \ldots, b_{m-1} + G_i \rangle$, so for some $n_{\ell,0}, \ldots, n_{\ell,m-1}$ the following equation holds $a_\ell - n_{\ell,0} b_0 - \ldots - n_{\ell,m-1} b_m \stackrel{def}{=} c_i \in G_i$.

Now use then induction hypothesis on $\{c_0, c_1, \cdots\}$.

10. Fact: If $P(\lambda,\kappa)$, $\lambda = \mu^+$ *then* every λ-free group of power λ is (μ,κ)-represented.

The following is a (strong) converse to 4,8,9 (so under suitable condition (μ,κ)-coverable $\equiv$ weakly (μ,κ)-coverable.)

11. Lemma : 1) Suppose $\lambda = \mu^+, |G| = \lambda$ and G is (μ,κ)-coverable *then* G is (μ,κ)-freely represented.

2) Then $\kappa > \aleph_0$ G weakly (μ,κ)-coverable is enough.

Proof: 1) Let $|G| = \lambda$ (i.e., the universe = the set of elements of G, is λ), $G = \bigcup_{\xi<\mu} H_\xi$, each H_ξ is a free pure subgroup of G, and $(\forall A \subseteq G)[|A| < \kappa \rightarrow (\exists \xi) A \subseteq H_\xi]$.

Let $G = \bigcup G_i$, G_i increasing continuous, $\|G_i\| < \lambda$ and let $S = \{i < \lambda$: G/G_i is not κ-free$\}$, we assume S is stationary and will arrive at contradiction thus finishing. For $i \in S$, let L_i be a pure subgroup of G of rank $< \kappa$, such that $L_i + G_i / L_i$ not free. Let $A_i \subseteq L_i$ be such that $|A_i| < \kappa$, and $L_i + G_i$ is the pure closure of $\langle G_i \cup A_i \rangle$.

So for every $i \in S$ for some $\xi(i) < \mu, A_i \subseteq H_{\xi(i)}$. So for some $\xi, T = \{i \in S: \xi(i) = \xi\}$ is stationary. Let N be an elementary submodel of an appropriate expansion of G, with universal $|G_i| = i \in T$. We shall prove that: (the pure closure of $G_i \cup A_i$ in $G)/G_i \cong$ pure closure of $(H_\xi \cap G_i) \cup A_i$ in $H_\xi / H_\xi \cap G_i$.

This suffices. So it suffices to show.

(*) if $a_1, \ldots, a_n \in A_i$, $0 < k < \omega$, $b \in G_i$, $b + \sum_{i=1}^{n} m^\ell a_\ell$ is divisible by k (in G), then we can find such $b \in H_\xi \cap G_i$ divisible by k even in H_ξ.

Proof of (*): As N is an elementary submodel, $b \in N$ as $b \in G_i = i$ we can find $a'_\ell \in H_\xi \cap N = H_\xi \cap G_i$ such that $b + \sum m^\ell a'_\ell$ is divisible by k (in G and even in G_i). Now let $b' = 0 - \sum m^\ell a'_\ell \in H_\xi \cap G_i$, and divisibility is in H_ξ using: $H_\xi \subset G$ purely.

2) Just take care that $A_i = L_i$, $L_i, G_i + L_i$ and $G \cap L_i$ will be pure subgroups of G.

We now restrict ourselves for a while to $\lambda = \aleph_2$ $\mu = \aleph_1$.

12. Lemma : The following are equivalent.

A) $P(\aleph_2,\aleph_1)$

B) every $\aleph_2$-free group of power $\aleph_2$ is $(\aleph_1,\aleph_1)$-coverable.

C) every $\aleph_2$-free group of power $\aleph_2$ is $(\aleph_1,2)$-coverable.

D) If $S \subset \{\delta : \delta < \aleph_2, cf\ \delta = \aleph_0\}$ is stationary. $A_\delta \subset \delta$ (is countable for $\delta \in S$ *then* there is a stationary $T \subset \aleph_1$ and $f : T \to S$ one-to-one such that $\{\xi \in T : A_{f(\xi)} \subset \bigcup_{\zeta < \xi} A_{f(\zeta)}\}$ is stationary.

Proof : (A) $\Longrightarrow$ (B) by 10, 4+6(1).

(B) $\Longrightarrow$ (C) trivial

(C) $\Longrightarrow$ (D). We prove $\neg$(D) $\Longrightarrow$ $\neg$(C).

Let $\{A_\delta : \delta \in S\}$ be a counterexample to (D). Let $A_\delta = \{a_{\delta,n} : n < \omega\}$. Let G be freely generated by $x_\eta (\eta \in {}^{\omega >}\aleph_2)$, $y_{\delta,n} (n < \omega, \delta \in S)$ except the relations (letting $\eta_\delta = \langle a_{\delta,0}, a_{\delta,1}, \cdots \rangle$)

$$p\ y_{\delta,n+1} = y_{\delta,n} - x_{\eta_\delta \restriction n}$$

(p a fixed prime but you can make it a natural number ≥ 1 depending on δ,n) Easily G is not $(\aleph_1,2)$-freely represented and by 1) we get a contradiction.

(D) $\Longrightarrow$ (A): See [Sh 2] for much more

13 Theorem : (D) is equi consistent with Mahlo..

Proof : See Harrington Shelah [H Sh].

14. Lemma: We can move the cardinals in 11, e.g. let $\mu = \mu^{<\kappa}$, $\kappa > \aleph_0$ then the following are equivalent.

(A)' $P(\mu^+, \kappa)$.

(B)' every μ^+-free group of power μ^+ is (μ, κ)-coverable.

(C)' every μ^+-free group of power μ^+ is weakly (μ, κ)-coverable.

(D)' for some regular χ, $\vartheta < \kappa + \aleph_1$, there are a stationary $S \subseteq \{\delta < \mu^+ : cf\ \delta = \vartheta\}$, and $A_\delta \subseteq \delta$ of order type $\vartheta\chi$, Sup $A_\delta = \delta$, $A_\delta = \bigcup_{i<\vartheta} A_{\delta,i}, A_{\delta,i} < A_{\delta,j}$ for $i < j$ otop $(A_{\delta,\alpha}) = \chi$, such that for every $i < \mu^+$ we can find pairwise disjoint $B_\delta \subseteq A_\delta$, such that $(\exists^{<\vartheta} i)(\exists^{<\chi} j \in A_{\delta,i}) j \not\in B_\delta$ (if $\kappa = \aleph_1$, (D)' can be replaced by "A_δ of order type ω , $|A_\delta - B_\delta| < \aleph_0$".

The consistency strength, for μ regular is as in 13.

Proof: As in [Sh 2].

However.

15 Observation: Suppose $\lambda_0 \leq \lambda$, $(\exists n)\lambda \leq \lambda_0^{+n}$ λ is regular, and

(A) for every χ, $\lambda < \chi^+ \leq \lambda$, every (χ^+)-free group of power χ^+ is χ-freely represented (i.e $P(\chi^+, \chi)$.

(B) every λ_0-free group of power λ_0 is (μ, κ)-freely represented.

Then every λ-free group of cardinality λ is (μ, κ)- freely represented.

Proof: By induction on λ. For $\lambda = \lambda_0$ this is (B) for λ a successor cardinal use (A).

Remark. We can phrase similar things for $\lambda \geq \lambda_0^{+\omega}$, but then for λ singular every λ-free group of power λ will by free be [Sh 1] so this is not an interesting case.

The consistency strength is much higher by Magidor [Ma].

Now by 14 and 15 and known set theory we can get positive results e.g. (using $\aleph_1$ for simplicity).

16 Theorem: 1) Suppose $2 < n < \omega$ and $P(\aleph_{m+1},\aleph_m)$ holds when $1 \leq m < m$. Then every $\aleph_n$-free group of cardinality $\aleph_n$ is $(\aleph_1,\aleph_1)$-freely represented hence in $(\aleph_1,\aleph_1)$-coverable.

2) From the consistency of $(n-1)$ supercompact cardinals we can get the consistency of $\bigwedge_{m=1}^{n-1} P(\aleph_{m+1},\aleph_m)$ and G.C.H. [$\aleph_0 < \kappa_1 < \cdots < \kappa_n$ are supercompact, w.l.o.g. satisfying Laver's conclusion [L], and use Levi collapse to make κ_ℓ to $\aleph_\ell$ ($\ell = 1,n$) and use Baumgartner [B] argument.]

Note

17. Lemma : 1) Let U be an abelian group, and let

$$F = \{(A,B): \langle A \cup B \rangle_G / \langle B \rangle_G \text{ is } (\mu,\kappa)\text{-represented}\}.$$

Then (in the context of [Sh 1], §1, or [Šh 2] §1 the following axioms holds) with χ there standing for μ here: II, III, IV, VI, VII.

18. Lemma: 1) If G is (μ,κ)-coverable then G is (μ,κ)-represented.

2) If $\kappa > \aleph_0$ weakly (μ,κ)-coverable suffice.

Proof: We can prove this by induction on $\|G\|$. If $\|G\| \leq \mu$ this is trivial. For $\|G\| > \mu$ a singular cardinal use the compactness theorem of [Sh 1] (where Lemma 17 shows the assumption holds. For $\|G\| > \mu$ a regular cardinal repeats the proof of 11.

19 Conclusion: Suppose $\kappa > \aleph_0$ and $\mathcal{P}_{<\kappa}(\mu)$ has a stationary subset of cardinality μ.

For any group G, G is (μ,κ)-represented *iff* G is (μ,κ)-coverable if G is weakly (μ,κ)-coverable.

Proof: The first implies the secod by Lemma 4, the second implies the third trivially, the third implies the first by Lemma 18.

References

[B] J. Baumgartner, A new class of order types, *Annals of Math. Logic*, 9 (1976), 187-222.

[BD] S. Ben-David, On Shelah compactness of cardinals, *Israel. J Math.* 31 (1978), 34-56.

[EK] R. Engelking and M. Karlowicz, Some theorems of set theory and their topological consequence, *Fund. Math.* 57 (1965), 275-285.

[F] M. Foreman, in preparation.

[HSh]
L. Harrington and S. Shelah, Equi-consistency results, *Notre Dame I. of Formal Logic*. Proc. of the Jerusalam Model Theory Year 1980/1, to appear.

[L] R. Laver, Making the supercompactness of κ indestructible under κ-directed closed forcing. *Israel J. Math*, 29 (1978), 385-388.

[Ma] M. Magidor,

[Sh1]
S. Shelah, A Compactness theorem in singular cardinals, free algebra, Whitehead problem an transversal. *Israel. J. Math.* 2(1975), 319-349.

[Sh2]
S. Shelah, Incompactness in regular cardinals. *Notre Dame J. of Formal Logic*, in press.

[Sh3]
S. Shelah, Proper forcing, Springer Lecture Notes, 940 (1982).

On countable theories with models - homogeneous models only

1. Theorem : Suppose every model of T of power λ is model homogeneous and $\lambda > |T|$ and T is countable. Then $\lambda \geq \lambda(T)$ (= the first cardinality in which T is stable), T is superstable, unidimensional and every model of T of power $\mu > \lambda(T)$ is model homogeneous.

Proof : We known that (see [Sh 2]):

(*) if M_1, M_2 are models homogeneous model of T of power λ and $\{N/\approx : N \prec M_1, \|N\| = |T|\} = \{N/\approx : N \prec M_2, \|N| = |T|\}$ then $M_1 \approx M_2$.

By [Sh 1], Ch. VIII, 4.2. if T is not superstable T has non-isomorphic models of power λ which contradict (*).

Suppose T is stable but not unidimensional. By V. 2.10 T has an $\mathbf{F}^a_{|T|^+}$-saturated model M of cardinality $> (2^\lambda)^+$, with a maximal indiscernible set $\mathbf{I} \subseteq M$ of power $\kappa_r(T)$. Let $\bar{c} \in \mathbf{I}$ (so $\bar{c} \in M$), $\mathbf{I}_1 \stackrel{def}{=} \mathbf{I} - \{\bar{c}\}$ and let $N \prec M$ be $\mathbf{F}^a_{|T|^+}$-primary over $\bigcup \mathbf{I}$ and $N_1 \prec N$ be $\mathbf{F}^a_{\kappa_r(T)}$-primary over $\bigcup \mathbf{I}_1$. By [Sh 1] IV 4.18 N_1 is isomorphic to N by an isomorphism f mapping $\mathbf{I}_1$ onto $\mathbf{I}$, and clearly M omit $Av(\mathbf{I}, \bigcup \mathbf{I})$, N_1 omit $Av(\mathbf{I}_1, \bigcup \mathbf{I}_1)$. So clearly we cannot extend f to an elementary mapping f^* from N into M (as then $f^*(\bar{c})$ realizes $Av(\mathbf{I}, \bigcup \mathbf{I})$). So if $\lambda > \|N\|$ we can finish (see below). Let N^* be such that $N^* \prec M, \|N^*\| = |T|$ and:

$$(N^*, N_1 \cap N^*, f \restriction (N^* \cap N_1), \mathbf{I} \cap N^*, \bar{c}) \prec (M, N_1, f, \mathbf{I}, \bar{c})$$

and $\mathbf{I} \subseteq N^*$ (possible as $|\mathbf{I}| = \kappa_r(T) = \aleph_0 = |T|$). Again $f \restriction (N^* \cap N_1)$ cannot be extended to an elementary mapping f^* from $N^* \cap N_1$ into M (as then $f^*(\bar{c})$ realizes $Av(\mathbf{I}, \bigcup \mathbf{I})$). So M is not $|T|^+$-model homogeneous and $\|M\| \geq \lambda$, so we can find $M^+, N^* \prec M^+ \prec M$, $\|M^+\| = \lambda$ and M^+ contradicts a hypothesis . So T is unidimensional.

The proof that $\lambda > \lambda(T)$ is similar (M will be $\mathbf{F}^a_{\kappa_r(T)}$-prime over $I, |I| = \kappa_r(T)$, so $\|M\| = \lambda(T)$, I maximal in M).

If T has the otop, we get contradiction as in the case of unsuperstable T. T cannot have the dop as it is unidimensional.

Now note:

2. Lemma: If T is superstable unidimensional (or even just with no $M, \varphi, \bar{a}$ such that $\aleph_0 \le |\varphi(M,\bar{a})| < \|M\|$) and with the $(<\infty,\lambda>)$-existence property, then (see [Sh 1] Ch. XI §2), $(\mathbf{F}^t_{\aleph_0}, \subseteq_t)$ satisfies Ax. A4-6, B1-6, C1$^+$, C2, C3$^+$, D2^{-1}, E1^1, E2$^+$ $\lambda(\mathbf{T}^t_{\aleph_0}, \subseteq_t) \le |T|$.

Proof: Suppose M is not model homogeneous, $\|M\| > \lambda(T)$ and we shall get a contradiction thus finishing. Clearly M is not saturated, hence by [Sh 1] IX 1.8 T is not $\aleph_0$ stable, [condition (3) fail hence (6) with λ, μ there standing for λ, $|T|$ here, now M is easily not μ^+-model homogeneous.] So $\lambda(T) = 2^{\aleph_0}$. As M is not model homogeneous there are $\mu < \|M\|$, $M_0 \prec M_1 \prec M$, $M^0 \prec M$ f an isomorphism from M_0 onto M^0 which cannot be extended to an elementary embedding of M_1 into M and $\|M_1\| \le \mu$. By [Sh 1] 2.6(2), and Lemma 2, M_0 has an $(\mathbf{F}^t_{\aleph_0}, \subseteq)$-decomposition $\langle N_\eta : \eta \in \{<>, <i> : i < \alpha_0\}\rangle$, N_η countable, and it can be extended to an $(\mathbf{F}^t_{\aleph_0}, \subseteq)$-decomposition $\langle N_\eta ; \eta \in \{<>, <i> : i < \alpha_1\}\rangle$ of M_1, $\|N_\eta\| = \aleph_0$. Clearly $\langle N^+_\eta : \eta \in \{<>, <i> : i < \alpha^0\}\rangle$ is an $(\mathbf{F}^t_{\aleph_0}, \subseteq)$-decomposition of M^0 when $N^+_\eta = f(N^+_\eta)$ and it can be extended to an $(\mathbf{F}^t_{\aleph_0}, \subseteq)$-decomposition of M : $\langle N^+_\eta : \eta \in \{<>, <i>, : i < \beta\rangle\}$, $\|N^+_\eta\| = \aleph_0$.

We can define by induction on $\alpha, \alpha_0 \le \alpha < \alpha_1$ a model $M_{0,\alpha}$: $M_{0,0} = M_0$, $M_{0,\alpha+1}$ is prime over $M_{0,\alpha} \cup N_{<\alpha>}$ (it exists as T has the $(<\infty,2)$-existence property), and for limit δ, $M_{0,\delta} = \bigcup_{\alpha_0 \le \alpha < \delta} M_{0,\alpha}$. We then can try to define by induction on $\alpha, \alpha_0 \le \alpha < \alpha_1$, an elementary embedding f_α of $M_{0,\alpha}$ into M, f_α extending f and f_β when $\alpha_0 \le \beta < \alpha$. If f_{α_1}, is defined this contradicts the choice of M_0, M_1, M^0, f. So for some α, f_α is defined but $f_{\alpha+1}$ is not, and (by renaming) W.l.o.g. $\alpha = \alpha_0$, $\alpha_1 = \alpha_0 + 1$. By the $(<\infty,2)$-existence property there is no N^1 realizing $f(tp_*(N_{<\alpha_0>}, N_{<>}))$, which is independent over $(M^0, f(N_{<>}))$.

Choose $N \prec M, \|N\| = |T|$, $|N_{<>}| \cup |N^+_{<>}| \cup |N_{<\alpha_0>}| \subseteq N$, f maps $N \cap M_0$ onto $N \cap M^1$, $tp_*(N, M_0)$ does not fork over $N \cap M_0$, $tp_*(N, M^0)$ does not fork over $M^0 \cap N, N \cap M_0 \prec M_0$. As $\|M\| > 2^{\aleph_0} + \mu$ w.l.o.g. $tp(N^+_{\alpha_0+i}, N^+_{<>})$ is constant for $i < \mu^+$ and $\{N^+_{\alpha_0+i} : i < \mu^+\}$ is independent over $(M_1 \cup M^0 \cup N, N^+_{<>})$.

If $\lambda \leq \mu^+$ let M_λ be prime over $N \cup \bigcup_{i<\lambda} N^+_{\alpha_0+i}$ (exists by the $(<\infty,2)$-existence property). So w.l.o.g. $M_\lambda \prec M$, and we shall show that M_λ is not $|T|^+$-model homogeneous, thus getting a contradiction hence every model of T of power $> \lambda(T)$ is model homogeneous thus finishing the proof . The non $|T|^+$-model homogeneity of M_λ is exemplified by $M_0 \cap N, M_1 \cap N$, and $f \restriction (M_0 \cap N)$. For this it suffices to prove that $f(tp_*(N_{<\alpha_0>}, M_0 \cap N))$ is not realized in M_λ, so suppose N^+ realizes it, $N^+ \prec M_\lambda$. So $N^+ \prec M$. Easily $tp_*(N^+, M_1 \cup M^0 \cup N)$ does not fork over N, (as M_λ is atomic over $N \cup \bigcup_{i<\lambda} N^+_{\alpha_0+i}$) and we have chosen N such that $tp_*(N, M^0)$ does not fork over $M^0 \cap N$ so by III 0.1, $tp_*(N^+ \cup N, M^0)$ does not fork over $M^0 \cap N$, so $tp_*(N^+, M^0)$ does not fork over $M^0 \cap N$. So N^+ realizes over M^0 the stationarization of $f(tp_*(N_{<\alpha_0>}, N \cap M_0))$ so we can show that it realizes $f(tp_*(N_{<\alpha_0>}, M_0))$, contradiction.

If $\lambda \geq \mu^+$ we can "lengthen" $\{N^+_{\alpha_0+1} : i < \mu^+\}$ and the proof is similar.

References

[Sh1]

S. Shelah, Classification Theory, completed for countable. North Holland Publ. Co.

[Sh2]

--------, Classification of non elementary classes II, Abstract Elementary classes, here.

On decomposable sentences for finite models

Saharon Shelah

A Definition : Suppose $\psi = \psi(\bar{P}, \bar{Q})$ (i.e. ψ is first order depending on the predicates $\bar{P} = \langle P_\ell : \ell < n \rangle$, $\bar{Q} = \langle Q_\ell : \ell < n \rangle$. If the truth value of $(A, \bar{P}, \bar{Q}) \models \psi(\bar{P}, \bar{Q})$ depend on the isomorphism types of $(A, \bar{P})$ and $(A, \bar{Q})$ only, we call $\psi(\bar{P}, \bar{Q})$ decomposable.

If this holds for all finite models we call $\psi(\bar{P}, \bar{Q})$ finitely decomposable.

Let $K_\psi = \{(A, \bar{Q});\ \exists \bar{P}$ such that $(A, \bar{P}, \bar{Q}) \models \psi\}$

B. Claim: If $\psi(\bar{P}, \bar{Q})$ is decomposable then there are $\psi_\ell(\bar{P})$, $\psi^\ell(\bar{Q})$ such that we can compute the truth value of $(A, \bar{P}, \bar{Q}) \models \psi$ from the truth values of $(A, \bar{P}) \models \psi_\ell(\bar{P})$ and $(A, \bar{Q}) \models \psi^\ell(\bar{Q})$.

Proof : Use saturated models.

C. **Conclusion:** If $\psi(\bar{P}, \bar{Q})$ is decomposable then there are $\vartheta_m(\bar{Q})(m < m_0)$ such that each $K_\psi^\lambda = \{M \in K_\psi : ||M|| = \lambda\}$ is the class of models of $\vartheta_m(\bar{Q})$ where m depends on λ (and is the same for all infinite cardinals).

1. **Example:** We deal with models with universe $n = \{0, 1, \ldots, n-1\}$, ($n < \omega$ arbitrary).

We shall find sentences $\psi(\bar{P}, \bar{Q}), \varphi(\bar{P})$ (not depending on n) such that

1) the truth value of $(n, \bar{P}, \bar{Q}) \models \psi(\bar{P}, \bar{Q})$ depend on the isomorphism type of $(n, \bar{P})$ and $(n, \bar{Q})$ only

2) $\psi(\bar{P}, \bar{Q}) \rightarrow \varphi(\bar{P})$

3) in each finite power $\varphi(\overline{P})$ has a unique model.

4) For $n < \omega$ (quite large), the set $K_n = \{(n,\overline{Q}) : (\exists \overline{P})[(n,\overline{P},\overline{Q}) \models \psi(\overline{P},\overline{Q})]\}$

is not definable (among models of the right signature and power n) by any first order sentence of size $= {}^{200}\sqrt{n}$ (and even such quantifier depth.)

Remark: We do not try to improve the bounds appearing here, clearly $n^{(1/2+\varepsilon)}$ suffices (for any positive ε).

2. Construction: Let $\varphi(\overline{P})$ just say that (n,P) is a model $(n,+,\times,0,1,<)$ satisfying the reasonable rules of arithmetic (addition , product) (but not necessarily the standard ones). Let $\psi = \psi_0(\overline{Q})$ be such that

$(A,Q_0,Q_1,Q_2,Q_3,F_1,F_2,+',\times',0',1') \models \psi_0$ *iff* Q_0,Q_1,Q_2 are monadic relations which form a partition of A, Q_3 a monadic relation, $Q_3 \subset Q_1$, also $\varphi^{Q_2}(+',\times' \cdots)$ hold , F_1,F_2 are one place function from Q_1 onto Q_3. (so $F_\ell(x)$ is undefined for $x \notin Q_1$), and:

$$(\forall x \in Q_3)[x = F_1(x) = F_2(x)]$$
$$(\forall x,y \in Q_1)[x = y \equiv (F_2(x) = F_1(y) \wedge F_2(x) = F_2(y))]$$
$$(\forall x,y \in Q_3)\ (\exists z \in Q_1)[\ F_1(z) = x \wedge F_2(z) = y]$$

Let $K_n = \{M : \|M\| = n,\ M \models \psi_0,\ |Q_0^M|^{100} < |Q_1|$ and $|Q_0|$ is even$\}$ (we can replace "even" by anything reasonable.

Before we shall define a ψ, such that $K_n = K_\psi^n$ we have to deal with

3. Question: If $(n,\overline{P}) \models \varphi(\overline{P})$, $Q \subseteq M$, can we define (by a short formula) $|Q|$ in $(n,\overline{P},Q)$, i.e. we want as formula $\vartheta(x,\overline{P},Q)$ such that:

$$(n,\overline{P}) \models \varphi(\overline{P}),\ Q \subset n \Longrightarrow (n,\overline{P},,Q) \models (\forall x)\,[\,|\{y : y < x\}| = |Q| \equiv \vartheta(x,\overline{P},Q)]$$

The following approximation (and more) for this appeared in Deneberg Gurevich and Shelah [2], and is included for completeness.

4. Fact: There is a formula $\vartheta(x,\overline{P},Q)$ such that for every n and $\overline{P}$, if $(n,\overline{P}) \models \varphi(\overline{P})$ and $Q \subset n$ then $(n,\overline{P},Q) \models (\exists x)\vartheta(x,\overline{P},Q)$ and $\models \vartheta(x,\overline{P},\overline{Q})$ implies

$$|Q| \leq |\{y : y < x\}| \leq |Q|^2 |\ln n|^2 + 10$$

Proof : Let $\vartheta_0(x,\overline{P},Q)$ says that x is the first prime number such that for every $y \neq z \in Q$: $y \not\equiv z \ mod \ x$ (all arithmetic statements are interpreted by $\overline{P}$).

Let $(n,\overline{P})$ be for notational simplicity the usual arithmetic. So clearly there is at most one such x and $|Q| \leq x$. Suppose that $T < n$ and for every prime $|Q| \leq p$ $|Q| \leq p \leq T$, there is a pair $y \neq z \in Q$ so that p divides $y-z$.

Then $A = \prod_{\substack{y,z \in Q \\ z>y}} (z-y)$ is divisible by $B = \Pi\{p:p \text{ prime}, |Q| \leq p < T\}$. Hence $B \leq A$; but $A \leq n^{|Q|^2}$, whereas $B \geq |Q|^{\pi}$, where π is the number of primes in ($|Q|,T)$. So $e^{|Q|^2 \ln n} = n^{|Q|^2} \geq |Q|^{T/\ln T - |Q|/\ln(Q)} = (e^{T \ln |Q| / \ln T})e^{-|Q|}$, hence

$$|Q|^2 \ln n + |Q| \geq T \ln |Q| / \ln T$$

Hence if e.g. $T = |Q|^2 (\ln n)^2$, $n \geq 10$ we get contradiction.

5. Fact: In 4) we can also define a one to one function from Q into $\{y:y < x\}$, and then we can do the same analysis on the image, replacing n by $\{y:y < x\}$) (or even if you want, $T = |Q|^2(\ln n)^2$); so we get a new bound

$$|Q| \leq |\{y:y < x'\}| \leq |Q|^2 (\ln T)^2$$

So if e.g. $|Q| \leq {}^3\sqrt{\ln n}$, we can find a one to one map from Q onto an initial segment: as by the previous analysis w.l.o.g. $Q \subseteq {}^2\sqrt{\ln n}$, the funcion $q: Q \to n$, $q(x) = |\{y \in Q: y < x\}|$ is represnted in $(n,\overline{P})$.

6. Fact: There is a formula $\vartheta(x,y,\overline{P},\overline{Q})$ such that if $(n,\overline{P},\overline{Q}) \models \varphi(\overline{P},\overline{Q})$, $\varphi(\overline{P}) \wedge \psi_0(\overline{P},\overline{Q})$, then $\vartheta(x,y,\overline{P},\overline{Q})$ defines an isomorphism from $(Q_2,+',\times',\cdots)$ onto an initial segment of $(n,\overline{P})$.

Proof : By (5) we can do this for large enough initial segment, of power $k = \sqrt{\ln n}$; then we know that in a model of finite arithmetic, 2^k is definable as well as the representation of every $\ell \leq 2^k$ by a subset of k (using binary representation). Doing it twice we finish.

7. The sentence ψ: So we have to describe the sentence ψ such that $K_n = K_\psi^n$ for every finite n. It will be the conjunction of $\varphi(\overline{P})$, $\psi_0(\overline{Q})$ and another sentence which we describe what it says, rather than write it down.

So let $M=(A,\bar{P},\bar{Q})\models\psi_0(\bar{Q})\wedge\varphi(\bar{P})$, $|A|=n$. For simplicity we ignore the case some Q_ℓ is empty. W.l.o.g. $(A,\bar{P})$ is the standard model. All considerations are uniform in the sense they do not depend on n.

By (6) we can define the number $|Q_2|$ hence the numbers $|Q_0|+|Q_1|=n-|Q_2|$. By (4) we can define an x such that:

$$|Q_0|\leq x\leq |Q_0|^2(\ln n)^2$$

We can also define the number $\ln n$. We recall that $|Q_1|$ is a perfect square (by the functions F_0,F_1). So there is a number $y<n$, $y^2=|Q_1|$. Can we define y in M?

It satisfies:

$$(*)\ n-|Q_2|-y^2\leq x\leq(n-|Q_2|-y^2)^2(\ln n)^2$$

We have already defined all numbers appearing here (by suitable formulas) except y. So it sufices to show that (*) has a unique solution when $M\in K_n$ (as then we can define it and write our demand on $|Q_0|$ which is $n-|Q_2|-y^2$); if however there are two solutions, then $M\notin K_n$).

Now if $M\in K_n$, $|Q_0|^{100}<|Q_1|$ and $y_1\neq y_2$ are solutions, we get a contradiction or $y\leq(\ln n)^{10}$, but then we can define $|Q_0|$ directly.

8. Non definability of K_n:

It is well known that two models of the theory of equality of power $>n$ satisfies the same first order sentence of quantifiers depth n. So by the Feferman Vaught theorem (see [CK]), if $M\restriction Q_2=N\restriction Q_2$, $M\restriction Q_1=N\restriction Q_1$ and $|Q_0^N|=|Q_0^M|+1$, (and M,N are finite) then M,N, satisfy the same first order sentences of quantifier depth $<|Q_0^N|$, but $M\in\bigcup_{n<\omega}K_n\Leftrightarrow N\notin\bigcup_{n,\omega}K_n$.

So we finish.

References

[1] C. C. Cang and H. J. Keisler, Model Theory, North Holland Publ. Co.

[2] L. Deneberg, Y. Gurevich and S. Shelah, Cardinalities definable by constant depth polynomial size circuits. Information and Control.

Remarks on Squares.

The combinatorial principle square (and some variants) was introduced by Jensen [J]. We have been interested in deriving weak forms of it from ZFC, plus possibly restrictions on cardinal arithmetic, see [Sh 1], [Sh 2], Magidor and Shelah [MS] and Abraham, Shelah and Solovay [ASS]. The modest remarks appearing here were first intented to appear in [ASS]. I thank Shai Ben-David for deleting inaccurances here.

Convention: λ will be a fixed regular uncountable cardinal, δ vary on limit ordinals.

1. Definition : 1) We call $\bar{C} = \langle C_\delta : \delta \in S \rangle$ a square (or S-square) if:

(i) $S \subseteq \lambda$ is a stationary set.

(ii) for $\delta \in S$, C_δ is a closed unbounded subset of δ.

(iii) if γ is a limit point of C_δ, where $(\delta \in S)$ then $\gamma \in S$ and $C_\gamma = C_\delta \cap \gamma$.

2) We say there is a diamond on $\bar{C}$ for χ where $\bar{C} = \langle C_\delta : \delta \in S \rangle$ is a square, *if* there are $A_\delta \subseteq \delta$ for $\delta \in S$ such that for every $A \subseteq \lambda$:

$\{\delta \in S : C_\delta$ has order type $\geq \chi$ and for every limit point γ of $C_\delta \cup \{\delta\}$, $A \cap \gamma = A_\gamma\}$.

It may be interesting to note that we can find square sequences on some S from cardinality hypothesis only.

2. Lemma : 1) Suppose $\lambda = \mu^+$, $\mu^{<\chi} = \mu$. Then we can find $S_\xi (\xi < \mu)$ such that :

a) $\bigcup_{\xi<\mu} S_\xi = \{\delta < \lambda : cf\ \delta < \chi\}$.

b) for each $\xi < \mu$, there is an S_ξ-square sequence $\langle C_i : i \in S_\xi \rangle$ (so $C_i \subseteq S_\xi$ for each i, $otp(C_i) < \chi$).

2) Suppose $\lambda = \mu^+$, μ singular, $(\forall \vartheta < \mu)[\vartheta^{<\chi} < \mu]$.

Then we can find S_ξ $(\xi < \mu)$ such that :

a) $\bigcup_{\xi<\mu} S_\xi = \{\delta < \lambda : cf\ \delta < \chi, cf\ \delta \neq cf\ \mu\} - S^*(\lambda)$. ($S^*(\lambda)$-the bad set, see [Sh 1]) and called it S^+.

b) for each $\xi < \mu$ there is a weak ($< \chi$)-square sequence $\langle C_i^\xi : i \in S_\xi \rangle$

c) if $\delta \in S_\xi$, $cf\ \delta < cf\ \mu$ then $C_\delta^\xi \cap S^+ \subseteq S_\xi$.

d) if $\delta \in S^+$, $cf\ \delta > cf\ \mu$ then there are $\xi_\gamma < \mu (\gamma < cf\ \mu)$, such that $C_\delta^{\xi_\gamma} = C_\delta^{\xi_0}$, and $C_\delta^{\xi_0} \cap S^+ \subseteq \bigcup_\gamma S_{\xi_\gamma}$.

Proof : 1) By Engelking and Karlowicz [EK] there are functions $f_i : \mu \to \mu$ for $i < 2^\mu$ such that for any distinct $i_\gamma < 2^\mu (\gamma < \gamma^* < \chi)$ and $\xi_\gamma < \mu$, for some $\zeta < \mu$, $f_{i_\gamma}(\zeta) = \xi_\gamma$ (for $\gamma < \gamma^*$). For each $\delta < \mu^+$ let $\langle B_\xi^\delta : \xi < \mu \rangle$ be a list of all subsets of δ of power $< \chi$ (possible as $\mu = \mu^{<\chi}$). Now define a function $g_\zeta : \mu^+ \to \mu$, by $g_\zeta(i) = f_i(\zeta)$.

Now for each $\zeta < \mu$ we define S_ζ:

(*) S_ζ is the set of limit ordinals $\delta < \mu$ of cofinality $< \chi$ such that $B_{g_\zeta(\delta)}^\delta$ is a closed unbounded subset of δ, moreover for each accumulation point γ of $B_{g_\zeta(\delta)}^\delta$, $B_{g_\zeta(\delta)}^\gamma = B_{g_\zeta(\delta)}^\delta \cap \gamma$.

Clealy for every γ, δ as in (*) $\gamma \in S_\zeta$. So condition b) is satisfied: $\langle B_{g_\zeta(\delta)}^\delta : \delta \in S_\zeta \rangle$ exemplify it.

Why condition a) holds? If $\delta < \lambda$, $cf\ \delta < \chi$, let C_δ be a closed unbounded subset of it of cardinality $< \chi$. Let for $\gamma \in C_\delta \cup \{\delta\}$, $\xi_\gamma < \mu$ be such that $B_{\xi_\gamma}^\gamma = C_\delta \cap \gamma$ (possible by the choice of $\langle B_\xi^\gamma : \xi < \mu \rangle$). So by the choice of the functions f_i, there is $\zeta < \mu$ such that for every $\gamma \in C_\delta \cup \{\delta\}, f_\gamma(\zeta) = \xi_\gamma$ hence $g_\zeta(\gamma) = \xi_\gamma$. So easily $\delta \in S_\zeta$.

2) Left to the reader (just see what proof of the theorem from [EK gives).

2. Conclusion: If for simplicity G.C.H., χ regular, $\mu > \chi^*, \lambda > \mu^+$ then there is a χ-square S with diamond on it. (see [ASS])

3. Question: Let $\lambda = \mu^+$, μ regular, $\diamondsuit_{\{\delta<\lambda: cf\ \delta=\mu\}}$, and assume G.C.H. Is there a μ-square with diamond on it.

4. Lemma: Let λ be regular uncountable cardinal, R a set of regular cardinals $< \lambda$, such that $|R| < \lambda$, and $(\forall \kappa \in R)\ \kappa^+ < \lambda$. Then we can find $S_\kappa (\kappa \in R)$ such that :

a) S_κ is a stationary subsets of λ.

b) for every $\delta \in S_\kappa$, $cf\ \delta = \kappa$.

c) if $\delta \in S_{\kappa_1}$, $\kappa_1 \neq \kappa_2$ then $S_{\kappa_2} \cap \delta$ is not stationary in δ.

Remark: In (d) only the case $\kappa_2 < \kappa_1$ is relevant.

Proof : For every κ choose pairwise disjoint stationary subsets $\{S(\kappa,i) : i < \lambda\}$ of $\{\delta < \lambda : cf\ \delta = \kappa\}$, such that $\kappa, i < Min\ S(\kappa,i)$ (exists by Solovay [So]). Suppose the lemma fails Now we define by induction on $\xi < \lambda$, $\kappa_\xi \in S$ and $\langle S^\xi_\kappa : \kappa < \kappa_\xi, \kappa \in R \rangle$, and $\gamma(\xi,\kappa)$ γ^ξ_κ such that

(i) $S^\xi_\kappa \subseteq S(\kappa, \gamma^\xi_\kappa)$ for $\kappa \in \kappa_\xi \cap R)$ (i.e. $\kappa < \kappa_\xi$, $\kappa \in R$

(ii) $\gamma^\xi_\kappa \neq \gamma^\zeta_\kappa$ for $\zeta < \xi$ (when both are defined).

(iii) if $\vartheta < \sigma < \kappa_\xi$, $\kappa \in R$, $\sigma \in R, \delta \in S^\xi_\sigma$ then $S^\xi_\vartheta \cap \delta$ is not stationary in δ.

(iv) the set $T_\xi = \{\delta : \delta \in \bigcup \{S(\kappa_\xi, i) : i \notin \{\gamma^\zeta_{\kappa_\xi} : \zeta < \xi\}\}$, and no $S^\xi_\kappa (\kappa \in R \cap \kappa_\xi)$ is stationary in $\delta\}$ is not stationary and so disjoint to some club C^ξ of λ.

There is no problem is the definition: for each ξ we define γ_κ^ξ, S_κ^ξ by indunction on $\kappa \in R$. If it impossible to choose S_κ^ξ then the set defined in (iv) for κ cannot be stationary (as then the lemma's conclusion holds - remember $\kappa, i < Min\ S(\kappa, i)$ and by Fodour Lemma for some γ, $S(\kappa,\gamma) \cap T$ is stationary and we could have choose $S_{\kappa_\xi}^\xi = S(\kappa_\xi,\gamma) \cap T$, $\gamma_\kappa^\xi = \gamma$, but we have assumed this is impossible.

Now as $|R| < \lambda$ for some κ_α, $A = \{\xi < \lambda : \kappa_\xi = \kappa_\alpha\}$ has power λ, and choose $B \subseteq A$, $|B| = \kappa_\alpha^+$ so $|B| < \lambda$. Let $B = \{\xi_\varepsilon : \varepsilon < \kappa^+\}$ and so $\xi^* = \bigcup_\varepsilon \xi_\varepsilon < \lambda$. Hence there is $\gamma < \lambda$ such that $\gamma \not\in \{\gamma_{\kappa_\alpha}^\zeta : \zeta < \xi^*\}$, and there is $\delta \in S(\kappa_\alpha,\gamma) \cap \bigcap_{\varepsilon < \kappa_\alpha^+} C^{\xi_\varepsilon}$. Working carefully with the choice of C^{ξ_ε} we see that for each $\varepsilon < \kappa_\alpha^+$, $\delta \cap (\bigcup_{\kappa < \kappa_\alpha} S_\kappa^{\xi_\varepsilon})$ is stationay in δ. So an ordinal of cofinality κ_α has κ_α^+ pairwise disjoint stationary subsets, contradiction.

References.

[ASS]
U. Avraham, S. Shelah and R. M. Solovay, Squares with diamonds and Souslin trees with special squares for successor of singulars, *Fund Math*, submitted.

[MS] M. Magidor and S. Shelah, Compactness in successor of singular, in preparation, see ***Abstracts of A.M.S.*** Aug (or. Oct.) 1983.

[Sh1]
S. Shelah, Successor of singular cardinals, Proc. of the ASL meeting Mons. 8/1978, *Logic Colloquium 78*. ed M. Boffa, D. van Dalem and K. McAlon, Studies in Logic and the Foundation of Math, Vol. 97, North Holland Pub. Co. 1979, 357-380.

[Sh2]
S. Shelah, Diamonds and uniformization, *J.Symb. Logic, in press.*

[So] R. M. Solovay. Real valued measurable cardinals, *Proc. Symp. in Pur Math.* Vol XIII 1 (ed. D. Scott) Amer. Math. Soc. (1971), 397-428.

Vol. 1090: Differential Geometry of Submanifolds. Proceedings, 1984. Edited by K. Kenmotsu. VI, 132 pages. 1984.

Vol. 1091: Multifunctions and Integrands. Proceedings, 1983. Edited by G. Salinetti. V, 234 pages. 1984.

Vol. 1092: Complete Intersections. Seminar, 1983. Edited by S. Greco and R. Strano. VII, 299 pages. 1984.

Vol. 1093: A. Prestel, Lectures on Formally Real Fields. XI, 125 pages. 1984.

Vol. 1094: Analyse Complexe. Proceedings, 1983. Edité par E. Amar, R. Gay et Nguyen Thanh Van. IX, 184 pages. 1984.

Vol. 1095: Stochastic Analysis and Applications. Proceedings, 1983. Edited by A. Truman and D. Williams. V, 199 pages. 1984.

Vol. 1096: Théorie du Potentiel. Proceedings, 1983. Edité par G. Mokobodzki et D. Pinchon. IX, 601 pages. 1984.

Vol. 1097: R.M. Dudley, H. Kunita, F. Ledrappier, École d'Éte de Probabilités de Saint-Flour XII – 1982. Edité par P.L. Hennequin. X, 396 pages. 1984.

Vol. 1098: Groups – Korea 1983. Proceedings. Edited by A.C. Kim and B.H. Neumann. VII, 183 pages. 1984.

Vol. 1099: C.M. Ringel, Tame Algebras and Integral Quadratic Forms. XIII, 376 pages. 1984.

Vol. 1100: V. Ivrii, Precise Spectral Asymptotics for Elliptic Operators Acting in Fiberings over Manifolds with Boundary. V, 237 pages. 1984.

Vol. 1101: V. Cossart, J. Giraud, U. Orbanz, Resolution of Surface Singularities. Seminar. VII, 132 pages. 1984.

Vol. 1102: A. Verona, Stratified Mappings – Structure and Triangulability. IX, 160 pages. 1984.

Vol. 1103: Models and Sets. Proceedings, Logic Colloquium, 1983, Part I. Edited by G.H. Müller and M.M. Richter. VIII, 484 pages. 1984.

Vol. 1104: Computation and Proof Theory. Proceedings, Logic Colloquium, 1983, Part II. Edited by M.M. Richter, E. Börger, W. Oberschelp, B. Schinzel and W. Thomas. VIII, 475 pages. 1984.

Vol. 1105: Rational Approximation and Interpolation. Proceedings, 1983. Edited by P.R. Graves-Morris, E.B. Saff and R.S. Varga. XII, 528 pages. 1984.

Vol. 1106: C.T. Chong, Techniques of Admissible Recursion Theory. IX, 214 pages. 1984.

Vol. 1107: Nonlinear Analysis and Optimization. Proceedings, 1982. Edited by C. Vinti. V, 224 pages. 1984.

Vol. 1108: Global Analysis – Studies and Applications I. Edited by Yu. G. Borisovich and Yu. E. Gliklikh. V, 301 pages. 1984.

Vol. 1109: Stochastic Aspects of Classical and Quantum Systems. Proceedings, 1983. Edited by S. Albeverio, P. Combe and M. Sirugue-Collin. IX, 227 pages. 1985.

Vol. 1110: R. Jajte, Strong Limit Theorems in Non-Commutative Probability. VI, 152 pages. 1985.

Vol. 1111: Arbeitstagung Bonn 1984. Proceedings. Edited by F. Hirzebruch, J. Schwermer and S. Suter. V, 481 pages. 1985.

Vol. 1112: Products of Conjugacy Classes in Groups. Edited by Z. Arad and M. Herzog. V, 244 pages. 1985.

Vol. 1113: P. Antosik, C. Swartz, Matrix Methods in Analysis. IV, 114 pages. 1985.

Vol. 1114: Zahlentheoretische Analysis. Seminar. Herausgegeben von E. Hlawka. V, 157 Seiten. 1985.

Vol. 1115: J. Moulin Ollagnier, Ergodic Theory and Statistical Mechanics. VI, 147 pages. 1985.

Vol. 1116: S. Stolz, Hochzusammenhängende Mannigfaltigkeiten und ihre Ränder. XXIII, 134 Seiten. 1985.

Vol. 1117: D.J. Aldous, J.A. Ibragimov, J. Jacod, Ecole d'Été de Probabilités de Saint-Flour XIII – 1983. Édité par P.L. Hennequin. IX, 409 pages. 1985.

Vol. 1118: Grossissements de filtrations: exemples et applications. Seminaire, 1982/83. Edité par Th. Jeulin et M. Yor. V, 315 pages. 1985.

Vol. 1119: Recent Mathematical Methods in Dynamic Programming. Proceedings, 1984. Edited by I. Capuzzo Dolcetta, W.H. Fleming and T. Zolezzi. VI, 202 pages. 1985.

Vol. 1120: K. Jarosz, Perturbations of Banach Algebras. V, 118 pages. 1985.

Vol. 1121: Singularities and Constructive Methods for Their Treatment. Proceedings, 1983. Edited by P. Grisvard, W. Wendland and J.R. Whiteman. IX, 346 pages. 1985.

Vol. 1122: Number Theory. Proceedings, 1984. Edited by K. Alladi. VII, 217 pages. 1985.

Vol. 1123: Séminaire de Probabilités XIX 1983/84. Proceedings. Edité par J. Azéma et M. Yor. IV, 504 pages. 1985.

Vol. 1124: Algebraic Geometry, Sitges (Barcelona) 1983. Proceedings. Edited by E. Casas-Alvero, G.E. Welters and S. Xambó-Descamps. XI, 416 pages. 1985.

Vol. 1125: Dynamical Systems and Bifurcations. Proceedings, 1984. Edited by B.L.J. Braaksma, H.W. Broer and F. Takens. V, 129 pages. 1985.

Vol. 1126: Algebraic and Geometric Topology. Proceedings, 1983. Edited by A. Ranicki, N. Levitt and F. Quinn. V, 523 pages. 1985.

Vol. 1127: Numerical Methods in Fluid Dynamics. Seminar. Edited by F. Brezzi, VII, 333 pages. 1985.

Vol. 1128: J. Elschner, Singular Ordinary Differential Operators and Pseudodifferential Equations. 200 pages. 1985.

Vol. 1129: Numerical Analysis, Lancaster 1984. Proceedings. Edited by P.R. Turner. XIV, 179 pages. 1985.

Vol. 1130: Methods in Mathematical Logic. Proceedings, 1983. Edited by C.A. Di Prisco. VII, 407 pages. 1985.

Vol. 1131: K. Sundaresan, S. Swaminathan, Geometry and Nonlinear Analysis in Banach Spaces. III, 116 pages. 1985.

Vol. 1132: Operator Algebras and their Connections with Topology and Ergodic Theory. Proceedings, 1983. Edited by H. Araki, C.C. Moore, Ş. Strătilă and C. Voiculescu. VI, 594 pages. 1985.

Vol. 1133: K.C. Kiwiel, Methods of Descent for Nondifferentiable Optimization. VI, 362 pages. 1985.

Vol. 1134: G.P. Galdi, S. Rionero, Weighted Energy Methods in Fluid Dynamics and Elasticity. VII, 126 pages. 1985.

Vol. 1135: Number Theory, New York 1983–84. Seminar. Edited by D.V. Chudnovsky, G.V. Chudnovsky, H. Cohn and M.B. Nathanson. V, 283 pages. 1985.

Vol. 1136: Quantum Probability and Applications II. Proceedings, 1984. Edited by L. Accardi and W. von Waldenfels. VI, 534 pages. 1985.

Vol. 1137: Xiao G., Surfaces fibrées en courbes de genre deux. IX, 103 pages. 1985.

Vol. 1138: A. Ocneanu, Actions of Discrete Amenable Groups on von Neumann Algebras. V, 115 pages. 1985.

Vol. 1139: Differential Geometric Methods in Mathematical Physics. Proceedings, 1983. Edited by H.D. Doebner and J.D. Hennig. VI, 337 pages. 1985.

Vol. 1140: S. Donkin, Rational Representations of Algebraic Groups. VII, 254 pages. 1985.

Vol. 1141: Recursion Theory Week. Proceedings, 1984. Edited by H.-D. Ebbinghaus, G.H. Müller and G.E. Sacks. IX, 418 pages. 1985.

Vol. 1142: Orders and their Applications. Proceedings, 1984. Edited by I. Reiner and K.W. Roggenkamp. X, 306 pages. 1985.

Vol. 1143: A. Krieg, Modular Forms on Half-Spaces of Quaternions. XIII, 203 pages. 1985.

Vol. 1144: Knot Theory and Manifolds. Proceedings, 1983. Edited by D. Rolfsen. V, 163 pages. 1985.

GPSR Compliance

The European Union's (EU) General Product Safety Regulation (GPSR) is a set of rules that requires consumer products to be safe and our obligations to ensure this.

If you have any concerns about our products, you can contact us on ProductSafety@springernature.com

In case Publisher is established outside the EU, the EU authorized representative is:

Springer Nature Customer Service Center GmbH
Europaplatz 3
69115 Heidelberg, Germany

Batch number: 10692380

Printed by Printforce, the Netherlands